T0339735

Environmental Metabolomics

Environmental Metabolomics
Applications in Field and Laboratory Studies to Understand from Exposome to Metabolome

Edited by

Diana Álvarez-Muñoz

Marinella Farré
Water and Soil Quality Research Group,
Department of Environmental Chemistry,
IDAEA-CSIC, Barcelona, Spain

ELSEVIER

Elsevier
Radarweg 29, PO Box 211, 1000 AE Amsterdam, Netherlands
The Boulevard, Langford Lane, Kidlington, Oxford OX5 1GB, United Kingdom
50 Hampshire Street, 5th Floor, Cambridge, MA 02139, United States

Notices

Knowledge and best practice in this field are constantly changing. As new research and experience broaden our
understanding, changes in research methods, professional practices, or medical treatment may become necessary.

Practitioners and researchers must always rely on their own experience and knowledge in evaluating and using any
information, methods, compounds, or experiments described herein. In using such information or methods they
should be mindful of their own safety and the safety of others, including parties for whom they have a professional
responsibility.

To the fullest extent of the law, neither the Publisher nor the authors, contributors, or editors, assume any liability
for any injury and/or damage to persons or property as a matter of products liability, negligence or otherwise, or
from any use or operation of any methods, products, instructions, or ideas contained in the material herein.

Library of Congress Cataloging-in-Publication Data
A catalog record for this book is available from the Library of Congress

British Library Cataloguing-in-Publication Data
A catalogue record for this book is available from the British Library

ISBN: 978-0-12-818196-6

For information on all Elsevier publications visit our website
at https://www.elsevier.com/books-and-journals

Publisher: Janco Candice
Acquisitions Editor: Louisa Munro
Editorial Project Manager: Sara Pianavilla
Production Project Manager: Kumar Anbazhagan
Cover Designer: Miles Hitchen

Typeset by TNQ Technologies

Working together
to grow libraries in
developing countries

www.elsevier.com • www.bookaid.org

Contents

Chapter 8: Mass spectrometry to explore exposome and metabolome of organisms exposed to pharmaceuticals and personal care products 235

Frédérique Courant, Hélène Fenet, Bénilde Bonnefille, Thibaut Dumas and Elena Gomez

Chapter 9: Metabolomics effects of nanomaterials: an ecotoxicological perspective 259

Marinella Farré and Awadhesh N. Jha

Contributors

Konstantinos A. Aliferis Pesticide Science Laboratory, Agriculture University of Athens, Athens, Greece; Department of Plant Science, McGill University, Sainte-Anne-de-Bellevue, QC, Canada

Diana Álvarez-Muñoz Water and Soil Quality Research Group, Department of Environmental Chemistry, IDAEA-CSIC, Barcelona, Spain

Francisca Arellano-Beltrán Department of Chemistry, Faculty of Experimental Sciences, University of Huelva, Campus El Carmen, Huelva, Spain; Research Center of Natural Resources, Health and the Environment (RENSMA), University of Huelva, Campus El Carmen, Huelva, Spain

Ana Arias-Borrego Department of Chemistry, Faculty of Experimental Sciences, University of Huelva, Campus El Carmen, Huelva, Spain; Research Center of Natural Resources, Health and the Environment (RENSMA), University of Huelva, Campus El Carmen, Huelva, Spain

Òscar Aznar-Alemany Water and Soil Quality Research Group, Department of Environmental Chemistry, IDAEA-CSIC, Barcelona, Spain

Julián Blasco Institute of Marine Sciences of Andalusia (CSIC), Campus Rio San Pedro, Cádiz, Spain

Bénilde Bonnefille HydroSciences, Univ Montpellier, CNRS, IRD, Montpellier, France

Belén Callejón-Leblic Department of Chemistry, Faculty of Experimental Sciences, University of Huelva, Campus El Carmen, Huelva, Spain; Research Center of Natural Resources, Health and the Environment (RENSMA), University of Huelva, Campus El Carmen, Huelva, Spain

Pedro Carriquiriborde Centro de Investigaciones del Medioambiente (CIM), Facultad de Ciencias Exactas, Universidad Nacional de la Plata — CONICET, La Plata, Buenos Aires, Argentina

Chien-Min Chen Department of Environmental Resources Management, Chia Nan University of Pharmacy & Science, Tainan, Taiwan

Frédérique Courant HydroSciences, Univ Montpellier, CNRS, IRD, Montpellier, France

Arthur David Univ Rennes, Inserm, EHESP, Irset (Institut de recherche en santé, environnement et travail), UMR_S 1085, Rennes, France

Xiaoping Diao State Key Laboratory of Marine Resource Utilization in South China Sea, Hainan University, Haikou, Hainan Province, China; Ministry of Education Key Laboratory of Tropical Island Ecology, Hainan Normal University, Haikou, Hainan Province, China

Thibaut Dumas HydroSciences, Univ Montpellier, CNRS, IRD, Montpellier, France

Marinella Farré Water and Soil Quality Research Group, Department of Environmental Chemistry, IDAEA-CSIC, Barcelona, Spain

Hélène Fenet HydroSciences, Univ Montpellier, CNRS, IRD, Montpellier, France

Tamara García-Barrera   Department of Chemistry, Faculty of Experimental Sciences, University of Huelva, Campus El Carmen, Huelva, Spain; Research Center of Natural Resources, Health and the Environment (RENSMA), University of Huelva, Campus El Carmen, Huelva, Spain

Ruben Gil-Solsona   Catalan Institute for Water Research (ICRA), Parc Científic i Tecnòlogic de la Universitat de Girona, Girona, Spain

Elena Gomez   HydroSciences, Univ Montpellier, CNRS, IRD, Montpellier, France

José Luis Gómez-Ariza   Department of Chemistry, Faculty of Experimental Sciences, University of Huelva, Campus El Carmen, Huelva, Spain; Research Center of Natural Resources, Health and the Environment (RENSMA), University of Huelva, Campus El Carmen, Huelva, Spain

Awadhesh N. Jha   University of Plymouth, Plymouth, Devon, United Kingdom

Vera Kovacevic   Department of Chemistry, University of Toronto, Toronto, ON, Canada; Environmental NMR Centre, Department of Physical and Environmental Sciences, University of Toronto Scarborough, Toronto, ON, Canada

Marta Llorca   Water and Soil Quality Research Group, Department of Environmental Chemistry, IDAEA-CSIC, Barcelona, Spain

Gema Rodríguez-Moro   Department of Chemistry, Faculty of Experimental Sciences, University of Huelva, Campus El Carmen, Huelva, Spain; Research Center of Natural Resources, Health and the Environment (RENSMA), University of Huelva, Campus El Carmen, Huelva, Spain

Sara Rodríguez-Mozaz   Catalan Institute for Water Research (ICRA), Parc Científic i Tecnòlogic de la Universitat de Girona, Girona, Spain

Pawel Rostkowski   NILU — Norwegian Institute for Air Research, Kjeller, Norway

Sara Ramírez-Acosta   Department of Chemistry, Faculty of Experimental Sciences, University of Huelva, Campus El Carmen, Huelva, Spain; Research Center of Natural Resources, Health and the Environment (RENSMA), University of Huelva, Campus El Carmen, Huelva, Spain

Albert Serra-Compte   Catalan Institute for Water Research (ICRA), Parc Científic i Tecnòlogic de la Universitat de Girona, Girona, Spain

Myrna J. Simpson   Department of Chemistry, University of Toronto, Toronto, ON, Canada; Environmental NMR Centre, Department of Physical and Environmental Sciences, University of Toronto Scarborough, Toronto, ON, Canada

Hailong Zhou   State Key Laboratory of Marine Resource Utilization in South China Sea, Hainan University, Haikou, Hainan Province, China; School of Life and Pharmaceutical Sciences, Hainan University, Haikou, Hainan Province, China

Preface

Metabolomics consists of the simultaneous characterization of the metabolites present in an organism and offers a "picture" of the biochemistry of the organism at any one time. Its application to the environment, known as Environmental Metabolomics, allows characterizing the interaction that occurs between organisms and the surrounding environment. Concretely, this book is focused on the interaction between organisms and contaminants that are present in the environment due to human activities and may have toxic effects.

The application of metabolomics in the environmental field for biomarkers discovery is relatively new. Scientific papers on this subject started being published about 10 years ago, but it has been in the last 5 years when the application of metabolomics for analyzing biological samples in environmental monitoring has strongly attracted the attention of researches. Consequently, an increasing number of papers are currently been published generating a high amount of data that need to be compiled and harmonized to get relevant information.

This book gathers information on environmental metabolomics when natural organisms are exposed to metals, persistent organic pollutants, and emerging pollutants. It shows the reader different experimental setups, analytical techniques, data processing, and data analysis. This book will lead you through the metabolomics workflow and will serve as a guide for implementation. Besides, it allows, for the first time, to have general biomarkers snapshot very useful for risk assessment. It also discusses the current limitations and future perspectives of environmental metabolomics.

The audience of this book is wide-ranging from undergraduate to graduate students interested in environmental research, researchers in the field of environmental toxicology and chemistry, legislators, and policy-makers.

We, the Editors, learned a lot from the authors and hope that you readers also do. We expect that the knowledge contained here will help to further gain insight and advance on environmental metabolomics science.

Diana Álvarez-Muñoz

Marinella Farré

Acknowledgments

Huge thanks to all the people who have been involved in this project, especially to the authors; without their hard work, this book would not have been possible. Thanks to Elsevier, particularly to the acquisitions editor, the editorial project manager, and the production project manager. Finally, we are thankful to our families for their support, for understanding our passion for science, and all the time dedicated to this subject.

CHAPTER 1

Fundamentals of environmental metabolomics

Vera Kovacevic[1,2], Myrna J. Simpson[1,2]

[1]*Department of Chemistry, University of Toronto, Toronto, ON, Canada;* [2]*Environmental NMR Centre, Department of Physical and Environmental Sciences, University of Toronto Scarborough, Toronto, ON, Canada*

Chapter Outline

1. Environmental stressors

Aquatic and terrestrial ecosystems are under constant threat from various environmental stressors that arise from natural or anthropogenic activities. Environmental stressors include biotic stressors such as pathogens and abiotic stressors such as drought, flood, extreme temperatures, and salinity (Nõges et al., 2016). The release of anthropogenic contaminants into the environment from urbanization, transportation, and industrial activities is also contributing to environmental stress (Schaeffer et al., 2016). Increases in the amount and variety of synthetic chemicals produced have caused contaminants to enter the environment at a very quick pace on a global scale (Bernhardt et al., 2017). The

Environmental Metabolomics. https://doi.org/10.1016/B978-0-12-818196-6.00001-7

transport and fate of contaminants is governed by their physical—chemical properties which include water solubility, *n*-octanol—water partition coefficients (K_{ow}), acid dissociation constants (pK_a) or base dissociation constants (pK_b), and vapor pressure (De Laender et al., 2015; Pereira et al., 2016). Environmental factors such as pH, sunlight intensity, temperature, and organic matter content also influence the transport and fate of contaminants in the environment (De Laender et al., 2015; Pereira et al., 2016). Some contaminants may be transported long distances to isolated regions and some contaminants may bioaccumulate in food webs (Gao et al., 2018; Xie et al., 2017). For instance, although metals are naturally occurring, anthropogenic activities have caused increased metal concentrations in the environment and metals often accumulate in organisms as they cannot be biodegraded (Peng et al., 2018a; Wise et al., 2018). Contaminants are frequently detected in all three environmental compartments of air, water, and soil (Gavrilescu et al., 2015; Net et al., 2015). Consequently, aquatic and terrestrial organisms are exposed to various classes of contaminants such as metals, persistent organic pollutants (POPs), pharmaceuticals and personal care products (PPCPs), industrial chemicals, plasticizers, flame retardants such as organophosphate esters, and pesticides (Peng et al., 2018b; van den Brink et al., 2016; Wilkinson et al., 2018).

1.1 Contaminant stressors

Contaminants of emerging concern (CECs) are not definitively defined, and there is no comprehensive list of CECs. Instead, CECs are thought to be any naturally occurring or anthropogenic compounds which are now detected or suspected to occur in soil, air, or water and whose persistence or toxicity may significantly alter the metabolism of an organism (Sauvé and Desrosiers, 2014). The United States Environmental Protection Agency's list of CECs includes POPs, PPCPs, veterinary medicines, endocrine-disrupting chemicals (EDCs), and nanomaterials (Ankley et al., 2008, Fig. 1.1). POPs are legacy pollutants that have existed and persisted in the environment for decades and include polychlorinated biphenyls, dibenzo-*p*-dioxins, dibenzofurans, and organochlorine pesticides such as dichlorodiphenyltrichloroethane (Nadal et al., 2015). The Stockholm Convention on POPs is an international environmental treaty that was initiated in 2001 to protect human health and the environment from POPs (Lallas, 2001). The screening criteria for POPs include persistence, bioaccumulation, potential for long range transport in the environment, and adverse impacts to organisms (McLachlan, 2018). Other organic contaminants such as PPCPs, veterinary medicines, and EDCs can more easily degrade in soil and water depending on their properties, but their extensive use has resulted in their frequent detection in the environment (Bártíková et al., 2016; Ebele et al., 2017; Song et al., 2018). Pharmaceuticals include over-the-counter medications and prescription drugs, and personal care products are in everyday products such as shampoos, hair dyes, toothpaste, and deodorants (Boxall et al., 2012). Veterinary medications include

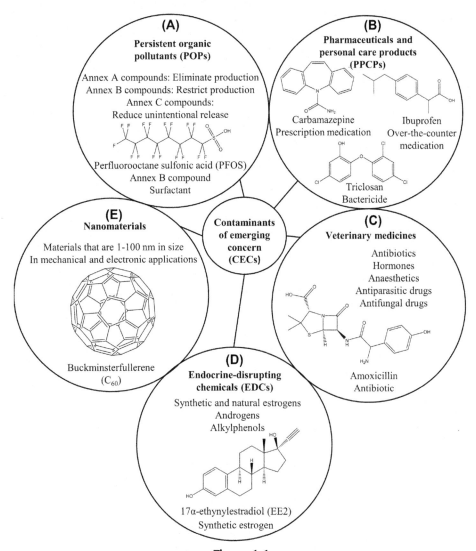

Figure 1.1

Names and chemical structures of some well-known contaminants of emerging concern (CECs): (A) persistent organic pollutants (POPs), (B) pharmaceuticals and personal care products (PPCPs), (C) veterinary medicines, (D) endocrine-disrupting chemicals (EDCs), and (E) nanomaterials.

antibiotics, hormones, anesthetics, and antiparasitic and antifungal drugs (Bártíková et al., 2016). The main entry route of veterinary medications into the environment is from treatment of livestock, aquaculture, and companion animals (Bártíková et al., 2016). EDCs include alkylphenol compounds, natural estrogens, natural androgens, synthetic hormones, and some pharmaceuticals and pesticides (Omar et al., 2016). For instance, one of the most potent endocrine disruptors is the synthetic estrogen 17α-ethynylestradiol found in birth control pills (Laurenson et al., 2014). There are also contaminants of industrial origin such as perfluorinated compounds which can be found in common products such as furniture, carpets, cookware, and firefighting materials (von der Trenck et al., 2018). Terrestrial and aquatic ecosystems are typically exposed to a mixture of pesticides from agricultural applications, and runoff water can have pesticide concentrations that are substantially above the legal limit (Lefrancq et al., 2017). Nanomaterials have electronic and mechanical applications and are being increasingly detected in the environment, most likely from sewage treatment plant sludge and solid waste (Sun et al., 2016). The common detection of organic contaminants in surface waters is mainly due to the inability of conventional wastewater treatment methods to efficiently remove compounds such as PPCPs, veterinary medications, and EDCs (Hernández et al., 2015; Rice and Westerhoff, 2017; Yang et al., 2017). Organic contaminants also enter soil and groundwater when soil is irrigated with wastewater or when soil is fertilized with sewage sludge, and from municipal soil waste landfills (Healy et al., 2017; Prosser and Sibley, 2015). The detection of even very low concentrations of CECs in the environment has raised concerns because organisms are constantly exposed to inputs of CECs which may bioaccumulate to concentrations that could cause deleterious impacts in biota (Meador et al., 2016). The potential harmful impacts of CECs on ecosystem and human health have resulted in the development of toxicity tests and the assessment and management of environmental contamination.

2. Ecotoxicology

Ecotoxicology has traditionally been described as the study of the toxic impacts of pollutants to ecosystem constituents, including animals, plants, and microorganisms (Truhaut, 1977). As such, environmental quality standards have been initiated to preserve ecosystems and human health by placing maximum permissible concentrations of contaminants that may be detected in water, soil, or biota (Lepper, 2005). Ecological risk assessment is generally done by comparing measured environmental concentrations (MECs) of a contaminant to its predicted no effect concentrations (PNECs) from ecotoxicological data which ideally represent the most sensitive species over several trophic levels (Papadakis et al., 2015). If the risk quotient calculated from the MEC/PNEC ratio is greater than one, then the contaminant is a concern and action should be undertaken to confirm the environmental risk, identify the sources of contamination, and reduce the release of contamination (Papadakis et al., 2015; Thomaidi et al., 2016).

Toxicologists have established and used acute toxicity tests on terrestrial and aquatic organisms based on mortality or immobilization rates with increasing contaminant concentrations (Bruce, 1985; Buckler et al., 2005). Most chronic toxicity tests also assess changes in development, growth, and reproduction parameters such as amount of offspring, time to first breeding, and number of broods (Thome et al., 2017; Toumi et al., 2013; Wang et al., 2006). Other tests have studied behavioral changes, mainly with fish, and these include changes in motor activity during light and dark photoperiods as well as the ability and length of time needed to catch prey (Gaworecki and Klaine, 2008; Kristofco et al., 2016). Since these approaches are not sensitive for very low sublethal concentrations of contaminants, new techniques were developed using biomarkers (Coppola et al., 2018; Valavanidis et al., 2006). A biomarker is defined as a biological response which includes any biochemical, physiological, histological, and morphological changes measured inside an organism that arise from contaminant exposure (Van Gestel and Van Brummelen, 1996). Substantive progress has been made in measuring biochemical impacts from contaminant exposure, and this includes measuring changes in biomarkers of oxidative stress, endocrine disruption, immunomodulation, xenobiotic detoxification systems, and DNA damage (Jasinska et al., 2015; Loughery et al., 2018; Valavanidis et al., 2006). Although biomarkers give important information about the potential deleterious impacts of contaminants on a biochemical level, they are not capable of evaluating the molecular mechanism of action of contaminants (Campos et al., 2012).

There is a need to study the mechanism or mode of action of contaminants to better understand the molecular processes of how organisms respond to contaminant stressors in the environment. Environmental omics research was initiated which involves the use of omics technologies to investigate the molecular-level responses of organisms to various environmental stressors (Martyniuk and Simmons, 2016). The omics techniques include genomics to study the structure, function, and expression of genes, transcriptomics to measure gene expression, proteomics to measure proteins and peptides, and metabolomics to measure metabolites which are the end products of cellular events (Loughery et al., 2018; Marjan et al., 2017; Revel et al., 2017). The data collected from omics techniques form a systems biology approach, and their integration into the field of ecotoxicology is referred to as ecotoxicogenomics (Snape et al., 2004). Also, omics technologies are sensitive and can detect changes in organism biochemistry at lower contaminant concentrations and more rapidly than conventional morality tests, reproduction tests, or cytotoxic evaluations (de Figueirêdo et al., 2019; Shin et al., 2018).

3. Metabolomics

Metabolomics is the use of advanced analytical techniques to identify and measure low-molecular weight metabolites that are generally less than 1000 Da in cells, tissues, biofluids, organs, or whole organisms (Gao et al., 2019; Lin et al., 2006). This includes primary metabolites which are involved in the development, growth, and reproduction of an organism as well as secondary metabolites which are produced by bacteria, fungi, and plants and have various ecological functions (Mazzei et al., 2016; Palazzotto and Weber, 2018). Metabolomics can be considered as the downstream process of genomics, transcriptomics, and proteomics, and the changes of metabolite levels directly relate to biochemical activity and the phenotype (Johnson et al., 2016). Fundamental metabolic pathways such as those involved in energy, carbohydrate, amino acid, and lipid metabolism are conserved from bacteria to eukaryotes (Peregrín-Alvarez et al., 2009). Metabolomics has applications in several fields including medicine (Wishart, 2016), pharmacology (Pang et al., 2018), toxicology (Ramirez et al., 2018), plant biochemistry (Skliros et al., 2018), and the environmental sciences (Zhang et al., 2018a).

The main analytical techniques that are used to collect metabolomics data sets are nuclear magnetic resonance (NMR) spectroscopy and mass spectrometry (MS) because of the ability for small molecule detection and the unique assets of each analytical instrument. Targeted metabolomics analysis detects a predetermined set of metabolites, usually chosen with regard to the biological sample to be analyzed or from metabolite libraries in software databases (Bingol, 2018). Nontargeted metabolomics analysis is the nonbiased analysis of as many metabolites that can be reliably identified and assigned by the analytical instrument and metabolomics databases (Bingol, 2018). Nontargeted metabolomics analysis frequently uses NMR spectroscopy or high resolution MS, while targeted metabolomics analysis often uses MS as the analytical method of choice (Emwas, 2015; Mullard et al., 2015). There are around 114,000 metabolites listed in the Human Metabolome Database (Wishart et al., 2017), but only around 1500 metabolites are identified in nontargeted analysis, 200–500 metabolites are identified in targeted analysis, and it is estimated that less than two dozen metabolites are regularly quantified in most metabolomics studies (Markley et al., 2017; Psychogios et al., 2011). Once metabolites are identified and quantified, online databases and tools may be used to aid in data interpretation and mechanistic understanding by relating the changes in metabolite levels to metabolic pathways that are likely impacted (Chong et al., 2018; Kanehisa et al., 2016). Through this process, metabolomics gives detailed information about the mode of action that may be occurring in the biological sample and may be used for high-throughput testing of individual contaminants and mixtures (Ahmed et al., 2019; Zampieri et al., 2018).

4. Environmental metabolomics

Environmental metabolomics involves applying metabolomic techniques to analyze the metabolic response of organisms as a result of their interactions with the environment (Bundy et al., 2009). Metabolomics is used to study various environmental stressors including UV light (Zhang et al., 2018b), elevated atmospheric carbon dioxide concentration (Creydt et al., 2019), ambient fine particulate matter (Xu et al., 2019), drought (Li et al., 2018b), extreme temperatures (Tomonaga et al., 2018), and contaminants (Roszkowska et al., 2018). Controlled laboratory exposures with target species are performed to acquire knowledge of the mode of action of abiotic stressors, biotic stressors, or contaminants (Garreta-Lara et al., 2016; Sivaram et al., 2019; Tang et al., 2017). Furthermore, field-based studies may be conducted to understand how the metabolome of organisms is impacted when exposed to an ecosystem under environmental stress (Gauthier et al., 2018). One of the goals of environmental metabolomics research is for utilization in environmental biomonitoring and risk assessment by applying metabolomics techniques to keystone organisms that play important roles in trophic levels and food webs (Bahamonde et al., 2016). To achieve these aims, environmental metabolomics studies have a workflow that includes study design, exposure, sample preparation, metabolite extraction, data collection, data analysis, and finally a biological interpretation of the analyzed data.

4.1 Study design

The experimental design of environmental metabolomics projects involves the selection of an environmental stressor and the target organism. Environmental metabolomics studies are done on a variety of organisms which span from microorganisms to plants and animals (Sivaram et al., 2019; Tian et al., 2017; Tomonaga et al., 2018). Microorganisms have been exposed to environmental contaminants, for instance, the yeast *Saccharomyces cerevisiae* has been exposed to copper (Farrés et al., 2016) and tetrachlorobisphenol A (Tian et al., 2017). Plant metabolomics has investigated the exposure to nanoparticles (Zhang et al., 2018a), how plants respond to polycyclic aromatic hydrocarbons or metals from remediation efforts (Pidatala et al., 2016; Sivaram et al., 2019), the impact of mineral deficiency (Sung et al., 2015), UV-B radiation (Zhang et al., 2018b), and drought (Khan et al., 2019). Common terrestrial organisms used in environmental metabolomics studies include nematodes (Ratnasekhar et al., 2015), earthworms (Tang et al., 2017), flies (Cox et al., 2017), and mice (Wang et al., 2018a). The choice of organism should reflect the environmental compartment under consideration, for instance, earthworms are commonly used to assess soil contamination in metabolomics studies due to their occurrence in a variety of soils worldwide (He et al., 2018; Tang et al., 2017). Aquatic organisms frequently used in environmental metabolomics studies include crustaceans (Garreta-Lara et al., 2016; Gómez-Canela et al., 2016) and fish such as medaka, rainbow trout, salmon, and fathead minnow (Kaneko et al., 2019). The studied aquatic organism should also reflect the research question, for instance, bivalves have a sessile lifestyle and can accumulate contaminants, and therefore an analysis of the metabolic profile of bivalves may reflect the contamination at the site of collection (Watanabe et al., 2015). Metabolomics studies may also be performed on a targeted selection of organisms that serve as research models. Model organisms may be represented by the rat and mouse for mammals, zebrafish *Danio rerio* for aquatic vertebrates, the water flea *Daphnia magna* for aquatic invertebrates, *Arabidopsis thaliana* for plants, the yeast *S. cerevisiae* for eukaryotes, and *Escherichia coli* for prokaryotes (Kim et al., 2015; Reed et al., 2017).

4.2 Collection of organisms and experimental exposure

Environmental metabolomics studies may be field-based where organisms are sampled directly from the environment or laboratory-based where organisms are cultured in the laboratory and then exposed under controlled conditions (Campillo et al., 2019; Davis et al., 2016). Both laboratory and field research should have careful planning of the number and type of control or reference site exposures and environmental stressor exposures, as well as the number of samples to be taken from each treatment group. This is important as there is natural variation in biological samples and adequate replication is needed for proper statistical analysis (Simmons et al., 2015). Regarding field-based work, there is the option of field trials which involves the sampling of free-living organisms in the environment (Gauthier et al., 2018; Melvin et al., 2018, 2019) or there is the option of field deployment which involves deploying organisms into environments that are under stress (Ekman et al., 2018). In field-based studies, the organisms that represent the stressor-exposed groups may be collected at contaminated locations and may be compared to organisms collected at more pristine locations which serve as the group from a reference site (Watanabe et al., 2015). Environmental metabolomics studies that sample free-living organisms at a field site are a step forward for validating this technique for use in environmental monitoring (Melvin et al., 2018, 2019). Also, using field-deployed organisms has a great value for environmental risk assessment. For instance, a study that used cage-deployed fathead minnows (*Pimephales promelas*) at sites across the Great Lakes basin noted that the profiles of endogenous polar metabolites had covariance with at most 49 contaminants (Davis et al., 2016). Taking organisms from the field takes into consideration the natural variation from different locations, for example, there are correlations between the metabolome of the pine tree (*Pinus pinaster*) and its original geographic origin even when grown in a common garden for 5 years (Meijón et al., 2016). Sampling organisms from the field also considers the individual variability, which may stem from genetic differences, in the metabolic profile of these organisms as they are not only one laboratory strain (Quina et al., 2019). Understanding the metabolic response of organisms which are sampled from the field may be challenging because field populations may be exposed to multiple stressors at once, including both abiotic and biotic stressors as well as contaminant stressors (Garreta-Lara et al., 2018; Mishra et al., 2019). The choice of organisms should be as close as possible to identical age and size to minimize natural variation in populations (Coppola et al., 2018). However, when sampling organisms from the field it may be difficult to distinguish gender, different life stages, and species, for instance, DNA barcoding is used to distinguish species of the genus *Atlantoscia*, a terrestrial isopod (Zimmermann et al., 2018). Organisms sampled from the environment have various factors that may impact the metabolome which should be recorded, such as season and geographic location (Wei et al., 2018), climate (Gargallo-Garriga et al., 2015), disease (Zacher et al., 2018), and habitat surroundings (Quina et al., 2019). Also, it is a challenge to be certain of all the current environmental stressors present at the time of collection and the history of organism exposure to environmental stressors (Olsvik et al., 2018). Field-based studies also face the difficultly of locating an ideal reference site that can act as a control which has minimal abiotic, biotic, and contaminant stressors but where the natural settings are analogous to the impacted site (Martyniuk, 2018). However, to aid in the metabolomics analysis and interpretation of field-based studies, there are standard reporting requirements for metabolomics experiments of biological samples taken from the environment (Morrison et al., 2007).

Laboratory populations are strictly controlled, and temperature, light, and diet are maintained constantly to limit any perturbations to the metabolome. Laboratory-based studies have the benefit of standard protocols that are available, for instance, the Organisation for Economic Co-operation and Development provides guidance documents for aquatic toxicity testing and soil toxicity testing that should be followed for laboratory work (OECD, 2002, 2004). Controlled laboratory conditions are optimal for determining the mode of action of contaminants as the metabolic perturbations can be directly accredited to the stressor (Gómez-Canela et al., 2016). For this reason there are standard reporting requirements for metabolomics experiments of mammalian/*in vivo* work (Griffin et al., 2007) and for metabolomics experiments of microbial and *in vitro* work (van der Werf et al., 2007). The route of exposure, such as dosing, air-borne inhalation, aqueous exposure, and others, should be carefully selected and documented (van der Werf et al., 2007). Although the metabolic disturbances from exposure to stressors in laboratory conditions can be extrapolated to predict the hazard in environmental settings, care must be taken when doing so to avoid an overestimation or an underestimation of risk (Murphy et al., 2018).

4.3 Sample collection, sample preparation, and metabolite extraction

The decision of which biological sample to analyze is important since different metabolic responses can be found across cells, tissues, organs, biological fluids, and the whole organism (Tavassoly et al., 2018). There are advantages of choosing each biological compartment. For instance, using cell cultures provides specific information on the metabolic pathways involving the changes of endogenous metabolites in cells, and different cell lines may have different metabolic alterations (Rodrigues et al., 2019). The metabolomics analysis of a certain tissue can give information about the state of an organ and different tissues of an organism contain different baseline metabolic profiles (Cappello et al., 2018). The collection of a biofluid such as blood or urine is minimally invasive and can provide information about the overall biological state of an organism (Liao et al., 2018).

Compared to DNA sequences which can be extracted from samples over 100,000 years old (Rohland et al., 2018), metabolite samples are highly unstable and there may be additional changes in metabolite levels from residual enzymatic activity (Bernini et al., 2011). Therefore, proper sample preparation is important to avoid any disturbances in the metabolome that may occur from preparing samples. Sample preparation procedures may need to quickly halt enzymatic activity with freezing in liquid nitrogen and samples must be kept cold throughout any storage (Bernini et al., 2011). Some sample types need to have water removed for analysis by NMR spectroscopy, for instance, tissue samples need to be lyophilized soon after collection and the solvent extracts of cell cultures need to be dried and reconstituted in an NMR buffer (Beckonert et al., 2007; Carrola et al., 2018). NMR-based metabolomics most frequently uses a deuterium oxide phosphate-based buffer which extracts polar metabolites (Beckonert et al., 2007). The phosphate buffer maintains a constant pH to minimize variations in chemical shift and a deuterated solvent allows for a lock signal in NMR spectroscopy (Schripsema, 2010). The metabolite extraction procedure for MS analysis typically uses solvents of varying polarity to extract polar and nonpolar metabolites and several extraction methods are based on the widely used Bligh and Dyer extraction (Bligh and Dyer, 1959; Wu et al., 2008). The metabolite extraction method must be developed and verified because one extraction protocol is usually not able to extract all the metabolites from the sample. Also, the metabolite extraction method depends on the biological matrix chosen to be analyzed, and there are well-developed protocols for metabolite extractions from mammalian and bacterial cell cultures (Dietmair et al., 2010; Winder et al., 2008), biological fluids such as blood and urine (Beckonert et al., 2007; Bruce et al., 2009), and plant- and animal-derived tissues (Valledor et al., 2014; Wu et al., 2008). To extract metabolites from cells, cold solvents such as aqueous methanol or aqueous acetonitrile are applied to cell cultures to quench the metabolism of cells (Dietmair et al., 2010; Rodrigues et al., 2019). For biological fluids such as blood, there is usually a protein precipitation step to isolate proteins from the metabolite sample by treatment with a solvent mixture such as methanol/ethanol or methanol/acetonitrile/acetone (Bruce et al., 2009). Urine samples typically have simple sample preparation and involve sample dilution with appropriate solvents or ultrafiltration (Khamis et al., 2017). For tissue samples or whole organism homogenates, there is manual grinding or homogenization with precooled extraction solvents such as methanol (Römisch-Margl et al., 2012). Tissue samples or small whole organism homogenates may be further sonicated to enhance the extraction efficiency (Wang et al., 2018b). Finally, centrifugation can separate the supernatant-containing metabolites from large molecules such as precipitated proteins and cellular and tissue debris which are pelleted from centrifugation

(Hamerly et al., 2015). Besides, the order of metabolite extraction and data collection from samples also should be randomized to avoid any bias or batch impacts.

4.4 Analytical techniques for data collection

The general analytical methods used in environmental metabolomics are MS and NMR spectroscopy. Fig. 1.2 illustrates the workflow in an environmental metabolomics study

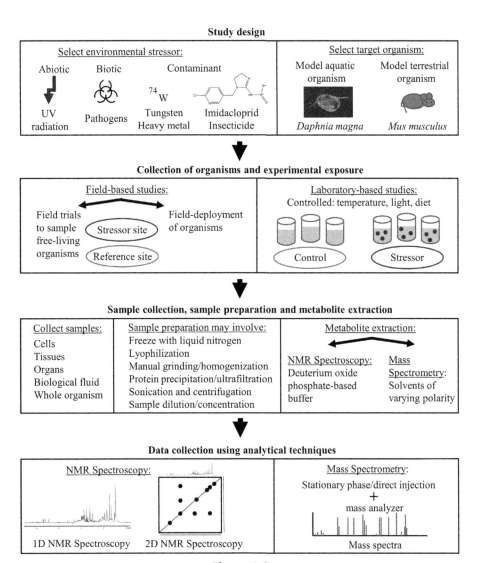

Figure 1.2

Workflow I of a generic environmental metabolomics study: from study design to data collection on the analytical instrument of choice.

beginning from study design to data collection on the analytical instrument of choice. Both NMR spectroscopy and MS analytical techniques encounter some challenges, and there is no single analytical technique that can identify and quantify the whole metabolome. Several studies propose direct infusion mass spectrometry (DIMS) coupled to high resolution mass analyzers such as Orbitrap mass analyzers or Fourier transform—ion cyclotron resonance (FT-ICR) mass analyzers because of the rapid and nontargeted analysis of a wide range of metabolites (Ghaste et al., 2016; Southam et al., 2017; Taylor et al., 2016). However, ion suppression and the complexity of the mass spectra without sample separation with a stationary phase are disadvantages of DIMS (Ghaste et al., 2016). Therefore, MS is usually coupled to liquid chromatography, gas chromatography, or capillary electrophoresis for analyte separation prior to mass detection (Beale et al., 2018; Gorrochategui et al., 2016; Sasaki et al., 2018). The stationary phase can aid in metabolite identification using the retention time of the analyte of interest (Chekmeneva et al., 2018). Gas chromatography—mass spectrometry (GC-MS) of polar metabolites typically requires chemical derivatization of metabolite extract samples to form volatile and thermally stable analytes and to also enhance the separation and detection of these metabolites (Miyagawa and Bamba, 2019). However, the quantitation of derivatized metabolites by GC-MS analysis may be difficult as the chemical derivatization efficiency may vary and the stability of the derivatized metabolites may not be constant (Miyagawa and Bamba, 2019). Metabolomics studies frequently use GC-MS with electron impact or chemical ionization (Li et al., 2015). Capillary electrophoresis—mass spectrometry (CE-MS) involves the separation of highly polar and ionic metabolites due to different migration speeds of the metabolites (Ramautar et al., 2019). CE-MS—based metabolomics studies often couple capillary electrophoresis to a time-of-flight (TOF) mass analyzer (Ramautar et al., 2019). Liquid chromatography—mass spectrometry (LC-MS) analysis includes the selection of an appropriate liquid chromatographic column, solvents for the mobile phases, potential additives to the mobile phase, and mobile-phase gradients for separation by liquid chromatography (Wang et al., 2018b). Most MS-based metabolomics studies use liquid chromatography with electrospray ionization (ESI) that is suitable for the ionization of a range of metabolites (Aszyk et al., 2018). Nontargeted metabolomics studies for global profiling frequently use high resolution and high accuracy mass analyzers such as FT-ICR mass analyzers or quadrupole-time-of-flight (QTOF) mass analyzers (Aszyk et al., 2018). MS-based metabolomics studies often use triple quadrupole mass spectrometers for targeted analysis due to the high sensitivity and selectivity (Nagana Gowda and Djukovic, 2014).

The main advantages of using MS for environmental metabolomics applications is that MS can detect and quantify metabolites at very low concentrations in the femtomolar to attomolar range and MS has a high dynamic range and resolution (Marshall and Powers, 2017). A disadvantage is that MS only detects metabolites that can promptly ionize, for instance, up to 60% of compounds from a reference chemical library were detectable by ESI-MS (Copeland et al., 2012). Ion suppression or ion enhancement may occur in metabolomics samples with coeluting compounds and a complex matrix and, as a result, MS has issues in reproducibility (Engskog et al., 2016). Another downside of MS-based metabolomics is that it may be difficult to know in advance the most optimal ionization technique and ionization polarity to select for if the composition of the sample is unknown, which may be a circumstance during nontargeted MS analysis (Panuwet et al., 2016). Also, metabolites may have retention time drifts during separation in the stationary phase and there may be instability in MS detection due to contamination of the ion source of the mass analyzer (Zelena et al., 2009).

NMR spectroscopy can identify unknown metabolites unambiguously, can distinguish isomers, and is excellent for structure elucidation of unknown compounds (Bingol and Bruschweiler, 2015). The most commonly used NMR method in environmental metabolomics is one-dimensional (1D) proton (^{1}H) NMR spectroscopy (Amberg et al., 2017). NMR spectroscopy is highly quantitative and reproducible, for example, the technical variation within metabolomics data sets using 1D ^{1}H NMR spectroscopy can be as low as 1.6% median spectral relative standard deviation (Bharti and Roy, 2012; Parsons et al., 2009). Water suppression is used with NMR spectroscopy to avoid baseline distortions and large errors in peak areas since water is at a substantially higher concentration than metabolites in sample extracts (Giraudeau et al., 2015). There are several water suppression techniques such as WATERGATE (Piotto et al., 1992), excitation sculpting (Hwang and Shaka, 1995), and presaturation utilizing relaxation gradients and echoes (Simpson and Brown, 2005). NMR spectroscopy has no need for sample separation by chromatography and often requires minimal sample preparation such as for urine samples (Takis et al., 2019). Only a single internal reference is needed for the absolute quantitation of metabolites using NMR spectroscopy (Nagana Gowda et al., 2018). Another advantage of NMR spectroscopy is that it has nondestructive data acquisition, for instance, NMR methods such as ^{1}H high resolution magic-angle spinning NMR (^{1}H HR-MAS NMR) can be performed on intact tissues (Battini et al., 2017). *In vivo* NMR spectroscopy using flow-based systems with solution-state NMR (Tabatabaei Anaraki et al., 2018), HR-MAS NMR (Bunescu et al., 2010), and comprehensive multiphase NMR (Mobarhan et al., 2016) is also emerging for environmental metabolomics applications of living organisms. A disadvantage of NMR spectroscopy is that only the most abundant metabolites at concentrations greater than 1 μM are detected (Nagana Gowda and Raftery, 2015). Also, the resolution in 1D ^{1}H NMR spectroscopy is low and although two-dimensional (2D) NMR methods increase resolution, the acquisition time of 2D NMR spectra is longer (Nagana Gowda and Raftery, 2015). However, there are developments that can increase the sensitivity and resolution of NMR spectroscopy for metabolomics analysis. These include technical improvements in pulse sequences such as *para*-H$_2$-induced hyperpolarization (Reile et al., 2016), cryoprobe technology, and microprobe technology (Saborano et al., 2019). Sample volumes are minimized from 500−600 µL using standard 5 mm NMR probes to 25−45 µL using 1.7 mm microprobes which is beneficial for mass-limited samples, and concentrating the sample can improve the signal-to-noise ratio (Martin, 2005; Nagato et al., 2015). The sensitivity gain can be up to 40-fold for samples of equal mass when transitioning from a 5 mm room temperature probe to a 1.7 mm microcryoprobe (Saborano et al., 2019). In addition, microcoil NMR has been used for the analysis of single animal eggs of the nematode *Heligmosomoides polygyrus bakeri* and the tardigrade *Richtersius coronifer* (Grisi et al., 2017) as well as the fish *Cypselurus poecilopterus* and the water flea *D. magna* (Fugariu et al., 2017). Microcoil NMR has the ability to analyze biological samples with a volume of approximately 0.1 nL (Grisi et al., 2017), and collecting an NMR spectrum on a 50 µm coil gives 3000 times the mass sensitivity compared to a 5 mm NMR probe (Fugariu et al., 2017).

4.5 *Raw data preprocessing and data analysis*

Raw data preprocessing of spectra collected from NMR spectroscopy or MS is done before statistical analysis and pattern detection with chemometrics models. Fourier transformation of the free induction decay signal produces the NMR spectrum and typically a line broadening of 0.3 Hz is used for ^{1}H NMR spectra (Kostidis et al., 2017). Each NMR spectrum should be manually phased for metabolomics studies since many automatic phasing methods may distort signals with small peak areas (Emwas et al., 2018). Next, baseline correction is performed, and chemical shifts are aligned to an internal calibrant such as 4,4-dimethyl-4-silapentane-1-sulfonic acid or 3-(trimethylsilyl)-2,2′,3,3′-tetradeuteropropionic acid (Dona et al., 2016). The NMR spectra undergo binning, also called bucketing, most commonly into equal widths of 0.02 or 0.04 ppm for ^{1}H NMR spectra and the signal in each bin is integrated (Vignoli et al., 2018). Binning the data allows multivariate methods to identify the regions of the NMR spectra that are most responsible for the variance within the data set (Karaman, 2017). With ^{1}H NMR spectra of polar metabolites, the region between 4.7 and 4.9 ppm is typically excluded from analysis because this region may add variance to the data due to residual water signals, and the regions outside of 0.5−10 ppm are excluded from analysis as these regions are comprised of mostly noise (Kim et al., 2016). Data scaling to reduce sample-to-sample variation is often carried out by normalization to total NMR spectral area, where the integrated value of one bin of the NMR spectrum is divided by the sum of the integrated values of all the spectral bins in one spectrum (Zacharias et al., 2018). Next, metabolites such as carbohydrates, amino acids, organic acids, alcohols, osmolytes, and nucleotides are identified by 1D ^{1}H and 2D NMR spectroscopy (Everett, 2015; Gu et al., 2019; Watanabe et al., 2015). The relative quantitation of metabolites can be completed as the signal intensity of an ^{1}H NMR peak is directly proportional to the concentration and number of protons that contribute to that particular resonance signal (Kumar et al., 2014). Absolute quantitation of ^{1}H NMR peaks can be performed by manually adding a specific concentration of one internal standard (Nagana Gowda et al., 2018), by adding an external standard in a coaxial NMR tube (Gardner et al., 2018), or using the Electronic REference To access In-vivo Concentration method to produce an artificial reference NMR signal (Akoka et al., 1999). MS analysis also has raw data preprocessing steps to simplify and convert a spectrum into peaks for further processing with multivariate tools. This includes centroiding to change each peak to one data point with a mass-to-charge and intensity value, as well as filtering methods to remove the baseline and measurement noise (Katajamaa and Orešič, 2007). Peak detection and deconvolution steps are implemented to assign different ions that belong to the same metabolite or to separate peaks from coeluting compounds (Ni et al., 2016). A preprocessing step for LC-MS or GC-MS data includes aligning the spectra along the chromatographic run time as shifts in retention time may occur due to column aging, the sample matrix, or changes in temperature, pressure, or the mobile phase (Watrous et al., 2017). Peak extraction and peak identification must be completed to quantify metabolites (Li et al., 2016). MS-based metabolomics also has data normalization methods to reduce sample-to-sample variability such as MS total useful signal normalization or probabilistic quotient normalization (Gagnebin et al., 2017).

The multivariate statistical analysis that is applied in environmental metabolomics studies is performed to simplify and facilitate data analysis in the often large data sets from NMR spectroscopy or MS. The commonly used method of principal component analysis (PCA) is an unsupervised statistical method that has no bias or knowledge about the treatment group of a sample (Lever et al., 2017). PCA is used to determine general patterns and visual differences in the metabolic profiles of different treatment groups (Gu et al., 2019). The outcome of a PCA model is a scores plot where each point in a two-dimensional or three-dimensional space is the spectrum of a single sample (Hu et al., 2017). The first few principal components explain most of the variation in the data set and most accurately reflect the content of the data (Lever et al., 2017). The clustering of samples in the PCA scores plot suggests that those samples have similar features in their metabolome (Hu et al., 2017). To easily identify group membership and visualize the data, an averaged PCA scores plot is made where the scores of all the spectra of one treatment group are averaged and plotted as average values with standard error (Wang et al., 2016). PCA loadings plots indicate the regions of the spectra that contribute to the variation in each principal component and hence indicate the metabolites that contribute to the metabolic variation between treatment groups (Lee et al., 2016).

Supervised statistical methods such as partial least squares (PLS), including PLS regression (PLS-r), PLS discriminant analysis (PLS-DA), and orthogonal (O)-PLS-DA models, use sample class as a factor to generate predictions about the data (Davis et al., 2017; Trygg et al., 2007). The data from classified samples are the training set, and these data are analyzed and used to predict the data from unclassified samples which are the test set (de Carvalho Rocha et al., 2018). PLS models require proper validation because these supervised techniques may overfit the data and result in the false appearance of group separation (Saccenti et al., 2014). For this reason, validation techniques are required to determine the optimal number of PLS components to avoid overfitting the data (Franitza et al., 2018; Li et al., 2018a). These include cross-validation methods such as leave-one-out cross validation, K-fold cross validation, Monte Carlo cross validation, and double cross validation (Li et al., 2018a). There is also an independent validation where a different data set is used to test the predictive ability of the PLS model (Franitza et al., 2018). To determine the statistical significance of the generated PLS model, a permutation test is frequently used which gives random group assignments to evaluate the prediction accuracy (Zhang et al., 2017b). For example, a PLS-DA model was created from data collected with ^{1}H NMR spectroscopy and underwent sevenfold cross validation and 200 permutation tests to distinguish the metabolomes of control rats from rats exposed to fine particulate matter (Zhang et al., 2017b).

In univariate statistical analysis, the individual metabolites are identified, and their absolute or relative quantification is completed. NMR spectroscopy has multiple methods for metabolite identification. In ^{1}H NMR-based metabolomics, many metabolites can be assigned from the chemical shift and peak multiplicity (Cappello et al., 2018). The Madison Metabolomics Consortium Database is a chemical shift database that is often used in metabolomics that has both theoretical and experimental and 1D and 2D NMR data as well as MS data (Cui et al., 2008). Metabolite identities can be validated with 2D NMR methods by using software packages to compare a chemical shift database of 2D NMR spectra to regions of 2D NMR spectra of biological samples (Nagato et al., 2015; Yuk et al., 2013). The 2D NMR spectra used for metabolite identification are mainly ^{1}H-^{1}H total correlation spectroscopy and ^{1}H-^{13}C heteronuclear single quantum coherence spectroscopy (HSQC) (Xia et al., 2008). Identification of unknowns using MS can be accomplished by matching high accurate mass experimental tandem mass spectra of an unknown metabolite to the best corresponding accurate mass and tandem mass spectral fragmentation pattern of all likely isomeric structures from a metabolite database (Boiteau et al., 2018; Duhrkop et al., 2015).

Metabolite data can be disclosed by reporting the absolute concentrations of metabolites across all treatment groups or by reporting the fold change of metabolites in an exposed group relative to a control group (Gómez-Canela et al., 2016). To determine if a metabolite level that was obtained from absolute quantification or relative quantification has a statistically significant change between two independent treatment groups, the main univariate statistical tests are the Student's *t*-test for data with a normality distribution (Mosley et al., 2018) and the Wilcoxon–Mann–Whitney test for data that are not normally distributed (Fay and Proschan, 2010; Pradhan et al., 2016). For example, a Student's *t*-test was applied to liquid chromatography tandem mass spectrometry (LC-MS/MS) data to determine significant changes in metabolites of the skin mucus of fathead minnows *P. promelas* after a 21 day exposure to treated wastewater effluent (Mosley et al., 2018). When more than two groups are being compared under the normality distribution, an analysis of variance (ANOVA) test with posthoc analysis can be used to determine significantly different metabolites for multiple group means (Fu et al., 2019; Kokushi et al., 2017; Ren et al., 2015). For instance, one-way ANOVA with Tukey's and Fisher's least significant difference tests were used to determine the statistical differences of metabolite changes in zebrafish *D. rerio* embryos from exposure to five concentrations of triclosan (Fu et al., 2019). The result of all these statistical tests is a *P*-value, and if this *P*-value is below a significance level that is generally taken to be 0.05, then the alternative hypothesis that there is a difference between two or more group means is accepted (Saccenti et al., 2014). Likewise, if the *P*-value is below the selected significance level of 0.05, then the null hypothesis that there is no significant difference between two groups is rejected (Saccenti et al., 2014).

4.6 Biological interpretation of the results

Understanding the implications of metabolite changes is challenging because metabolites are interconnected and can be part of several metabolic pathways (Caspi et al., 2015). Online databases of metabolic data include the Kyoto Encyclopedia of Genes and Genomes (KEGG) Pathway Database which shows many metabolic pathways and the individual metabolites in each step of a pathway (Kanehisa et al., 2016). The online database MetaboAnalyst 4.0 has several data interpretation tools for both NMR spectroscopy and MS data, such as metabolite pathway analysis (MetPA) where the changes of the individual metabolites are correlated to metabolic pathway information (Chong et al., 2018). The user inputs a list of metabolites and their abundances from two treatment groups and selects one of the 21 model organisms which MetaboAnalyst 4.0 has metabolite pathway library information from a database such as the KEGG (Xia and Wishart, 2010). Next, the parameters for pathway analysis are specified, which includes the algorithm for pathway enrichment analysis and the algorithm for topological analysis (Xia and Wishart, 2010). The output is presented as a map-style network visualization system and as a list of metabolic pathways in order of their statistical significance between the two treatment groups (Xia and Wishart, 2011). For instance, metabolic pathway analysis performed using MetPA determined that nine metabolic pathways were significantly altered in the soil nematode *Caenorhabditis elegans* from exposure to sublethal concentrations of titanium dioxide nanoparticles (Ratnasekhar et al., 2015).

Raw data preprocessing and data analysis

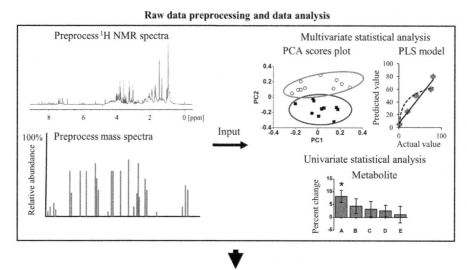

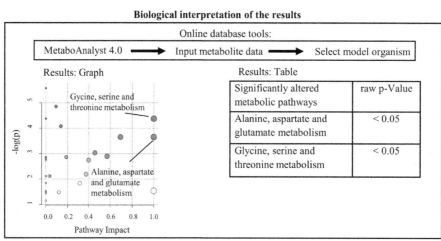

Figure 1.3

Workflow II of a generic environmental metabolomics study: from data preprocessing to biological interpretation with metabolite pathway analysis.

Fig. 1.3 illustrates the workflow involved in an environmental metabolomics study, beginning from data preprocessing to biological interpretation of the data using metabolite pathway analysis tools.

5. Exposure to contaminant mixtures

Contaminants exist in the environment as complex mixtures, and therefore efforts are being made to apply environmental metabolomics techniques to study exposures to contaminant mixtures or environmental samples such as wastewater treatment plant effluent (Van Meter et al., 2018; Wagner et al., 2019; Zhen et al., 2018). Environmental metabolomics studies of contaminant mixtures investigate the metabolic response from the combined sublethal exposure of contaminants (Ahmed et al., 2019; Van Meter et al., 2018). It is beneficial to evaluate the mode of action of the individual contaminants that compose a mixture of two or more contaminants to more effectively assess any interactive impacts on the metabolome (Altenburger et al., 2018). For example, the juvenile green frogs *Lithobates clamitans* were exposed to five pesticides as single, double, or triple pesticide mixtures, and metabolomics analysis with GC-MS revealed different changes in metabolites and metabolic pathways between the treatments (Van Meter et al., 2018). The risk assessment of a complex mixture that may have unknown constituents or that has tens or hundreds of contaminants is likely better achieved by evaluating the toxicity of the mixture as a whole (Kienzler et al., 2016). For this reason environmental metabolomics studies have examined the toxicity of environmental samples such as wastewater treatment plant effluent, ambient fine particulate matter, and contaminated sediments, and this is considered to be a whole mixture approach (Berlioz-Barbier et al., 2018; Chiu et al., 2017; Mosley et al., 2018; Song et al., 2019; Zhen et al., 2018). For example, nanoliquid chromatography coupled to quadrupole-time-of-flight-mass spectrometry (QTOF-MS) showed that the metabolome of the benthic invertebrate *Chironomus riparius* was impacted with mostly changes in lipid metabolism from both field and laboratory exposure to wastewater treatment plant effluents (Berlioz-Barbier et al., 2018). Understanding an organism's metabolic response after exposure to complex environmental samples is a challenge because the molecular-level responses to the individual contaminants are not always known. Also, there are other environmental factors that may impact the metabolic response and the bioavailability of contaminants such as dissolved oxygen concentration, food, temperature, salinity, pH, and organic matter content (Campillo et al., 2019; Garreta-Lara et al., 2018; Kovacevic et al., 2019; Mishra et al., 2019).

6. Summary and environmental biomonitoring efforts

Environmental metabolomics has made great progress since one of its first applications in 1997 using ^{1}H NMR spectroscopy to analyze changes in endogenous metabolites of earthworm exposed to copper (Gibb et al., 1997). Since then, there have been detailed characterizations of the mode of action of individual contaminants in target species using advanced analytical techniques such as 2D gas chromatography coupled to time-of-flight-mass spectrometry (TOF-MS) (Liu et al., 2018) and 2D ^{1}H-^{13}C HSQC NMR (Yuk et al., 2013). Metabolomics techniques are now being used to detect molecular-level changes in target organisms from exposure to environmentally relevant concentrations of contaminants in the ng/L to low µg/L range (Flores-Valverde et al., 2010; Ussery et al., 2018). The molecular-level responses of organisms exposed to sublethal contaminant mixtures are also being studied to understand the combined action of contaminants on the metabolome (Melvin et al., 2017). In addition, metabolomics approaches are applied to assess exposure to complex environmental samples such as soils contaminated with crude oil (Whitfield Åslund et al., 2013) or ambient fine particulate matter (Xu et al., 2019). Combining NMR spectroscopy and MS techniques for nontargeted and targeted analysis can provide a holistic view of the impacted metabolic pathways and the mode of action of contaminants (Zhang et al., 2017a). One of the end goals of environmental metabolomics research is to be applied as an early and sensitive warning signal for environmental biomonitoring and ecological risk assessment (Bahamonde et al., 2016). Progress toward this has been made using field studies where target organisms are cage deployed to contaminated sites and their metabolic profiles are compared with organisms which are cage deployed to less contaminated reference sites (Cappello et al., 2017; Ekman et al., 2018). Also, organisms such as mice and fish have been collected after lifelong existence in contaminated sites, and their metabolic profiles have been analyzed and compared to the metabolic profiles of organisms collected from less contaminated sites (Gauthier et al., 2018; Melvin et al., 2019; Quina et al., 2019). The concentrations of legacy and more recent synthetic organic contaminants in a field site may also be measured and related to the metabolic profiles of free dwelling organisms in that area (Morris et al., 2019). The aim of environmental metabolomics research is to be used as a bottom-up approach where changes in molecular parameters indicate the physiological function of organisms, which may then be related to the health of a population and ecosystem (Maloney, 2019). Eventually, the bioindicators of environmental stressors may be more reliably known as research in environmental metabolomics progresses. As a result, metabolite changes and biomarkers may be tracked with metabolomics techniques to allow for the rapid and routine monitoring of ecosystem health.

References

Ahmed, K.E.M., Frøysa, H.G., Karlsen, O.A., Blaser, N., Zimmer, K.E., Berntsen, H.F., Verhaegen, S., Ropstad, E., Kellmann, R., Goksøyr, A., 2019. Effects of defined mixtures of POPs and endocrine disruptors on the steroid metabolome of the human H295R adrenocortical cell line. Chemosphere 218, 328−339.

Akoka, S., Barantin, L., Trierweiler, M., 1999. Concentration measurement by proton NMR using the ERETIC method. Anal. Chem. 71, 2554−2557.

Altenburger, R., Scholze, M., Busch, W., Escher, B.I., Jakobs, G., Krauss, M., Krüger, J., Neale, P.A., Ait-Aissa, S., Almeida, A.C., Seiler, T.B., Brion, F., Hilscherova, K., Hollert, H., Novak, J., Schlichting, R., Serra, H., Shao, Y., Tindall, A., Tollefsen, K.E., Umbuzeiro, G., Williams, T.D., Kortenkamp, A., 2018.

Mixture effects in samples of multiple contaminants—An inter-laboratory study with manifold bioassays. Environ. Int. 114, 95—106.

Amberg, A., Riefke, B., Schlotterbeck, G., Ross, A., Senn, H., Dieterle, F., Keck, M., 2017. NMR and MS methods for metabolomics. In: Gautier, J.C. (Ed.), Drug Safety Evaluation. Methods in Molecular Biology. Humana Press, New York, NY, pp. 229—258.

Ankley, G.T., Erickson, R.J., Hoff, D.J., Mount, D.R., Lazorchak, J., Beaman, J., Linton, T.K., 2008. Aquatic Life Criteria for Contaminants of Emerging Concern: Part 1 General Challenges and Recommendations. OW/ORD Emerging Contaminants Workgroup EPA, pp. 1—46.

Aszyk, J., Byliński, H., Namieśnik, J., Kot-Wasik, A., 2018. Main strategies, analytical trends and challenges in LC-MS and ambient mass spectrometry-based metabolomics. Trends Anal. Chem. 108, 278—295.

Bahamonde, P.A., Feswick, A., Isaacs, M.A., Munkittrick, K.R., Martyniuk, C.J., 2016. Defining the role of omics in assessing ecosystem health: perspectives from the Canadian environmental monitoring program. Environ. Toxicol. Chem. 35, 20—35.

Bártíková, H., Podlipná, R., Skálová, L., 2016. Veterinary drugs in the environment and their toxicity to plants. Chemosphere 144, 2290—2301.

Battini, S., Faitot, F., Imperiale, A., Cicek, A.E., Heimburger, C., Averous, G., Bachellier, P., Namer, I.J., 2017. Metabolomics approaches in pancreatic adenocarcinoma: tumor metabolism profiling predicts clinical outcome of patients. BMC Med. 15, 56.

Beale, D.J., Pinu, F.R., Kouremenos, K.A., Poojary, M.M., Narayana, V.K., Boughton, B.A., Kanojia, K., Dayalan, S., Jones, O.A.H., Dias, D.A., 2018. Review of recent developments in gc—ms approaches to metabolomics-based research. Metabolomics 14, 152.

Beckonert, O., Keun, H.C., Ebbels, T.M., Bundy, J., Holmes, E., Lindon, J.C., Nicholson, J.K., 2007. Metabolic profiling, metabolomic and metabonomic procedures for NMR spectroscopy of urine, plasma, serum and tissue extracts. Nat. Protoc. 2, 2692—2703.

Berlioz-Barbier, A., Buleté, A., Fildier, A., Garric, J., Vulliet, E., 2018. Non-targeted investigation of benthic invertebrates (*Chironomus riparius*) exposed to wastewater treatment plant effluents using nanoliquid chromatography coupled to high-resolution mass spectrometry. Chemosphere 196, 347—353.

Bernhardt, E.S., Rosi, E.J., Gessner, M.O., 2017. Synthetic chemicals as agents of global change. Front. Ecol. Environ. 15, 84—90.

Bernini, P., Bertini, I., Luchinat, C., Nincheri, P., Staderini, S., Turano, P., 2011. Standard operating procedures for pre-analytical handling of blood and urine for metabolomic studies and biobanks. J. Biomol. NMR 49, 231—243.

Bharti, S.K., Roy, R., 2012. Quantitative ^{1}H NMR spectroscopy. Trends Anal. Chem. 35, 5—26.

Bingol, K., 2018. Recent advances in targeted and untargeted metabolomics by NMR and MS/NMR methods. High Throughput 7, 9.

Bingol, K., Bruschweiler, R., 2015. Two elephants in the room: new hybrid nuclear magnetic resonance and mass spectrometry approaches for metabolomics. Curr. Opin. Clin. Nutr. Metab. Care 18, 471—477.

Bligh, E.G., Dyer, W.J., 1959. A rapid method of total lipid extraction and purification. Can. J. Biochem. Physiol. 37, 911—917.

Boiteau, R.M., Hoyt, D.W., Nicora, C.D., Kinmonth-Schultz, H.A., Ward, J.K., Bingol, K., 2018. Structure elucidation of unknown metabolites in metabolomics by combined NMR and MS/MS prediction. Metabolites 8, 8.

Boxall, A.B., Rudd, M.A., Brooks, B.W., Caldwell, D.J., Choi, K., Hickmann, S., Innes, E., Ostapyk, K., Staveley, J.P., Verslycke, T., Ankley, G.T., Beazley, K.F., Belanger, S.E., Berninger, J.P., Carriquiriborde, P., Coors, A., Deleo, P.C., Dyer, S.D., Ericson, J.F., Gagne, F., Giesy, J.P., Gouin, T., Hallstrom, L., Karlsson, M.V., Larsson, D.G., Lazorchak, J.M., Mastrocco, F., McLaughlin, A., McMaster, M.E., Meyerhoff, R.D., Moore, R., Parrott, J.L., Snape, J.R., Murray-Smith, R., Servos, M.R., Sibley, P.K., Straub, J.O., Szabo, N.D., Topp, E., Tetreault, G.R., Trudeau, V.L., Van Der Kraak, G., 2012. Pharmaceuticals and personal care products in the environment: What are the big questions? Environ. Health Perspect. 120, 1221—1229.

Bruce, R.D., 1985. An up-and-down procedure for acute toxicity testing. Fund. Appl. Toxicol. 5, 151–157.

Bruce, S.J., Tavazzi, I., Parisod, V., Rezzi, S., Kochhar, S., Guy, P.A., 2009. Investigation of human blood plasma sample preparation for performing metabolomics using ultrahigh performance liquid chromatography/mass spectrometry. Anal. Chem. 81, 3285–3296.

Buckler, D.R., Mayer, F.L., Ellersieck, M.R., Asfaw, A., 2005. Acute toxicity value extrapolation with fish and aquatic invertebrates. Arch. Environ. Contam. Toxicol. 49, 546–558.

Bundy, J.G., Davey, M.P., Viant, M.R., 2009. Environmental metabolomics: a critical review and future perspectives. Metabolomics 5, 3–21.

Bunescu, A., Garric, J., Vollat, B., Canet-Soulas, E., Graveron-Demilly, D., Fauvelle, F., 2010. In vivo proton HR-MAS NMR metabolic profile of the freshwater cladoceran *Daphnia magna*. Mol. Biosyst. 6, 121–125.

Copeland, J.C., Zehr, L.J., Cerny, R.L., Powers, R., 2012. The applicability of molecular descriptors for designing an electrospray ionization mass spectrometry compatible library for drug discovery. Comb. Chem. High Throughput Screen. 15, 806–815.

Campillo, J.A., Sevilla, A., González-Fernández, C., Bellas, J., Bernal, C., Cánovas, M., Albentosa, M., 2019. Metabolomic responses of mussel *Mytilus galloprovincialis* to fluoranthene exposure under different nutritive conditions. Mar. Environ. Res. 144, 194–202.

Campos, A., Tedesco, S., Vasconcelos, V., Cristobal, S., 2012. Proteomic research in bivalves: towards the identification of molecular markers of aquatic pollution. J. Proteomics 75, 4346–4359.

Cappello, T., Giannetto, A., Parrino, V., Maisano, M., Oliva, S., De Marco, G., Guerriero, G., Mauceri, A., Fasulo, S., 2018. Baseline levels of metabolites in different tissues of mussel *Mytilus galloprovincialis* (*Bivalvia: mytilidae*). Comp. Biochem. Physiol. Genom. Proteonomics 26, 32–39.

Cappello, T., Maisano, M., Mauceri, A., Fasulo, S., 2017. ^{1}H NMR-based metabolomics investigation on the effects of petrochemical contamination in posterior adductor muscles of caged mussel *Mytilus galloprovincialis*. Ecotoxicol. Environ. Saf. 142, 417–422.

Carrola, J., Pinto, R.J.B., Nasirpour, M., Freire, C.S.R., Gil, A.M., Santos, C., Oliveira, H., Duarte, I.F., 2018. NMR metabolomics reveals metabolism-mediated protective effects in liver (HepG2) cells exposed to subtoxic levels of silver nanoparticles. J. Proteome Res. 17, 1636–1646.

Caspi, R., Billington, R., Ferrer, L., Foerster, H., Fulcher, C.A., Keseler, I.M., Kothari, A., Krummenacker, M., Latendresse, M., Mueller, L.A., Ong, Q., Paley, S., Subhraveti, P., Weaver, D.S., Karp, P.D., 2015. The MetaCyc database of metabolic pathways and enzymes and the BioCyc collection of pathway/genome databases. Nucleic Acids Res. 44, D471–D480.

Chekmeneva, E., dos Santos Correia, G., Gómez-Romero, M., Stamler, J., Chan, Q., Elliott, P., Nicholson, J.K., Holmes, E., 2018. Ultra-performance liquid chromatography–high-resolution mass spectrometry and direct infusion–high-resolution mass spectrometry for combined exploratory and targeted metabolic profiling of human urine. J. Proteome Res. 17, 3492–3502.

Chiu, K.H., Dong, C.D., Chen, C.F., Tsai, M.L., Ju, Y.R., Chen, T.M., Chen, C.W., 2017. NMR-based metabolomics for the environmental assessment of Kaohsiung Harbor sediments exemplified by a marine amphipod (*Hyalella azteca*). Mar. Pollut. Bull. 124, 714–724.

Chong, J., Soufan, O., Li, C., Caraus, I., Li, S., Bourque, G., Wishart, D.S., Xia, J., 2018. MetaboAnalyst 4.0: towards more transparent and integrative metabolomics analysis. Nucleic Acids Res. 46, W486–W494.

Coppola, F., Almeida, Â., Henriques, B., Soares, A.M.V.M., Figueira, E., Pereira, E., Freitas, R., 2018. Biochemical responses and accumulation patterns of *Mytilus galloprovincialis* exposed to thermal stress and Arsenic contamination. Ecotoxicol. Environ. Saf. 147, 954–962.

Cox, J.E., Thummel, C.S., Tennessen, J.M., 2017. Metabolomic studies in *Drosophila*. Genetics 206, 1169–1185.

Creydt, M., Vuralhan-Eckert, J., Fromm, J., Fischer, M., 2019. Effects of elevated CO_2 concentration on leaves and berries of black elder (*Sambucus nigra*) using UHPLC-ESI-QTOF-MS/MS and gas exchange measurements. J. Plant Physiol. 234, 71–79.

Cui, Q., Lewis, I.A., Hegeman, A.D., Anderson, M.E., Li, J., Schulte, C.F., Westler, W.M., Eghbalnia, H.R., Sussman, M.R., Markley, J.L., 2008. Metabolite identification via the madison metabolomics consortium database. Nat. Biotechnol. 26, 162−164.

Davis, J.M., Ekman, D.R., Skelton, D.M., LaLone, C.A., Ankley, G.T., Cavallin, J.E., Villeneuve, D.L., Collette, T.W., 2017. Metabolomics for informing adverse outcome pathways: androgen receptor activation and the pharmaceutical spironolactone. Aquat. Toxicol. 184, 103−115.

Davis, J.M., Ekman, D.R., Teng, Q., Ankley, G.T., Berninger, J.P., Cavallin, J.E., Jensen, K.M., Kahl, M.D., Schroeder, A.L., Villeneuve, D.L., Jorgenson, Z.G., Lee, K.E., Collette, T.W., 2016. Linking field-based metabolomics and chemical analyses to prioritize contaminants of emerging concern in the Great Lakes basin. Environ. Toxicol. Chem. 35, 2493−2502.

de Carvalho Rocha, W.F., Sheen, D.A., Bearden, D.W., 2018. Classification of samples from NMR-based metabolomics using principal components analysis and partial least squares with uncertainty estimation. Anal. Bioanal. Chem. 410, 6305−6319.

de Figueirêdo, L.P., Daam, M.A., Mainardi, G., Mariën, J., Espíndola, E.L.G., van Gestel, C.A.M., Roelofs, D., 2019. The use of gene expression to unravel the single and mixture toxicity of abamectin and difenoconazole on survival and reproduction of the springtail *Folsomia candida*. Environ. Pollut. 244, 342−350.

De Laender, F., Morselli, M., Baveco, H., Van den Brink, P.J., Di Guardo, A., 2015. Theoretically exploring direct and indirect chemical effects across ecological and exposure scenarios using mechanistic fate and effects modelling. Environ. Int. 74, 181−190.

Dietmair, S., Timmins, N.E., Gray, P.P., Nielsen, L.K., Krömer, J.O., 2010. Towards quantitative metabolomics of mammalian cells: development of a metabolite extraction protocol. Anal. Biochem. 404, 155−164.

Dona, A.C., Kyriakides, M., Scott, F., Shephard, E.A., Varshavi, D., Veselkov, K., Everett, J.R., 2016. A guide to the identification of metabolites in NMR-based metabonomics/metabolomics experiments. Comput. Struct. Biotechnol. J. 14, 135−153.

Duhrkop, K., Shen, H., Meusel, M., Rousu, J., Bocker, S., 2015. Searching molecular structure databases with tandem mass spectra using CSI:FingerID. Proc. Natl. Acad. Sci. U.S.A. 112, 12580−12585.

Ebele, A.J., Abdallah, M.A.E., Harrad, S., 2017. Pharmaceuticals and personal care products (PPCPs) in the freshwater aquatic environment. Emerg. Contam. 3, 1−16.

Ekman, D.R., Keteles, K., Beihoffer, J., Cavallin, J.E., Dahlin, K., Davis, J.M., Jastrow, A., Lazorchak, J.M., Mills, M.A., Murphy, M., Nguyen, D., Vajda, A.M., Villeneuve, D.L., Winkelman, D.L., Collette, T.W., 2018. Evaluation of targeted and untargeted effects-based monitoring tools to assess impacts of contaminants of emerging concern on fish in the South Platte River, CO. Environ. Pollut. 239, 706−713.

Emwas, A.H.M., 2015. The strengths and weaknesses of NMR spectroscopy and mass spectrometry with particular focus on metabolomics research. In: Bjerrum, J. (Ed.), Metabonomics. Methods in Molecular Biology. Humana Press, New York, NY, pp. 161−193.

Emwas, A.H., Saccenti, E., Gao, X., McKay, R.T., dos Santos, V.A.P.M., Roy, R., Wishart, D.S., 2018. Recommended strategies for spectral processing and post-processing of 1D ^{1}H-NMR data of biofluids with a particular focus on urine. Metabolomics 14, 31.

Engskog, M.K.R., Haglöf, J., Arvidsson, T., Pettersson, C., 2016. LC−MS based global metabolite profiling: the necessity of high data quality. Metabolomics 12, 114.

Everett, J.R., 2015. A new paradigm for known metabolite identification in metabonomics/metabolomics: metabolite identification efficiency. Comput. Struct. Biotechnol. J. 13, 131−144.

Farrés, M., Piña, B., Tauler, R., 2016. LC-MS based metabolomics and chemometrics study of the toxic effects of copper on *Saccharomyces cerevisiae*. Metallomics 8, 790−798.

Fay, M.P., Proschan, M.A., 2010. Wilcoxon-Mann-Whitney or t-test? On assumptions for hypothesis tests and multiple interpretations of decision rules. Stat. Surv. 4, 1−39.

Flores-Valverde, A.M., Horwood, J., Hill, E.M., 2010. Disruption of the steroid metabolome in fish caused by exposure to the environmental estrogen 17α-ethinylestradiol. Environ. Sci. Technol. 44, 3552−3558.

Franitza, L., Nicolotti, L., Granvogl, M., Schieberle, P., 2018. Differentiation of rums produced from sugar cane juice (rhum agricole) from rums manufactured from sugar cane molasses by a metabolomics approach. J. Agric. Food Chem. 66, 3038−3045.

Fu, J., Gong, Z., Kelly, B.C., 2019. Metabolomic profiling of zebrafish (*Danio rerio*) embryos exposed to the antibacterial agent triclosan. Environ. Toxicol. Chem. 38, 240−249.

Fugariu, I., Soong, R., Lane, D., Fey, M., Maas, W., Vincent, F., Beck, A., Schmidig, D., Treanor, B., Simpson, A.J., 2017. Towards single egg toxicity screening using microcoil NMR. Analyst 142, 4812−4824.

Gagnebin, Y., Tonoli, D., Lescuyer, P., Ponte, B., de Seigneux, S., Martin, P.Y., Schappler, J., Boccard, J., Rudaz, S., 2017. Metabolomic analysis of urine samples by UHPLC-QTOF-MS: impact of normalization strategies. Anal. Chim. Acta 955, 27−35.

Gao, K., Fu, J., Xue, Q., Fu, J., Fu, K., Zhang, A., Jiang, G., 2019. Direct determination of free state low molecular weight compounds in serum by online TurboFlow SPE HPLC-MS/MS and its application. Talanta 194, 960−968.

Gao, X., Huang, C., Rao, K., Xu, Y., Huang, Q., Wang, F., Ma, M., Wang, Z., 2018. Occurrences, sources, and transport of hydrophobic organic contaminants in the waters of Fildes Peninsula, Antarctica. Environ. Pollut. 241, 950−958.

Gardner, A., Parkes, H.G., Carpenter, G.H., So, P.W., 2018. Developing and standardizing a protocol for quantitative proton nuclear magnetic resonance (^{1}H NMR) spectroscopy of saliva. J. Proteome Res. 17, 1521−1531.

Gargallo-Garriga, A., Sardans, J., Pérez-Trujillo, M., Oravec, M., Urban, O., Jentsch, A., Kreyling, J., Beierkuhnlein, C., Parella, T., Peñuelas, J., 2015. Warming differentially influences the effects of drought on stoichiometry and metabolomics in shoots and roots. New Phytol. 207, 591−603.

Garreta-Lara, E., Campos, B., Barata, C., Lacorte, S., Tauler, R., 2018. Combined effects of salinity, temperature and hypoxia on *Daphnia magna* metabolism. Sci. Total Environ. 610, 602−612.

Garreta-Lara, E., Campos, B., Barata, C., Lacorte, S., Tauler, R., 2016. Metabolic profiling of *Daphnia magna* exposed to environmental stressors by GC−MS and chemometric tools. Metabolomics 12, 86.

Gauthier, P.T., Evenset, A., Christensen, G.N., Jorgensen, E.H., Vijayan, M.M., 2018. Lifelong exposure to PCBs in the remote Norwegian arctic disrupts the plasma stress metabolome in arctic charr. Environ. Sci. Technol. 52, 868−876.

Gavrilescu, M., Demnerová, K., Aamand, J., Agathos, S., Fava, F., 2015. Emerging pollutants in the environment: present and future challenges in biomonitoring, ecological risks and bioremediation. N. Biotechnol. 32, 147−156.

Gaworecki, K.M., Klaine, S.J., 2008. Behavioral and biochemical responses of hybrid striped bass during and after fluoxetine exposure. Aquat. Toxicol. 88, 207−213.

Ghaste, M., Mistrik, R., Shulaev, V., 2016. Applications of fourier transform ion cyclotron resonance (FT-ICR) and Orbitrap based high resolution mass spectrometry in metabolomics and lipidomics. Int. J. Mol. Sci. 17, 816.

Gibb, J.O.T., Svendsen, C., Weeks, J.M., Nicholson, J.K., 1997. ^{1}H NMR spectroscopic investigations of tissue metabolite biomarker response to Cu(II) exposure in terrestrial invertebrates: identification of free histidine as a novel biomarker of exposure to copper in earthworms. Biomarkers 2, 295−302.

Giraudeau, P., Silvestre, V., Akoka, S., 2015. Optimizing water suppression for quantitative NMR-based metabolomics: a tutorial review. Metabolomics 11, 1041−1055.

Gómez-Canela, C., Miller, T.H., Bury, N.R., Tauler, R., Barron, L.P., 2016. Targeted metabolomics of *Gammarus pulex* following controlled exposures to selected pharmaceuticals in water. Sci. Total Environ. 562, 777−788.

Gorrochategui, E., Jaumot, J., Lacorte, S., Tauler, R., 2016. Data analysis strategies for targeted and untargeted LC-MS metabolomic studies: overview and workflow. Trends Anal. Chem. 82, 425−442.

Griffin, J.L., Nicholls, A.W., Daykin, C.A., Heald, S., Keun, H.C., Schuppe-Koistinen, I., Griffiths, J.R., Cheng, L.L., Rocca-Serra, P., Rubtsov, D.V., Robertson, D., 2007. Standard reporting requirements for biological samples in metabolomics experiments: mammalian/in vivo experiments. Metabolomics 3, 179—188.

Grisi, M., Vincent, F., Volpe, B., Guidetti, R., Harris, N., Beck, A., Boero, G., 2017. NMR spectroscopy of single sub-nL ova with inductive ultra-compact single-chip probes. Sci. Rep. 7, 44670.

Gu, J., Su, F., Hong, P., Zhang, Q., Zhao, M., 2019. ^{1}H NMR-based metabolomic analysis of nine organophosphate flame retardants metabolic disturbance in Hep G2 cell line. Sci. Total Environ. 665, 162—170.

Hamerly, T., Tripet, B.P., Tigges, M., Giannone, R.J., Wurch, L., Hettich, R.L., Podar, M., Copié, V., Bothner, B., 2015. Untargeted metabolomics studies employing NMR and LC—MS reveal metabolic coupling between *Nanoarcheum equitans* and its archaeal host *Ignicoccus hospitalis*. Metabolomics 11, 895—907.

He, Z., Wang, Y., Zhang, Y., Cheng, H., Liu, X., 2018. Stereoselective bioaccumulation of chiral PCB 91 in earthworm and its metabolomic and lipidomic responses. Environ. Pollut. 238, 421—430.

Healy, M.G., Fenton, O., Cormican, M., Peyton, D.P., Ordsmith, N., Kimber, K., Morrison, L., 2017. Antimicrobial compounds (triclosan and triclocarban) in sewage sludges, and their presence in runoff following land application. Ecotoxicol. Environ. Saf. 142, 448—453.

Hernández, F., Ibáñez, M., Botero-Coy, A.M., Bade, R., Bustos-López, M.C., Rincón, J., Moncayo, A., Bijlsma, L., 2015. LC-QTOF MS screening of more than 1,000 licit and illicit drugs and their metabolites in wastewater and surface waters from the area of Bogotá, Colombia. Anal. Bioanal. Chem. 407, 6405—6416.

Hu, W.T., Guo, W.L., Meng, A.Y., Sun, Y., Wang, S.F., Xie, Z.Y., Zhou, Y.C., He, C., 2017. A metabolomic investigation into the effects of temperature on *Streptococcus agalactiae* from Nile tilapia (*Oreochromis niloticus*) based on UPLC—MS/MS. Vet. Microbiol. 210, 174—182.

Hwang, T.L., Shaka, A.J., 1995. Water suppression that works. Excitation sculpting using arbitrary wave-forms and pulsed-field gradients. J. Magn. Reson. 112, 275—279.

Jasinska, E.J., Goss, G.G., Gillis, P.L., Van Der Kraak, G.J., Matsumoto, J., de Souza Machado, A.A., Giacomin, M., Moon, T.W., Massarsky, A., Gagné, F., Servos, M.R., Wilson, J., Sultana, T., Metcalfe, C.D., 2015. Assessment of biomarkers for contaminants of emerging concern on aquatic organisms downstream of a municipal wastewater discharge. Sci. Total Environ. 530, 140—153.

Johnson, C.H., Ivanisevic, J., Siuzdak, G., 2016. Metabolomics: beyond biomarkers and towards mechanisms. Nat. Rev. Mol. Cell Biol. 17, 451—459.

Kanehisa, M., Furumichi, M., Tanabe, M., Sato, Y., Morishima, K., 2016. KEGG: new perspectives on genomes, pathways, diseases and drugs. Nucleic Acids Res. 45, D353—D361.

Kaneko, G., Ushio, H., Ji, H., 2019. Application of magnetic resonance technologies in aquatic biology and seafood science. Fish. Sci. 85, 1—17.

Karaman, I., 2017. Preprocessing and pretreatment of metabolomics data for statistical analysis. In: Sussulini, A. (Ed.), Metabolomics: from Fundamentals to Clinical Applications. Advances in Experimental Medicine and Biology. Springer, Cham, pp. 145—161.

Katajamaa, M., Orešič, M., 2007. Data processing for mass spectrometry-based metabolomics. J. Chromatogr. A 1158, 318—328.

Khamis, M.M., Adamko, D.J., El-Aneed, A., 2017. Mass spectrometric based approaches in urine metabolomics and biomarker discovery. Mass Spectrom. Rev. 36, 115—134.

Khan, N., Bano, A., Rahman, M.A., Rathinasabapathi, B., Babar, M.A., 2019. UPLC-HRMS-based untargeted metabolic profiling reveals changes in chickpea (*Cicer arietinum*) metabolome following long-term drought stress. Plant Cell Environ. 42, 115—132.

Kienzler, A., Bopp, S.K., van der Linden, S., Berggren, E., Worth, A., 2016. Regulatory assessment of chemical mixtures: requirements, current approaches and future perspectives. Regul. Toxicol. Pharmacol. 80, 321—334.

Kim, H.J., Koedrith, P., Seo, Y.R., 2015. Ecotoxicogenomic approaches for understanding molecular mechanisms of environmental chemical toxicity using aquatic invertebrate, *Daphnia* model organism. Int. J. Mol. Sci. 16, 12261−12287.

Kim, S., Yoon, D., Lee, M., Yoon, C., Kim, S., 2016. Metabolic responses in zebrafish (*Danio rerio*) exposed to zinc and cadmium by nuclear magnetic resonance-based metabolomics. Chem. Ecol. 32, 136−148.

Kokushi, E., Shintoyo, A., Koyama, J., Uno, S., 2017. Evaluation of 2,4-dichlorophenol exposure of Japanese medaka, *Oryzias latipes*, using a metabolomics approach. Environ. Sci. Pollut. Res. 24, 27678−27686.

Kostidis, S., Addie, R.D., Morreau, H., Mayboroda, O.A., Giera, M., 2017. Quantitative NMR analysis of intra- and extracellular metabolism of mammalian cells: a tutorial. Anal. Chim. Acta 980, 1−24.

Kovacevic, V., Simpson, A.J., Simpson, M.J., 2019. The concentration of dissolved organic matter impacts the metabolic response in *Daphnia magna* exposed to 17α-ethynylestradiol and perfluorooctane sulfonate. Ecotoxicol. Environ. Saf. 170, 468−478.

Kristofco, L.A., Cruz, L.C., Haddad, S.P., Behra, M.L., Chambliss, C.K., Brooks, B.W., 2016. Age matters: developmental stage of *Danio rerio* larvae influences photomotor response thresholds to diazinion or diphenhydramine. Aquat. Toxicol. 170, 344−354.

Kumar, V., Dwivedi, D.K., Jagannathan, N.R., 2014. High-resolution NMR spectroscopy of human body fluids and tissues in relation to prostate cancer. NMR Biomed. 27, 80−89.

Lallas, P.L., 2001. The stockholm convention on persistent organic pollutants. Am. J. Int. Law 95, 692−708.

Laurenson, J.P., Bloom, R.A., Page, S., Sadrieh, N., 2014. Ethinyl estradiol and other human pharmaceutical estrogens in the aquatic environment: a review of recent risk assessment data. AAPS J. 16, 299−310.

Lee, S.H., Wang, T.Y., Hong, J.H., Cheng, T.J., Lin, C.Y., 2016. NMR-based metabolomics to determine acute inhalation effects of nano- and fine-sized ZnO particles in the rat lung. Nanotoxicology 10, 924−934.

Lefrancq, M., Jadas-Hécart, A., La Jeunesse, I., Landry, D., Payraudeau, S., 2017. High frequency monitoring of pesticides in runoff water to improve understanding of their transport and environmental impacts. Sci. Total Environ. 587, 75−86.

Lepper, P., 2005. Manual on the Methodological Framework to Derive Environmental Quality Standards for Priority Substances in Accordance with Article 16 of the Water Framework Directive (2000/60/EC). Fraunhofer-Institute Molecular Biology and Applied Ecology, Schmallenberg, Germany.

Lever, J., Krzywinski, M., Altman, N., 2017. Points of significance: principal component analysis. Nat. Methods 14, 641−642.

Li, D.X., Gan, L., Bronja, A., Schmitz, O.J., 2015. Gas chromatography coupled to atmospheric pressure ionization mass spectrometry (GC-API-MS): review. Anal. Chim. Acta 891, 43−61.

Li, H., Cai, Y., Guo, Y., Chen, F., Zhu, Z.J., 2016. MetDIA: targeted metabolite extraction of multiplexed MS/MS spectra generated by data-independent acquisition. Anal. Chem. 88, 8757−8764.

Li, H.D., Xu, Q.S., Liang, Y.Z., 2018a. libPLS: an integrated library for partial least squares regression and linear discriminant analysis. Chemometr. Intell. Lab. Syst. 176, 34−43.

Li, M., Li, Y., Zhang, W., Li, S., Gao, Y., Ai, X., Zhang, D., Liu, B., Li, Q., 2018b. Metabolomics analysis reveals that elevated atmospheric CO_2 alleviates drought stress in cucumber seedling leaves. Anal. Biochem. 559, 71−85.

Liao, Y., Hu, R., Wang, Z., Peng, Q., Dong, X., Zhang, X., Zou, H., Pu, Q., Xue, B., Wang, L., 2018. Metabolomics profiling of serum and urine in three beef cattle breeds revealed different levels of tolerance to heat stress. J. Agric. Food Chem. 66, 6926−6935.

Lin, C.Y., Viant, M.R., Tjeerdema, R.S., 2006. Metabolomics: methodologies and applications in the environmental sciences. J. Pestic. Sci. 31, 245−251.

Liu, Y., Wang, X., Li, Y., Chen, X., 2018. Metabolomic analysis of short-term sulfamethazine exposure on marine medaka (*Oryzias melastigma*) by comprehensive two-dimensional gas chromatography-time-of-flight mass spectrometry. Aquat. Toxicol. 198, 269−275.

Loughery, J.R., Kidd, K.A., Mercer, A., Martyniuk, C.J., 2018. Part B: morphometric and transcriptomic responses to sub-chronic exposure to the polycyclic aromatic hydrocarbon phenanthrene in the fathead minnow (*Pimephales promelas*). Aquat. Toxicol. 199, 77−89.

Maloney, E.M., 2019. How do we take the pulse of an aquatic ecosystem? Current and historical approaches to measuring ecosystem integrity. Environ. Toxicol. Chem. 38, 289–301.

Marjan, P., Martyniuk, C.J., Fuzzen, M.L.M., MacLatchy, D.L., McMaster, M.E., Servos, M.R., 2017. Returning to normal? Assessing transcriptome recovery over time in male rainbow darter (*Etheostoma caeruleum*) liver in response to wastewater-treatment plant upgrades. Environ. Toxicol. Chem. 36, 2108–2122.

Markley, J.L., Brüschweiler, R., Edison, A.S., Eghbalnia, H.R., Powers, R., Raftery, D., Wishart, D.S., 2017. The future of NMR-based metabolomics. Curr. Opin. Biotechnol. 43, 34–40.

Marshall, D.D., Powers, R., 2017. Beyond the paradigm: combining mass spectrometry and nuclear magnetic resonance for metabolomics. Prog. Nucl. Magn. Reson. Spectrosc. 100, 1–16.

Martin, G.E., 2005. Small-volume and high-sensitivity NMR probes. Ann. Rep. NMR Spectrosc. 56, 1–96.

Martyniuk, C.J., 2018. Are we closer to the vision? A proposed framework for incorporating omics into environmental assessments. Environ. Toxicol. Pharmacol. 59, 87–93.

Martyniuk, C.J., Simmons, D.B., 2016. Spotlight on environmental omics and toxicology: a long way in a short time. Comp. Biochem. Physiol. Genom. Proteonomics 19, 97–101.

Mazzei, P., Vinale, F., Woo, S.L., Pascale, A., Lorito, M., Piccolo, A., 2016. Metabolomics by proton high-resolution magic-angle-spinning nuclear magnetic resonance of tomato plants treated with two secondary metabolites isolated from *trichoderma*. J. Agric. Food Chem. 64, 3538–3545.

McLachlan, M.S., 2018. Can the Stockholm Convention address the spectrum of chemicals currently under regulatory scrutiny? advocating a more prominent role for modeling in POP screening assessment. Environ. Sci. Process. Impacts 20, 32–37.

Meador, J.P., Yeh, A., Young, G., Gallagher, E.P., 2016. Contaminants of emerging concern in a large temperate estuary. Environ. Pollut. 213, 254–267.

Meijón, M., Feito, I., Oravec, M., Delatorre, C., Weckwerth, W., Majada, J., Valledor, L., 2016. Exploring natural variation of *Pinus pinaster* Aiton using metabolomics: is it possible to identify the region of origin of a pine from its metabolites? Mol. Ecol. 25, 959–976.

Melvin, S.D., Habener, L.J., Leusch, F.D., Carroll, A.R., 2017. ^{1}H NMR-based metabolomics reveals sub-lethal toxicity of a mixture of diabetic and lipid-regulating pharmaceuticals on amphibian larvae. Aquat. Toxicol. 184, 123–132.

Melvin, S.D., Lanctôt, C.M., Doriean, N.J.C., Bennett, W.W., Carroll, A.R., 2019. NMR-based lipidomics of fish from a metal(loid) contaminated wetland show differences consistent with effects on cellular membranes and energy storage. Sci. Total Environ. 654, 284–291.

Melvin, S.D., Lanctôt, C.M., Doriean, N.J.C., Carroll, A.R., Bennett, W.W., 2018. Untargeted NMR-based metabolomics for field-scale monitoring: temporal reproducibility and biomarker discovery in mosquitofish (*Gambusia holbrooki*) from a metal(loid)-contaminated wetland. Environ. Pollut. 243, 1096–1105.

Mishra, P., Gong, Z., Kelly, B.C., 2019. Assessing pH-dependent toxicity of fluoxetine in embryonic zebrafish using mass spectrometry-based metabolomics. Sci. Total Environ. 650, 2731–2741.

Miyagawa, H., Bamba, T., 2019. Comparison of sequential derivatization with concurrent methods for GC/MS-based metabolomics. J. Biosci. Bioeng. 127, 160–168.

Mobarhan, Y.L., Fortier-McGill, B., Soong, R., Maas, W.E., Fey, M., Monette, M., Stronks, H.J., Schmidt, S., Heumann, H., Norwood, W., Simpson, A.J., 2016. Comprehensive multiphase NMR applied to a living organism. Chem. Sci. 7, 4856–4866.

Morris, A.D., Letcher, R.J., Dyck, M., Chandramouli, B., Cosgrove, J., 2019. Concentrations of legacy and new contaminants are related to metabolite profiles in Hudson Bay polar bears. Environ. Res. 168, 364–374.

Morrison, N., Bearden, D., Bundy, J.G., Collette, T., Currie, F., Davey, M.P., Haigh, N.S., Hancock, D., Jones, O.A.H., Rochfort, S., Sansone, S.A., Štys, D., Teng, Q., Field, D., Viant, M.R., 2007. Standard reporting requirements for biological samples in metabolomics experiments: environmental context. Metabolomics 3, 203–210.

Mosley, J.D., Ekman, D.R., Cavallin, J.E., Villeneuve, D.L., Ankley, G.T., Collette, T.W., 2018. High-resolution mass spectrometry of skin mucus for monitoring physiological impacts and contaminant biotransformation products in fathead minnows exposed to wastewater effluent. Environ. Toxicol. Chem. 37, 788−796.

Mullard, G., Allwood, J.W., Weber, R., Brown, M., Begley, P., Hollywood, K.A., Jones, M., Unwin, R.D., Bishop, P.N., Cooper, G.J.S., Dunn, W.B., 2015. A new strategy for MS/MS data acquisition applying multiple data dependent experiments on Orbitrap mass spectrometers in non-targeted metabolomic applications. Metabolomics 11, 1068−1080.

Murphy, C.A., Nisbet, R.M., Antczak, P., Garcia-Reyero, N., Gergs, A., Lika, K., Mathews, T., Muller, E.B., Nacci, D., Peace, A., Remien, C.H., Schultz, I.R., Stevenson, L.M., Watanabe, K.H., 2018. Incorporating suborganismal processes into dynamic energy budget dodels for ecological risk assessment. Integrated Environ. Assess. Manag. 14, 615−624.

Nadal, M., Marquès, M., Mari, M., Domingo, J.L., 2015. Climate change and environmental concentrations of POPs: a review. Environ. Res. 143, 177−185.

Nagana Gowda, G.A., Djukovic, D., Bettcher, L.F., Gu, H., Raftery, D., 2018. NMR-guided mass spectrometry for absolute quantitation of human blood metabolites. Anal. Chem. 90, 2001−2009.

Nagana Gowda, G.A., Djukovic, D., 2014. Overview of mass spectrometry-based metabolomics: opportunities and challenges. In: Raftery, D. (Ed.), Mass Spectrometry in Metabolomics. Methods in Molecular Biology (Methods and Protocols). Humana Press, New York, NY, pp. 3−12.

Nagana Gowda, G.A., Raftery, D., 2015. Can NMR solve some significant challenges in metabolomics? J. Magn. Reson. 260, 144−160.

Nagato, E.G., Lankadurai, B.P., Soong, R., Simpson, A.J., Simpson, M.J., 2015. Development of an NMR microprobe procedure for high-throughput environmental metabolomics of *Daphnia magna*. Magn. Reson. Chem. 53, 745−753.

Net, S., Delmont, A., Sempéré, R., Paluselli, A., Ouddane, B., 2015. Reliable quantification of phthalates in environmental matrices (air, water, sludge, sediment and soil): a review. Sci. Total Environ. 515, 162−180.

Ni, Y., Su, M., Qiu, Y., Jia, W., Du, X., 2016. ADAP-GC 3.0: Improved peak detection and deconvolution of Co-eluting metabolites from GC/TOF-MS data for metabolomics studies. Anal. Chem. 88, 8802−8811.

Nõges, P., Argillier, C., Borja, Á., Garmendia, J.M., Hanganu, J., Kodeš, V., Pletterbauer, F., Sagouis, A., Birk, S., 2016. Quantified biotic and abiotic responses to multiple stress in freshwater, marine and ground waters. Sci. Total Environ. 540, 43−52.

OECD, 2002. Guidance Document on Aquatic Toxicity Testing of Difficult Substances and Mixtures. OECD Publishing, Paris, France, p. 53.

OECD, 2004. OECD Guideline for Testing of Chemicals No. 222, Earthworm Reproduction Test (*Eisenia fetida/Eisenia andrei*). OECD Publishing, Paris, France.

Olsvik, P.A., Aulin, M., Samuelsen, O.B., Hannisdal, R., Agnalt, A.L., Lunestad, B.T., 2018. Whole-animal accumulation, oxidative stress, transcriptomic and metabolomic responses in the pink shrimp (*Pandalus montagui*) exposed to teflubenzuron. J. Appl. Toxicol. 39, 485−497.

Omar, T.F.T., Ahmad, A., Aris, A.Z., Yusoff, F.M., 2016. Endocrine disrupting compounds (EDCs) in environmental matrices: review of analytical strategies for pharmaceuticals, estrogenic hormones, and alkylphenol compounds. Trends Anal. Chem. 85, 241−259.

Palazzotto, E., Weber, T., 2018. Omics and multi-omics approaches to study the biosynthesis of secondary metabolites in microorganisms. Curr. Opin. Microbiol. 45, 109−116.

Pang, H.Q., Yue, S.J., Tang, Y.P., Chen, Y.Y., Tan, Y.J., Cao, Y.J., Shi, X.Q., Zhou, G.S., Kang, A., Huang, S.L., Shi, Y.J., Sun, J., Tang, Z.S., Duan, J.A., 2018. Integrated metabolomics and network pharmacology approach to explain possible action mechanisms of Xin-Sheng-Hua Granule for treating anemia. Front. Pharmacol. 9, 165.

Panuwet, P., Hunter Jr., R.E., D'Souza, P.E., Chen, X., Radford, S.A., Cohen, J.R., Marder, M.E., Kartavenka, K., Ryan, P.B., Barr, D.B., 2016. Biological matrix effects in quantitative tandem mass spectrometry-based analytical methods: advancing biomonitoring. Crit. Rev. Anal. Chem. 46, 93−105.

Papadakis, E.N., Vryzas, Z., Kotopoulou, A., Kintzikoglou, K., Makris, K.C., Papadopoulou-Mourkidou, E., 2015. A pesticide monitoring survey in rivers and lakes of northern Greece and its human and ecotoxicological risk assessment. Ecotoxicol. Environ. Saf. 116, 1—9.

Parsons, H.M., Ekman, D.R., Collette, T.W., Viant, M.R., 2009. Spectral relative standard deviation: a practical benchmark in metabolomics. Analyst 134, 478—485.

Peng, W., Li, X., Xiao, S., Fan, W., 2018a. Review of remediation technologies for sediments contaminated by heavy metals. J. Soils Sediments 18, 1701—1719.

Peng, Y., Fang, W., Krauss, M., Brack, W., Wang, Z., Li, F., Zhang, X., 2018b. Screening hundreds of emerging organic pollutants (EOPs) in surface water from the Yangtze River Delta (YRD): occurrence, distribution, ecological risk. Environ. Pollut. 241, 484—493.

Peregrín-Alvarez, J.M., Sanford, C., Parkinson, J., 2009. The conservation and evolutionary modularity of metabolism. Genome Biol. 10, R63.

Pereira, V.J., da Cunha, J.P.A.R., de Morais, T.P., de Oliveira, J.P.R., de Morais, J.B., 2016. Physical-chemical properties of pesticides: concepts, applications, and interactions with the environment. Biosci. J. 32, 627—641.

Pidatala, V.R., Li, K., Sarkar, D., Ramakrishna, W., Datta, R., 2016. Identification of biochemical pathways associated with lead tolerance and detoxification in *Chrysopogon zizanioides* L. Nash (Vetiver) by metabolic profiling. Environ. Sci. Technol. 50, 2530—2537.

Piotto, M., Saudek, V., Sklenár, V., 1992. Gradient-tailored excitation for single-quantum NMR spectroscopy of aqueous solutions. J. Biomol. NMR 2, 661—665.

Pradhan, S.N., Das, A., Meena, R., Nanda, R.K., Rajamani, P., 2016. Biofluid metabotyping of occupationally exposed subjects to air pollution demonstrates high oxidative stress and deregulated amino acid metabolism. Sci. Rep. 6, 35972.

Prosser, R.S., Sibley, P.K., 2015. Human health risk assessment of pharmaceuticals and personal care products in plant tissue due to biosolids and manure amendments, and wastewater irrigation. Environ. Int. 75, 223—233.

Psychogios, N., Hau, D.D., Peng, J., Guo, A.C., Mandal, R., Bouatra, S., Sinelnikov, I., Krishnamurthy, R., Eisner, R., Gautam, B., 2011. The human serum metabolome. PLoS One 6 e16957.

Quina, A.S., Durão, A.F., Muñoz-Muñoz, F., Ventura, J., da Luz Mathias, M., 2019. Population effects of heavy metal pollution in wild Algerian mice (*Mus spretus*). Ecotoxicol. Environ. Saf. 171, 414—424.

Ramautar, R., Somsen, G.W., de Jong, G.J., 2019. CE-MS for metabolomics: developments and applications in the period 2016—2018. Electrophoresis 40, 165—179.

Ramirez, T., Strigun, A., Verlohner, A., Huener, H.A., Peter, E., Herold, M., Bordag, N., Mellert, W., Walk, T., Spitzer, M., Jiang, X., Sperber, S., Hofmann, T., Hartung, T., Kamp, H., van Ravenzwaay, B., 2018. Prediction of liver toxicity and mode of action using metabolomics in vitro in HepG2 cells. Arch. Toxicol. 92, 893—906.

Ratnasekhar, C., Sonane, M., Satish, A., Mudiam, M.K.R., 2015. Metabolomics reveals the perturbations in the metabolome of *Caenorhabditis elegans* exposed to titanium dioxide nanoparticles. Nanotoxicology 9, 994—1004.

Reed, L.K., Baer, C.F., Edison, A.S., 2017. Considerations when choosing a genetic model organism for metabolomics studies. Curr. Opin. Chem. Biol. 36, 7—14.

Reile, I., Eshuis, N., Hermkens, N.K.J., van Weerdenburg, B.J.A., Feiters, M.C., Rutjes, F.P.J.T., Tessari, M., 2016. NMR detection in biofluid extracts at sub-μM concentrations via para-H_2 induced hyperpolarization. Analyst 141, 4001—4005.

Ren, S., Hinzman, A.A., Kang, E.L., Szczesniak, R.D., Lu, L.J., 2015. Computational and statistical analysis of metabolomics data. Metabolomics 11, 1492—1513.

Revel, M., Châtel, A., Mouneyrac, C., 2017. Omics tools: new challenges in aquatic nanotoxicology? Aquat. Toxicol. 193, 72—85.

Rice, J., Westerhoff, P., 2017. High levels of endocrine pollutants in US streams during low flow due to insufficient wastewater dilution. Nat. Geosci. 10, 587—591.

Rodrigues, D., Pinto, J., Araújo, A.M., Jerónimo, C., Henrique, R., Bastos, M.d.L., Guedes de Pinho, P., Carvalho, M., 2019. GC-MS metabolomics reveals distinct profiles of low-and high-grade bladder cancer cultured cells. Metabolites 9, 18.

Rohland, N., Glocke, I., Aximu-Petri, A., Meyer, M., 2018. Extraction of highly degraded DNA from ancient bones, teeth and sediments for high-throughput sequencing. Nat. Protoc. 13, 2447−2461.

Römisch-Margl, W., Prehn, C., Bogumil, R., Röhring, C., Suhre, K., Adamski, J., 2012. Procedure for tissue sample preparation and metabolite extraction for high-throughput targeted metabolomics. Metabolomics 8, 133−142.

Roszkowska, A., Yu, M., Bessonneau, V., Bragg, L., Servos, M., Pawliszyn, J., 2018. Metabolome profiling of fish muscle tissue exposed to benzo[*a*]pyrene using in vivo solid-phase microextraction. Environ. Sci. Technol. Lett. 5, 431−435.

Saborano, R., Eraslan, Z., Roberts, J., Khanim, F.L., Lalor, P.F., Reed, M.A.C., Günther, U.L., 2019. A framework for tracer-based metabolism in mammalian cells by NMR. Sci. Rep. 9, 2520.

Saccenti, E., Hoefsloot, H.C.J., Smilde, A.K., Westerhuis, J.A., Hendriks, M.M.W.B., 2014. Reflections on univariate and multivariate analysis of metabolomics data. Metabolomics 10, 361−374.

Sasaki, K., Sagawa, H., Suzuki, M., Yamamoto, H., Tomita, M., Soga, T., Ohashi, Y., 2018. Metabolomics platform with capillary electrophoresis coupled with high-resolution mass spectrometry for plasma analysis. Anal. Chem. 91, 1295−1301.

Sauvé, S., Desrosiers, M., 2014. A review of what is an emerging contaminant. Chem. Cent. J. 8, 15.

Schaeffer, A., Amelung, W., Hollert, H., Kaestner, M., Kandeler, E., Kruse, J., Miltner, A., Ottermanns, R., Pagel, H., Peth, S., Poll, C., Rambold, G., Schloter, M., Schulz, S., Streck, T., Roß-Nickoll, M., 2016. The impact of chemical pollution on the resilience of soils under multiple stresses: a conceptual framework for future research. Sci. Total Environ. 568, 1076−1085.

Schripsema, J., 2010. Application of NMR in plant metabolomics: techniques, problems and prospects. Phytochem. Anal. 21, 14−21.

Shin, T.H., Lee, D.Y., Lee, H.S., Park, H.J., Jin, M.S., Paik, M.J., Manavalan, B., Mo, J.S., Lee, G., 2018. Integration of metabolomics and transcriptomics in nanotoxicity studies. BMB Rep. 51, 14−20.

Simmons, D.B.D., Benskin, J.P., Cosgrove, J.R., Duncker, B.P., Ekman, D.R., Martyniuk, C.J., Sherry, J.P., 2015. Omics for aquatic ecotoxicology: control of extraneous variability to enhance the analysis of environmental effects. Environ. Toxicol. Chem. 34, 1693−1704.

Simpson, A.J., Brown, S.A., 2005. Purge NMR: effective and easy solvent suppression. J. Magn. Reson. 175, 340−346.

Sivaram, A.K., Subashchandrabose, S.R., Logeshwaran, P., Lockington, R., Naidu, R., Megharaj, M., 2019. Metabolomics reveals defensive mechanisms adapted by maize on exposure to high molecular weight polycyclic aromatic hydrocarbons. Chemosphere 214, 771−780.

Skliros, D., Kalloniati, C., Karalias, G., Skaracis, G.N., Rennenberg, H., Flemetakis, E., 2018. Global metabolomics analysis reveals distinctive tolerance mechanisms in different plant organs of lentil (*Lens culinaris*) upon salinity stress. Plant Soil 429, 451−468.

Snape, J.R., Maund, S.J., Pickford, D.B., Hutchinson, T.H., 2004. Ecotoxicogenomics: the challenge of integrating genomics into aquatic and terrestrial ecotoxicology. Aquat. Toxicol. 67, 143−154.

Song, X., Wen, Y., Wang, Y., Adeel, M., Yang, Y., 2018. Environmental risk assessment of the emerging EDCs contaminants from rural soil and aqueous sources: analytical and modelling approaches. Chemosphere 198, 546−555.

Song, Y., Li, R., Zhang, Y., Wei, J., Chen, W., Chung, C.K.A., Cai, Z., 2019. Mass spectrometry-based metabolomics reveals the mechanism of ambient fine particulate matter and its components on energy metabolic reprogramming in BEAS-2B cells. Sci. Total Environ. 651, 3139−3150.

Southam, A.D., Weber, R.J.M., Engel, J., Jones, M.R., Viant, M.R., 2017. A complete workflow for high-resolution spectral-stitching nanoelectrospray direct-infusion mass-spectrometry-based metabolomics and lipidomics. Nat. Protoc. 12, 310−328.

Sun, T.Y., Bornhöft, N.A., Hungerbühler, K., Nowack, B., 2016. Dynamic probabilistic modeling of environmental emissions of engineered nanomaterials. Environ. Sci. Technol. 50, 4701−4711.

Sung, J., Lee, S., Lee, Y., Ha, S., Song, B., Kim, T., Waters, B.M., Krishnan, H.B., 2015. Metabolomic profiling from leaves and roots of tomato (*Solanum lycopersicum* L.) plants grown under nitrogen, phosphorus or potassium-deficient condition. Plant Sci. 241, 55−64.

Tabatabaei Anaraki, M., Dutta Majumdar, R., Wagner, N., Soong, R., Kovacevic, V., Reiner, E.J., Bhavsar, S.P., Ortiz Almirall, X., Lane, D., Simpson, M.J., Heumann, H., Schmidt, S., Simpson, A.J., 2018. Development and application of a low-volume flow system for solution-state in vivo NMR. Anal. Chem. 90, 7912−7921.

Takis, P.G., Ghini, V., Tenori, L., Turano, P., Luchinat, C., 2019. Uniqueness of the NMR approach to metabolomics. Trends Anal. Chem. 120, 115300.

Tang, R., Ding, C., Ma, Y., Wang, J., Zhang, T., Wang, X., 2017. Time-dependent responses of earthworms to soil contaminated with low levels of lead as detected using ^{1}H NMR metabolomics. RSC Adv. 7, 34170−34181.

Tavassoly, I., Goldfarb, J., Iyengar, R., 2018. Systems biology primer: the basic methods and approaches. Essays Biochem. 62, 487−500.

Taylor, N.S., Kirwan, J.A., Yan, N.D., Viant, M.R., Gunn, J.M., McGeer, J.C., 2016. Metabolomics confirms that dissolved organic carbon mitigates copper toxicity. Environ. Toxicol. Chem. 35, 635−644.

Thomaidi, V.S., Stasinakis, A.S., Borova, V.L., Thomaidis, N.S., 2016. Assessing the risk associated with the presence of emerging organic contaminants in sludge-amended soil: a country-level analysis. Sci. Total Environ. 548, 280−288.

Thome, C., Mitz, C., Sreetharan, S., Mitz, C., Somers, C.M., Manzon, R.G., Boreham, D.R., Wilson, J.Y., 2017. Developmental effects of the industrial cooling water additives morpholine and sodium hypochlorite on lake whitefish (*Coregonus clupeaformis*). Environ. Toxicol. Chem. 36, 1955−1965.

Tian, J., Ji, Z., Wang, F., Song, M., Li, H., 2017. The toxic effects of tetrachlorobisphenol A in *Saccharomyces cerevisiae* cells via metabolic interference. Sci. Rep. 7, 2655.

Tomonaga, S., Okuyama, H., Tachibana, T., Makino, R., 2018. Effects of high ambient temperature on plasma metabolomic profiles in chicks. Anim. Sci. J. 89, 448−455.

Toumi, H., Boumaiza, M., Millet, M., Radetski, C.M., Felten, V., Fouque, C., Férard, J.F., 2013. Effects of deltamethrin (pyrethroid insecticide) on growth, reproduction, embryonic development and sex differentiation in two strains of *Daphnia magna* (Crustacea, Cladocera). Sci. Total Environ. 458, 47−53.

Truhaut, R., 1977. Ecotoxicology: objectives, principles and perspectives. Ecotoxicol. Environ. Saf. 1, 151−173.

Trygg, J., Holmes, E., Lundstedt, T., 2007. Chemometrics in metabonomics. J. Proteome Res. 6, 469−479.

Ussery, E., Bridges, K.N., Pandelides, Z., Kirkwood, A.E., Bonetta, D., Venables, B.J., Guchardi, J., Holdway, D., 2018. Effects of environmentally relevant metformin exposure on Japanese medaka (*Oryzias latipes*). Aquat. Toxicol. 205, 58−65.

Valavanidis, A., Vlahogianni, T., Dassenakis, M., Scoullos, M., 2006. Molecular biomarkers of oxidative stress in aquatic organisms in relation to toxic environmental pollutants. Ecotoxicol. Environ. Saf. 64, 178−189.

Valledor, L., Escandón, M., Meijón, M., Nukarinen, E., Cañal, M.J., Weckwerth, W., 2014. A universal protocol for the combined isolation of metabolites, DNA, long RNA s, small RNA s, and proteins from plants and microorganisms. Plant J. 79, 173−180.

van den Brink, N.W., Arblaster, J.A., Bowman, S.R., Conder, J.M., Elliott, J.E., Johnson, M.S., Muir, D.C.G., Natal-da-Luz, T., Rattner, B.A., Sample, B.E., Shore, R.F., 2016. Use of terrestrial field studies in the derivation of bioaccumulation potential of chemicals. Integrated Environ. Assess. Manag. 12, 135−145.

van der Werf, M.J., Takors, R., Smedsgaard, J., Nielsen, J., Ferenci, T., Portais, J.C., Wittmann, C., Hooks, M., Tomassini, A., Oldiges, M., Fostel, J., Sauer, U., 2007. Standard reporting requirements for biological samples in metabolomics experiments: microbial and in vitro biology experiments. Metabolomics 3, 189−194.

Van Gestel, C.A.M., Van Brummelen, T.C., 1996. Incorporation of the biomarker concept in ecotoxicology calls for a redefinition of terms. Ecotoxicology 5, 217–225.

Van Meter, R.J., Glinski, D.A., Purucker, S.T., Henderson, W.M., 2018. Influence of exposure to pesticide mixtures on the metabolomic profile in post-metamorphic green frogs (*Lithobates clamitans*). Sci. Total Environ. 624, 1348–1359.

Vignoli, A., Ghini, V., Meoni, G., Licari, C., Takis, P.G., Tenori, L., Turano, P., Luchinat, C., 2018. High-throughput metabolomics by 1D NMR. Angew. Chem. Int. Ed. 58, 968–994.

von der Trenck, K.T., Konietzka, R., Biegel-Engler, A., Brodsky, J., Hädicke, A., Quadflieg, A., Stockerl, R., Stahl, T., 2018. Significance thresholds for the assessment of contaminated groundwater: perfluorinated and polyfluorinated chemicals. Environ. Sci. Eur. 30, 19.

Wagner, N.D., Helm, P.A., Simpson, A.J., Simpson, M.J., 2019. Metabolomic responses to pre-chlorinated and final effluent wastewater with the addition of a sub-lethal persistent contaminant in *Daphnia magna*. Environ. Sci. Pollut. Res. 1–13.

Wang, A., Holladay, S.D., Wolf, D.C., Ahmed, S.A., Robertson, J.L., 2006. Reproductive and developmental toxicity of arsenic in rodents: a review. Int. J. Toxicol. 25, 319–331.

Wang, D., Zhu, W., Chen, L., Yan, J., Teng, M., Zhou, Z., 2018a. Neonatal triphenyl phosphate and its metabolite diphenyl phosphate exposure induce sex- and dose-dependent metabolic disruptions in adult mice. Environ. Pollut. 237, 10–17.

Wang, L., Huang, X., Laserna, A.K.C., Li, S.F.Y., 2016. [1]H nuclear magnetic resonance-based metabolomics study of earthworm *Perionyx excavatus* in vermifiltration process. Bioresour. Technol. 218, 1115–1122.

Wang, P., Ng, Q.X., Zhang, H., Zhang, B., Ong, C.N., He, Y., 2018b. Metabolite changes behind faster growth and less reproduction of *Daphnia similis* exposed to low-dose silver nanoparticles. Ecotoxicol. Environ. Saf. 163, 266–273.

Watanabe, M., Meyer, K.A., Jackson, T.M., Schock, T.B., Johnson, W.E., Bearden, D.W., 2015. Application of NMR-based metabolomics for environmental assessment in the Great Lakes using zebra mussel (*Dreissena polymorph*a). Metabolomics 11, 1302–1315.

Watrous, J.D., Henglin, M., Claggett, B., Lehmann, K.A., Larson, M.G., Cheng, S., Jain, M., 2017. Visualization, quantification, and alignment of spectral drift in population scale untargeted metabolomics data. Anal. Chem. 89, 1399–1404.

Wei, F., Sakata, K., Asakura, T., Date, Y., Kikuchi, J., 2018. Systemic homeostasis in metabolome, ionome, and microbiome of wild yellowfin goby in estuarine ecosystem. Sci. Rep. 8, 3478.

Whitfield Åslund, M., Stephenson, G.L., Simpson, A.J., Simpson, M.J., 2013. Comparison of earthworm responses to petroleum hydrocarbon exposure in aged field contaminated soil using traditional ecotoxicity endpoints and ^{1}H NMR-based metabolomics. Environ. Pollut. 182, 263–268.

Wilkinson, J.L., Hooda, P.S., Swinden, J., Barker, J., Barton, S., 2018. Spatial (bio)accumulation of pharmaceuticals, illicit drugs, plasticisers, perfluorinated compounds and metabolites in river sediment, aquatic plants and benthic organisms. Environ. Pollut. 234, 864–875.

Winder, C.L., Dunn, W.B., Schuler, S., Broadhurst, D., Jarvis, R., Stephens, G.M., Goodacre, R., 2008. Global metabolic profiling of *Escherichia coli* cultures: an evaluation of methods for quenching and extraction of intracellular metabolites. Anal. Chem. 80, 2939–2948.

Wise Jr., J.P., Wise, J.T.F., Wise, C.F., Wise, S.S., Gianios Jr., C., Xie, H., Walter, R., Boswell, M., Zhu, C., Zheng, T., Perkins, C., Wise Sr, J.P., 2018. A three year study of metal levels in skin biopsies of whales in the Gulf of Mexico after the Deepwater Horizon oil crisis. Comp. Biochem. Physiol. C Toxicol. Pharmacol. 205, 15–25.

Wishart, D.S., 2016. Emerging applications of metabolomics in drug discovery and precision medicine. Nat. Rev. Drug Discov. 15, 473–484.

Wishart, D.S., Feunang, Y.D., Marcu, A., Guo, A.C., Liang, K., Vázquez-Fresno, R., Sajed, T., Johnson, D., Li, C., Karu, N., Sayeeda, Z., Lo, E., Assempour, N., Berjanskii, M., Singhal, S., Arndt, D., Liang, Y., Badran, H., Grant, J., Serra-Cayuela, A., Liu, Y., Mandal, R., Neveu, V., Pon, A., Knox, C., Wilson, M., Manach, C., Scalbert, A., 2017. HMDB 4.0: the human metabolome database for 2018. Nucleic Acids Res. 46, D608–D617.

Wu, H., Southam, A.D., Hines, A., Viant, M.R., 2008. High-throughput tissue extraction protocol for NMR- and MS-based metabolomics. Anal. Biochem. 372, 204−212.

Xia, J., Bjorndahl, T.C., Tang, P., Wishart, D.S., 2008. MetaboMiner - semi-automated identification of metabolites from 2D NMR spectra of complex biofluids. BMC Bioinformatics 9, 507.

Xia, J., Wishart, D.S., 2010. MetPA: a web-based metabolomics tool for pathway analysis and visualization. Bioinformatics 26, 2342−2344.

Xia, J., Wishart, D.S., 2011. Web-based inference of biological patterns, functions and pathways from metabolomic data using MetaboAnalyst. Nat. Protoc. 6, 743−760.

Xie, Z., Lu, G., Yan, Z., Liu, J., Wang, P., Wang, Y., 2017. Bioaccumulation and trophic transfer of pharmaceuticals in food webs from a large freshwater lake. Environ. Pollut. 222, 356−366.

Xu, Y., Wang, W., Zhou, J., Chen, M., Huang, X., Zhu, Y., Xie, X., Li, W., Zhang, Y., Kan, H., Ying, Z., 2019. Metabolomics analysis of a mouse model for chronic exposure to ambient $PM_{2.5}$. Environ. Pollut. 247, 953−963.

Yang, Y., Ok, Y.S., Kim, K.H., Kwon, E.E., Tsang, Y.F., 2017. Occurrences and removal of pharmaceuticals and personal care products (PPCPs) in drinking water and water/sewage treatment plants: a review. Sci. Total Environ. 596, 303−320.

Yuk, J., Simpson, M.J., Simpson, A.J., 2013. 1-D and 2-D NMR-based metabolomics of earthworms exposed to endosulfan and endosulfan sulfate in soil. Environ. Pollut. 175, 35−44.

Zacharias, H.U., Altenbuchinger, M., Gronwald, W., 2018. Statistical analysis of NMR metabolic fingerprints: established methods and recent advances. Metabolites 8, 47.

Zacher, L.S., Horstmann, L., Hardy, S.M., 2018. A field-based study of metabolites in sacculinized king crabs *Paralithodes camtschaticus* (Tilesius, 1815) and *Lithodes aequispinus* Benedict, 1895 (Decapoda: anomura: Lithodidae). J. Crustac Biol. 38, 794−803.

Zampieri, M., Szappanos, B., Buchieri, M.V., Trauner, A., Piazza, I., Picotti, P., Gagneux, S., Borrell, S., Gicquel, B., Lelievre, J., Papp, B., Sauer, U., 2018. High-throughput metabolomic analysis predicts mode of action of uncharacterized antimicrobial compounds. Sci. Transl. Med. 10 eaal3973.

Zelena, E., Dunn, W.B., Broadhurst, D., Francis-McIntyre, S., Carroll, K.M., Begley, P., O'Hagan, S., Knowles, J.D., Halsall, A., Consortium, H., Wilson, I.D., Kell, D.B., 2009. Development of a robust and repeatable UPLC− MS method for the long-term metabolomic study of human serum. Anal. Chem. 81, 1357−1364.

Zhang, H., Du, W., Peralta-Videa, J.R., Gardea-Torresdey, J.L., White, J.C., Keller, A., Guo, H., Ji, R., Zhao, L., 2018a. Metabolomics reveals how cucumber (*Cucumis sativus*) reprograms metabolites to cope with silver ions and silver nanoparticle-induced oxidative stress. Environ. Sci. Technol. 52, 8016−8026.

Zhang, P., Zhu, W., Wang, D., Yan, J., Wang, Y., Zhou, Z., He, L., 2017a. A combined NMR- and HPLC-MS/ MS-based metabolomics to evaluate the metabolic perturbations and subacute toxic effects of endosulfan on mice. Environ. Sci. Pollut. Res. 24, 18870−18880.

Zhang, X., Ding, X., Ji, Y., Wang, S., Chen, Y., Luo, J., Shen, Y., Peng, L., 2018b. Measurement of metabolite variations and analysis of related gene expression in Chinese liquorice (*Glycyrrhiza uralensis*) plants under UV-B irradiation. Sci. Rep. 8, 6144.

Zhang, Y., Hu, H., Shi, Y., Yang, X., Cao, L., Wu, J., Asweto, C.O., Feng, L., Duan, J., Sun, Z., 2017b. ^{1}H NMR-based metabolomics study on repeat dose toxicity of fine particulate matter in rats after intratracheal instillation. Sci. Total Environ. 589, 212−221.

Zhen, H., Ekman, D.R., Collette, T.W., Glassmeyer, S.T., Mills, M.A., Furlong, E.T., Kolpin, D.W., Teng, Q., 2018. Assessing the impact of wastewater treatment plant effluent on downstream drinking water-source quality using a zebrafish (*Danio Rerio*) liver cell-based metabolomics approach. Water Res. 145, 198−209.

Zimmermann, B.L., Palaoro, A.V., Bouchon, D., Almerão, M.P., Araujo, P.B., 2018. How coexistence may influence life history: the reproductive strategies of sympatric congeneric terrestrial isopods (Crustacea, Oniscidea). Can. J. Zool. 96, 1214−1220.

CHAPTER 2

Analytical techniques in metabolomics

Arthur David[1],[a], Pawel Rostkowski[2],[a]

[1]Univ Rennes, Inserm, EHESP, Irset (Institut de recherche en santé, environnement et travail), UMR_S 1085, Rennes, France; [2]NILU — Norwegian Institute for Air Research, Kjeller, Norway

Chapter Outline

1. Introduction

Metabolomics is a highly interdisciplinary field and is defined as an approach for the identification and semiquantification of all known and unknown metabolites (small molecules

---

[a] These authors contributed equally to this work.

Environmental Metabolomics. https://doi.org/10.1016/B978-0-12-818196-6.00002-9

with a weight below 1500 Da) within a biological system or systems (i.e., cell, tissue, organ, biological fluid, or organism) (Viant, 2007). In several cases, the small metabolites present in biological fluids and tissues, as well as environmental samples, may be exogenous in the source (e.g., food additives, plasticizers, cosmetics, and personal care products). In this case, the combination of the metabolome and xenobiotics as well as their products from phase I and II metabolism can be referred to as the xenometabolome (Al-Salhi et al., 2012; David et al., 2014; Holmes et al., 2007; Johnson et al., 2012a; Vineis et al., 2017).

Properly designed and conducted, metabolomics experiments and their associated data can provide valuable information on the biological response to internal and external stressors and allows for the potential to uncover biomarkers of effect (i.e., changes in endogenous metabolite profiles) as well as biomarkers of exposure (i.e., identification of xenobiotic mixtures). Hence, untargeted and targeted metabolomics has wide applicability in medicine, toxicology, plant sciences, environmental sciences, and more recently in exposomics (Balog et al., 2013; Bonvallot et al., 2018; David et al., 2014, 2017; Johnson et al., 2012b; Scalbert et al., 2014; Sumner et al., 2003).

Two main analytical techniques are currently being used in metabolomics: nuclear magnetic resonance (NMR) spectroscopy and high-resolution mass spectrometry (HRMS). Modern separation techniques, such as liquid chromatography (LC), gas chromatography (GC), or capillary electrophoresis (CE), are often coupled with HRMS. Among all these techniques, LC-HRMS typically utilizing an electrospray ionization (ESI) source is probably the most popular approach. One of the main reasons behind it is its improved sensitivity compared to NMR analysis and the soft ionization process in the ESI which allows structural elucidation.

Literature survey recently conducted by Miggiels et al. (2019), shows the increasing adaptation of metabolomics,—a trend that was previously observed by Kuehnbaum and Britz-McKibbin (2013) (Fig. 2.1A). A further search conducted on Web of Science in March 2019 shows that there were 2204 papers specifically mentioning the main technique published in 2018 and already 518 published as for March 24, 2019. The trends visualized in Fig. 2.1B show that LC-MS is recently a dominant technique, while the CE-MS is rather a niche technique. Fig. 2.1C shows the increasing number of applications of fast sample introduction and emerging ion mobility separations.

Metabolomics experiments can usually be divided into two categories: untargeted and targeted. Untargeted metabolomics aims to detect and characterize as much metabolome as possible without bias (Patti et al., 2012). In targeted studies, a specified set of compounds is being quantified with the use of authentic standards (Soule et al., 2015). Targeted approaches cover a relatively small number of metabolites, while untargeted metabolomics can measure up to thousands of features at a time (Bowen and Northen, 2010; Naz et al., 2014; Yanes et al., 2011). Therefore, this approach provides a large amount of information and the potential for biomarker discovery.

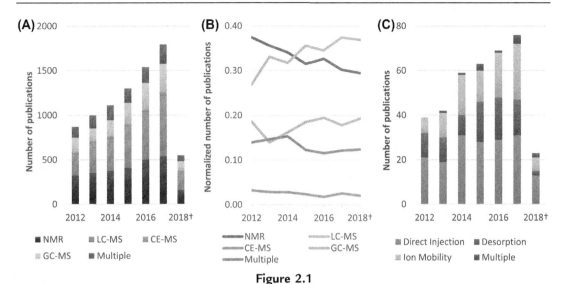

Figure 2.1

Literature survey of publications with metabolom* OR metabonom* in the title, abstract, or keywords in the period 2012—18 (data extracted on October 31, 2018) on the Web of Science: (A) number of publications mentioning the main techniques in metabolomics; (B) relative number of publications of the main techniques in metabolomics; and (C) number of mass spectrometry—related publications mentioning the use of fast sample introduction methods. *Reprint from Miggiels P., Wouters B., van Westen G.J.P., Dubbelman A.-C. and Hankemeier T., Novel technologies for metabolomics: more for less, Trends Anal. Chem. 120, 2019, 115323.*

Development and application of a mass spectrometry (MS)-based method pose few challenges: (I) metabolites can vary a lot in their chemical properties and concentrations which makes the selection of a comprehensive extraction method a problematic task; (II) chromatography that enables the resolution of large numbers of molecules makes high-throughput analyses challenging (Chetwynd and David, 2018); and (III) another challenge could be caused by the selection of ionization coverage and efficiency. Commonly used techniques like ESI, atmospheric pressure chemical ionization (APCI), and electron ionization (EI) work effectively for analytes with certain physical and chemical properties. ESI is, for example, better suited for polar metabolites, while EI (without sample derivatization) will be suitable for more nonpolar compounds. In addition, confident metabolite identification is difficult due to the existence of compounds with identical masses, like isomers and isobar.

This chapter aims to give an overview of currently available analytical techniques for metabolomics and provide insight regarding their respective advantages and drawbacks in terms of metabolite coverage, their simplicity, as well as their potential to be used in a high-throughput manner. Metabolomics consists of several steps (generation of chemical profiles using analytical methods, mining of data using bioinformatics and chemometrics, and interpretation of data), and all these steps are interrelated and will impact upon each other. It is therefore essential at the stage of the experiment design not only to ensure that the right analytical techniques are selected to detect classes of metabolites of interest but also to provide a high level of repeatability to minimize experimental errors.

2. Sample preparation methods for metabolomics

Metabolomics studies often use minimal sample preparation associated with very short chromatographic runs or direct infusion so that a maximum number of samples can be analyzed in a very short time. In addition to reducing the time of analysis, minimal sample preparation also contributes to reducing the costs of analysis. Hence, it has often been argued that an ideal sample preparation method for global metabolomics has to be simple, fast, and nonselective to ensure adequate depth of metabolite coverage (Chen et al., 2016; Vuckovic, 2012).

Minimal sample preparation is relevant when the study is focused on endogenous metabolites that are present at sufficiently high concentrations. However, for targeted metabolomics focusing of low abundant metabolites (e.g., steroidomics) or even untargeted metabolomics—based approach aiming to detect low levels of contaminants in complex biological matrices, selective extractions can be more favorable. Selective sample preparations such as solid-phase extraction (SPE) are more efficient to remove properly matrix components such as residual proteins and salts (David et al., 2014; Michopoulos et al., 2009; Vuckovic, 2012) and therefore reduce ion suppression and improve column reproducibility/prolongation (David et al., 2014; Michopoulos et al., 2009; Vuckovic, 2012).

This section will focus mainly on extraction and purification steps used in sample preparation for the analysis of human or animal biofluids and tissues. A particular interest will be paid to minimal and more selective sample preparation categories (Table 2.1). Minimal sample preparation is generally recommended for global metabolomics application, and selective sample preparation should be favored for targeted approach or when an increase in analytical sensitivity is needed to detect lower abundant metabolites. Finally, recommendations regarding the evaluation of the sample preparation are given to ensure that the most appropriate sample preparation is selected in accordance with the research question.

2.1 Minimal sample preparations

The most commonly used simple sample preparation methods are dilution, solvent protein precipitation (PPT) for biofluids, and solvent-based extraction for tissues.

Table 2.1: Overview of examples of minimal and more selective sample preparation used in metabolomics, their advantages, and limitations.

Sample preparation type	Recommended matrix	Main advantages	Limitations
Minimal			
Dilution	Biofluids such as urine	Fast and high throughput	Not adapted for biofluids with high content of proteins (e.g., blood)
		No method development	Mild centrifugation might be needed
		High metabolite coverage	Higher detection limits
Solvent protein precipitation for biofluids	Blood (plasma and serum) Cerebrospinal fluid		
Simple extraction		Fast and high throughput	Not very efficient to remove proteins
		Organic solvent disrupt binding	Repeatability and column lifetime can be affected
		Between proteins and metabolites	
		Good metabolite recovery	
Biphasic extraction		Complementary detection method	Time-consuming
		Can be used for polar and nonpolar fractions	Contamination of polar phase by lipids
Solvent extraction for tissues	Any animal or plant tissue	Metabolite concentration can be higher in tissues	Requires a step to physically break up the tissue
			Less high throughput
Selective			
Solid-phase extraction	Biofluids or tissues	Efficient removal of proteins and salts	Can be too selective
		Sample clean-up and enrichment	Time-consuming
		A large diversity of sorbents can be used	
Solid-phase microextraction	biofluids	Can be used in vivo in awake animal	Lower coverage of the metabolome
		Minimizes the preparation steps	Extract only the free fraction
		A large diversity of sorbents can be used	

Continued

Table 2.1: Overview of examples of minimal and more selective sample preparation used in metabolomics, their advantages, and limitations.—cont'd

Sample preparation type	Recommended matrix	Main advantages	Limitations
Filtration (delipidation)	biofluids and tissues (After extraction)	Efficient removal of proteins Efficient removal of lysophospholipids Sample clean-up and enrichment Fast and high throughput	Lower recoveries for some analytes
Electromigration based Extraction	biofluids	Fast and high throughput Sample clean-up and enrichment remove proteins	Too selective

Although ultrafiltration can be considered as simple sample preparation, this method will be discussed in the selective sample preparation section since many recent innovations have been made to remove specifically some classes of endogenous metabolites such as the lyso- and phospholipids during this step.

2.1.1 Dilution

Dilution is very popular for urine samples in LC-MS—based metabolomics applications and sometimes referred to as "dilute-and-shoot"-LC-MS (DS-LC-MS) (Fernández-Peralbo and Luque de Castro, 2012). Typical dilution factors with purified water before LC-MS analysis range between 1:1 and 1:10, and preservatives or filtration are usually required in order to prevent bacterial growth during storage (Vuckovic, 2012). Dilution is well adapted for urine (as opposed to plasma or serum) since this matrix contains lower protein amounts. However, a mild centrifugation step can be recommended after sample collection to remove any large particles before LC-MS analysis. The increase in MS sensitivity observed in recent years will probably encourage metabolomics application of DS-LC-MS, in particular when substances of interest are present at relatively high levels or with good ionization efficiency (Fernández-Peralbo and Luque de Castro, 2012).

2.1.2 Solvent protein precipitation for biofluids

As mentioned earlier, plasma and serum from blood (or plasma from cerebrospinal fluid) contain higher proportions of proteins and therefore require a step for protein removal. One of the most commonly used methods for protein removal is organic solvent—based PPT followed by centrifugation (Raterink et al., 2014; Vuckovic, 2012). The use of an organic solvent is preferred over protein denaturation using heat or inorganic acid as it has been shown to improve metabolite coverage (Want et al., 2006). Organic solvents will also help to disrupt binding between metabolites and proteins and will, therefore, better reflect the total metabolite concentrations. Furthermore, the use of a cold solvent (or storage of sample with organic solvent in the freezer before centrifugation) is often recommended to minimize the extent of enzymatic conversion of metabolites and to improve the precipitation of proteins.

In past years, the performance of various PPT methods for metabolomics, and in particular the choice of the solvent of extraction, has been evaluated in terms of protein removal efficiency, metabolite coverage, and precision (Raterink et al., 2014; Sitnikov et al., 2016; Vuckovic, 2012). No general consensus has been reached regarding the best method, which is inherent in the nature of metabolome which consists of a wide variety of compounds with very different polarities and no single solvent can extract simultaneously all metabolites. Nevertheless, many studies have shown that acetonitrile or acetone gave better results in terms of protein removal, while methanol, ethanol, or a mixture of both results in improved metabolic coverage and method precision (Vuckovic, 2012). A recent study where a side-by-side comparison of different sample preparation strategies including different solvent of extraction confirmed that a consensus is difficult to reach because all methods showed high overlap and redundancy; however, it was suggested that PPT using a mix of methanol/ethanol was offering the best compromise (Sitnikov et al., 2016). As for the ratio of biofluid to organic solvent, a ratio of 1−4 is often recommended (Bruce et al., 2009; Michopoulos et al., 2009).

Another very popular solvent-based protocol for biofluids such as serum and also urine uses biphasic solvent systems to extract both polar and nonpolar compounds separately. The original biphasic method is called the Bligh and Dyer method. It utilizes chloroform for the nonpolar fraction and methanol/water for the polar fraction, each fraction being analyzed separately. This method was initially intended to extract lipids and involves multiple steps that include solvent additions, mixing, and centrifugation. More recently, chloroform was replaced by methyl tert-butyl ether in the Matyash method as it is noncarcinogenic and the method was claimed to be as efficient as the Bligh and Dyer method (Matyash et al., 2008). Over the years, many modifications, such as solvent ratios and solvent addition strategies (e.g., stepwise, two-step, or simultaneously), have been tested for both methods to optimize their efficiency. Recently, Sostare et al. (Sostare et al., 2018) reoptimized the solvent ratios in the Matyash method (2.6/2.0/2.4, v/v/v) and compared it to the original Matyash method (10/3/2.5, v/v/v) and the conventional Bligh and Dyer method (stepwise, 2.0/2.0/1.8, v/v/v) using two biofluids (human serum and urine). They concluded that, overall, the modified Matyash method yielded a higher number of peaks and putatively annotated metabolites compared to the original Matyash method and the Bligh and Dyer method (Matyash et al., 2008; Sostare et al., 2018). In the end, analyzing the lipid fraction separately can be advantageous to decrease analytical interferences such as ion suppressions since lipids such as lyso- and phospholipids are usually present at high concentrations in biofluids like blood and tissues. Another advantage of biphasic solvent methods lies in the fact that both polar and nonpolar fractions can be analyzed separately using complementary analytical methods (e.g., GC and LC or normal and reversed-phase for LC-HRMS) to improve the coverage of the metabolome. However, this is disadvantageous in terms of being time-consuming, and contamination by lipids in the methanol/water phase (in particular polar phospholipids) can be problematic (Wu et al., 2008).

2.1.3 Solvent extraction for tissues

Compared to biofluids, analyzing tissue samples requires a first step to physically break up the tissue and homogenize it. This first step aiming at disrupting tissues is usually time-consuming and from this point of view, metabolomics analyses of tissue are usually performed in a less high-throughput manner compared with biofluids. One of the most traditional methods to disrupt tissues consists of grinding the frozen tissue in a liquid nitrogen—cooled mortar and pestle. However, this method is considered labor-intensive, is more susceptible to cross contamination, and requires the samples to be transferred carefully before solvent extraction (Wu et al., 2008). More recent techniques include extraction solvent using an electric tissue homogenizer, a sonication probe, or a bead-based homogenizer (Wu et al., 2008). Bead beating has been recommended in several studies by many authors because it allows many samples to be extracted at the same time. It is relatively fast (from few seconds to few minutes depending on the size of the tissue), and there is less risk of carryover between samples (Roemisch-Margl et al., 2011; Vuckovic and Pawliszyn, 2011; Wu et al., 2008).

Like for biofluids, the choice of the solvent of extraction is a critical issue for tissue metabolomics, and this aspect has been the subject of several studies in the past years. For instance, Wu et al. (2008) optimized tissue (fish liver) extraction methods compatible with high-throughput NMR spectroscopy and MS-based metabolomics using bead beating and the classic Bligh—Dyer protocol (Wu et al., 2008). They mainly focused on the solvent addition strategies and concluded overall that a two-step method provided good quality data based on lipid partitioning, reproducibility, yield, and throughput. Another study reported the development of a standardized high-throughput method using a bead-based homogenizer in combination with a simple extraction protocol for rapid and reproducible extraction of metabolites from multiple animal tissue samples such as liver, kidney, muscle, brain, and fat tissue from mouse and bovine (Roemisch-Margl et al., 2011). The authors showed that a simple methanolic extraction was the most efficient method for reliable results and good coverage of the metabolome (Roemisch-Margl et al., 2011). More recently, Sostare et al. (2018) showed that their modified Matyash method (see Section 2.1.2) tended to provide a higher yield and reproducibility for most sample types (including tissues from whole *Daphnia magna*) compared to the original Matyash method and the conventional Bligh and Dyer method (Sostare et al., 2018). As highlighted in the previous section, biphasic and simple solvent extractions have both their advantages and drawbacks which need to be carefully considered in relation to the original biological hypothesis. In any case, the automatization of the sample extraction procedure is needed to increase the overall sample throughput for tissue metabolomics studies (Wu et al., 2008).

2.2 Selective sample preparations

Minimal sample preparations are not very efficient to remove properly matrix components such as proteins and salts. As already mentioned, residual proteins or salts have significant detrimental effects on ion suppression for LC-ESI-MS but it can also cause detector saturation and affect chromatographic performance and column reproducibility/prolongation, which is a problem for high-throughput analysis (David et al., 2014; Michopoulos et al., 2009; Vuckovic, 2012). Hence, more selective sample preparations can be used to address this issue of matrix effects, increase column lifetime and overall method robustness by injecting much cleaner extracts (Vuckovic, 2012). Improving the clean-up procedure is also a necessary step to improve the detection limits of trace xenobiotics when the aim is to identify biomarkers of exposure. These selective sample preparations can be used as a second step or as a substitute for minimal sample preparation methods.

2.2.1 Solid-phase extraction and solid-phase microextraction

SPE is a very efficient sample-pretreatment method for the removal of interfering substances and the enrichment of analytes (Raterink et al., 2014). The principle of SPE consists of using a sorbent to extract one or several analytes present in aqueous or organic solutions. One of the main advantages of SPE lies in the fact that the sorbent can be washed before the elution (using ultrapure H_2O, for instance) to remove matrix interferences such as salts and thus provide a better sample clean-up compared to simple solvent PPT methods (Vuckovic, 2012). Significant preconcentrations of the sample can be done using SPE. A large variety of sorbents with different retention mechanisms are available in order to cover a wide range of metabolites and can be used for untargeted and targeted metabolomics. Sorbents are typically housed in cartridge format but can be used as well in 96-well plates, which can allow automatization of the procedure. Moreover, besides off-line SPE procedures, online SPE can also be utilized with LC-ESI-HRMS for nontargeted screening (Stravs et al., 2016).

Several metabolomics studies have already utilized SPE to analyze directly biofluids such as blood plasma (Michopoulos et al., 2009) or urine (Chetwynd et al., 2015) and as clean-up procedure after solvent extraction of tissues (David et al., 2017).

More specifically, SPE, such as C18 reversed-phase, has been utilized with UHPLC-MS to perform global metabolomics on human plasma (Michopoulos et al., 2009). The authors reported that the repeatability of the methodology was improved significantly when the sample preparation was performed using SPE in comparison with a methanol PPT (Michopoulos et al., 2009). More recently, Yang et al. (2013) developed a method combining liquid—liquid extraction and normal-phase SPE (NH$_2$ phase) to profile human plasma using UHPLC-MS (Yang et al., 2013). This method allowed for the detection of a significantly higher number of metabolites in comparison with a methanol PPT.

To date, metabolomics applications using SPE for the sample preparation are limited because it has been argued that obtaining high metabolite coverage may be challenging. However, uses of sorbent with polymeric resins which contain more polar groups embedded in a nonpolar polymeric backbone have a great potential for metabolomics (Mitra, 2003; Steehler, 2004). Moreover, the use of mixed-mode polymer materials with either cation- or anion-exchange sites also offers the possibility to enhance the coverage of the metabolome (Mitra, 2003). Recently, a metabolomics study where a comparison of SPE methods was performed on blood plasma showed that sample concentration by either polymer or mixed-mode ion-exchange SPEs gave comprehensive metabolite coverage of plasma extracts. Still, the use of cation-exchange SPE significantly increased detection of many metabolites and xenobiotics in the plasma extracts (David et al., 2014).

More recently, it has been shown that solid-phase microextraction (SPME) could be a promising alternative to SPE. Studies have shown that comprehensive metabolite profiles of human plasma can be reliably obtained using different SPME coatings based on known SPE sorbents (Vuckovic, 2012; Vuckovic and Pawliszyn, 2011). In particular, in vivo SPME, which allows performing in situ metabolite extraction in living animals using an SPME device directly inserted in the bloodstream of an animal, seems very attractive to perform metabolomics on circulating blood directly collected in awake animals (Vuckovic, 2012). The advantage of in vivo SPME lies in the fact that it combines extraction and metabolism quenching of analytes (which minimizes the number of preparation steps prior to analysis), and the technique is directly compatible with LC-MS injection (Bojko et al., 2014). Some limitations of the method include the fact that SPME coating equilibrates with only the free fraction of the analytes, and that it seems that there is a lower coverage of the analytes when compared to standard solvent-extraction protocols (Bojko et al., 2014). The use of SPME in metabolomics is still limited, but the applicability of the method was already proved with its application toward various complex matrices (i.e., biofluids, soft tissues, and plants) (Bojko et al., 2014).

2.2.2 Filtration and delipidation/deproteinization techniques

Filtration or ultrafiltration is a simple sample preparation method often used for biological fluids such as urine or blood, but it can also be used as the second step after solvent extraction. With this method, samples are filtered through a special filter using a vacuum or centrifugation which only allows passage of molecules of specific molecular weight. Hence, the main advantage of filtration is that it can physically separate small metabolites from proteins or other macromolecules present in a sample (Vuckovic, 2012). However, these methods seem more adapted to recover polar metabolites since hydrophobic compounds may be adversely retained on the filter unless organic solvent is used during (or after) the filtration (Tiziani et al., 2008).

Among other filtration techniques, deproteinization and simultaneous delipidation (mainly lyso- and phospholipids) of the sample has gained increased attention because it can reduce ion-suppression effects by removing highly abundant species and can improve the repeatability of the chromatographic method. Several lipid depletion plates from different manufacturers (e.g., Hybrid SPE: Sigma Aldrich, Ostro: Waters, Captiva ND: Agilent, ISOLUTE PLD+: Biotage, and Phree: Phenomenex), 96-well plates, are now commercially available and can remove phospholipids as well as proteins from a sample in a single step (Raterink et al., 2014). Samples are loaded with organic solvents such as acetonitrile or methanol (ratio of sample to organic solvent are usually 1/3 or 1/4) on the cartridge or 96-well plates and are subsequently filtered through the sorbent using a vacuum. The main advantage of this method is that it requires little to no method development and can be used in a high-throughput manner compared to PPT and SPE. However, even if recoveries are usually acceptable for most of the analytes, some of them can be more or less retained on the sorbent (David et al., 2014). However, injections of more concentrated extracts can make up for these lower recoveries.

Applications of these recent innovations in metabolomics studies have shown that they are effective for removing phospholipids and when used, in combination to solvent deproteinization, can enhance analyte detection of nonlipid species in comparison to extractions with organic solvents and a membrane-based solvent-free technique (Tulipani et al., 2013). Another application of this sample preparation technique in untargeted metabolomics—based approaches includes a study where plasma samples were extracted using phospholipid filtration plates in combination with polymeric or mixed-mode exchange SPE (David et al., 2014). In this study, the use of delipidation plates in combination with SPE allowed for the injection of more concentrated plasma extracts onto a nanoflow LC—nanoESI-MS platform without blockage of the nanocolumn or nanospray, thus resulting in a wider coverage of the (xeno)metabolome compared to PPT (David et al., 2014).

2.2.3 Electromigration-based extraction techniques

In this section, we describe an innovative sample preparation technique based on electromigration sample extraction. Although applications of electromigration-based extraction in metabolomics has been very limited so far, it offers interesting prospects for future metabolomics studies, in particular for targeted metabolomics. The principle of electromigration-based extraction is to induce selective migration of charged compounds using an electric field. Several techniques such as electromembrane extraction and electroextraction can be used and have been reviewed from a metabolomics viewpoint by Raterink et al. (2014). Both techniques are based on sample-enrichment processes, i.e., analytes are extracted from a sample to an acceptor compartment. The main advantages of these techniques lies in the fact they can offer speed since the electric field enhances the extraction rate of analytes, selectivity (proteins are excluded from blood extraction), and excellent potential for both miniaturization and high-throughput analysis using, for instance, nanoESI-DI-MS (Raterink et al., 2014). Hence, electromigration-based extraction offers perspectives for targeted metabolomics analyses of amino acids or acetylcarnitines; however, further work is needed to apply these methods to a wider range of metabolites (Raterink et al., 2014).

2.2.4 Derivatization methods (nonvolatile metabolites)

Derivatization of polar compounds (to reduce their polarity and improve thermal stability and volatility) is necessary for GC-MS—based metabolomics. Functional groups of molecules containing carboxylic acids (-COOH), alcohols (-OH), amines (-NH$_2$), and thiols can be derivatized by silylation, acylation, and alkylation (Dettmer et al., 2007). Commonly used reagents are listed in Table 2.2.

Several derivatization approaches are being used: (1) offline, (2) microwave-assisted (MAD), (3) in-time (in-line), and (4) in-liner.

Offline derivatization is so far the most common approach in GC-MS—based metabolomics (Beale et al., 2018; Hyötyläinen, 2013). Typically, it refers to a procedure where the extracts are dried down and then reconstituted in the derivatizing reagent, which is then heated for a prescribed amount of time and afterward transferred to a clean vial prior to GC-MS analysis.

MAD utilizes microwave heating to improve the efficiency of reactions comprising silylation, acylation, and alkylation (Söderholm et al., 2010) and has primarily been applied for the preparation of environmental, clinical, forensic, and industrial samples (Beale et al., 2018). MAD allows to substantially reduce the time needed to derivatize the samples compared to conventional approaches and is even being implemented by some autosampler providers into in-line derivatization modules. In-time (in-line) derivatization protocols allow to reduce the costs and improve the throughput of the research through shortening (saving) the time for sample preparation, freeing up lab human resources, and working toward elimination of the human error that can occur during manual transfer of extract and reagents. This procedure enables also the maximum utilization of the GC-MS instrument that can be operated continuously.

Table 2.2: Common derivatization reagents for GC-MS analysis after Beale et al. (2018), Lai and Fiehn (2018), and https://www.sigmaaldrich.com/content/dam/sigma-aldrich/docs/Supelco/Application_Notes/4537.pdf.

Derivatization procedure	Reagents	Function group derivatized									
		Amides	Amines	Carbohydrates	Carbonyls	Carboxyls	Esters	Hydroxyls	Nitriles	Thiols	Sulphonamides
Alkylation	Diazomethane					✓					
	Methyl chloroformate		✓			✓					
	Ethyl chloroformate		✓			✓					
Acylation	Acetic anhydride		✓	✓				✓		✓	✓
	N-methyl-bis(trifluoroacetamide) (MBTFA)		✓	✓				✓		✓	
	Trifluoroacetic anhydride	✓	✓					✓			✓
	Heptafluorobutyric anhydride	✓						✓			
	Pentafluoropropionic anhydride	✓	(Secondary)					✓			
Esterification	Methanolic HCl					✓					
	Pentafluorobenzyl bromide					✓					
	Trimethylanilinium hydroxide					✓					
Silylation	N, O-bis(trimethylsilyl) acetamide (BSA)	✓	✓	✓	✓	✓	✓	✓			
	O-trimethylsilyl (TMS): N-methyl-N-(trimethylsilyl) trifluoroacetamide (MSTFA)	✓	✓	✓	✓	✓	✓	✓			
	O-trimethylsilyl (TMS): bis(trimethylsilyl) trifluoroacetamide (BSTFA)	✓	✓	✓	✓	✓	✓	✓			
	O-trimethylsilyl (TMS): trimethylchlorosilane (TMCS)	✓	✓	✓	✓	✓	✓	✓			
	Tert-butyldimethylsilyl (TBDMS): N-methyl-N-tert-butyldi-methylsilyltrifluoroacetamide (MTBSTFA)	✓				✓					
	Hexamethyldisilazane (HMDS)							✓			

Continued

Table 2.2: Common derivatization reagents for GC-MS analysis after Beale et al. (2018), Lai and Fiehn (2018), and https://www.sigmaaldrich.com/content/dam/sigma-aldrich/docs/Supelco/Application_Notes/4537.pdf.—cont'd

Derivatization procedure	Reagents	Function group derivatized									
		Amides	Amines	Carbohydrates	Carbonyls	Carboxyls	Esters	Hydroxyls	Nitriles	Thiols	Sulphonamides
	HMDS + TMCS			✓				✓			
	HMDS + TMCS + pyridine			✓				✓			
	Trifluoroacetic acid (TFA)			✓				✓			
	N-trimethylsilyl-imidazole (TMSI)					✓					
	TMSI + pyridine			✓	✓						
Oximation	Methoxyamine HCl				✓						
	Ethoxyamine HCl				✓						

Advancements in MAD and in-line derivatization addressed a number of issues related to the extended sample preparation time (lower throughput, greater cost, and use of lab resources).

In cases where such technologies are not available, in-liner derivatization inside the GC inlet could be applied. To achieve this, the extract and the reagents are drawn into the GC syringe as multilayers split with an air gap between solutions and then being injected into a hot inlet (Ferreira et al., 2013).

Without doubts in-line and in-liner derivatization procedures are beneficial as they limit the steps in the whole process. However, the latter requires more frequent maintenance of the inlet and replacements of the liner. Dirty inlet could lead to some side reactions producing artifacts and overestimating relative response ratio or concentrations of metabolites (Beale et al., 2018).

To ensure that a maximum number of analytes is being detected with an acceptable recovery, it is important to optimize extraction and derivatization procedure. It can be achieved by investigating several derivatization reagents, with both authentic standards (containing several classes of metabolites) and spiked sample matrices. A sufficient amount of derivatizing reagent and optimum conditions are essential for the efficient derivatization of metabolites. Low efficiency of the derivatization of compounds with multiple functional groups may result in eluting multiple peaks for the same metabolite. Moreover, the stability of the derivatized extract and metabolite degradation during storage or their decomposition in the analytical system requires careful evaluation and validation in different matrices prior to the quantification of metabolites.

2.3 Recommendations for the evaluation of sample preparation

Metabolomics consists of several successive steps, and all of them are interrelated and will impact upon each other. It is therefore essential to test the specificity of a sample preparation method before launching large metabolomics studies. All sample preparation methods described in this chapter present their specificity. The choice of a particular method (solvent of extraction, for instance) can favor the detection of specific classes of metabolites or, on the contrary, prevent the discovery of others.

Evaluation and comparison of sample preparation methods for metabolomics often focus on metabolite coverage (i.e., the number of detected features) and repeatability (Sitnikov et al., 2016). However, without proper metabolite identification, it is difficult to determine which method is better than another (David et al., 2014; Vuckovic, 2012). Relying only on the number of metabolite features without proper identification can be highly misleading. Besides metabolite annotation, it is also recommended to assess the analytical sensitivity of the method, in particular when the study aims at detecting very low abundant metabolites. It is therefore suggested to perform a spiking experiment using realistic concentrations and quantitative recovery studies across different metabolite classes. So far, a very limited number of studies have provided proper metabolite identification to compare sample preparation methods and perform quantitative recovery studies or evaluation of matrix effects (David et al., 2014; Sitnikov et al., 2016). More care should be provided for the evaluation of sample preparation in order to improve the whole metabolomics workflow.

3. Separation and identification methods

3.1 Nuclear magnetic resonance spectroscopy

NMR was reported as a used technique in over 30% of the recent peer-reviewed publications in metabolomics (Fig. 2.1). This technique is highly reproducible, cost-efficient, and currently more established for high-throughput than MS. NMR can be considered as a complementary technique to GC and LC-MS, as it can provide information about the more abundant but difficult to analyze metabolites. The major drawback of NMR is its significantly lower sensitivity compared to MS. A recent review presents the future of application NMR in metabolomics (Markley et al., 2017). A number of different strategies aiming to increase the sensitivity of NMR have been published so far and include established techniques such as introduction of higher field magnets (Moser et al., 2017), cryogenically cooled probes (Jézéquel et al., 2015), and emerging techniques like high-temperature superconducting oils (Ramaswamy et al., 2013), microcoil-NMR probes (Saggiomo and Velders, 2015), and hyperpolarization.

3.2 Mass spectrometry

Due to the complexity of the metabolome, analysis of the complete chemical pattern of a biological system is currently not doable with a single generic method. It is necessary to employ several complementary techniques for this purpose. In the majority of cases, MS, especially high-resolution (HRMS), coupled with different separation techniques, such as GC, LC, CE or standalone, is an excellent choice due to its selectivity and sensitivity.

A range of different mass analyzers have been installed in mass spectrometers by various suppliers. They are an essential part of the instrument needed to distinguish one mass peak from another. To achieve this, the mass analyzer must be able to resolve individual ions of similar mass. Mass resolution and resolving power are important parameters describing to what extent the resolution of individual ions could be achieved. Mass resolution is defined as a degree of separation between two mass spectral peaks of equal height and width observed in a spectrum (Δm) at full-width half mass. Mass resolving power is then defined as ($m_1/(m_2-m_1)$) (G Marshall et al., 2013). The performance of mass analyzers is then typically defined by mass resolving power and accuracy, dynamic range, tandem analysis capabilities, and acquisition speed. When focusing on mass resolving power and mass accuracy, low- and high-resolution mass spectrometry (LRMS and HRMS, respectively) can be differentiated (Table 2.3).

The mass detecting error is typically below 5ppm for (Q)TOFs and below 2ppm for (Q)-Orbitraps. Mass resolution is a factor of great importance as it could significantly affect the peak picking process and therefore marker identification.

Mass accuracy is dependent on the calibration of a mass spectrometer. The process itself is dependent on the MS supplier that usually obliges the users to follow instrument specifications. In general, continuous and noncontinuous calibration forms are in use. A continuous mass calibration requires calibrating solution to be continuously infused so that peaks of calibrants are present in each spectrum. Since it causes the risk of potential interferences from the infused calibrations, calibration before and after injection (noncontinuous) could be considered. Another option is to use common background ion(s) as a calibrant improving mass accuracy.

When comparing high-resolution mass spectrometers, Orbitrap seems to be more stable than QTOF systems (Kaufmann, 2014) which have to be recalibrated continuously or with

Table 2.3: List of common mass analyzers and instrumental capabilities.

Mass analyzer	Mass resolution	Mass range	MS/MS	MS^E	Acquisition speed
Quadrupole	~1000	50–6000	Yes (only on triple quadrupole systems)	No	Medium
Ion trap (IT)	~1000	50–4000	Yes	Yes	Medium
TOF	2,500 –40,000	20–500000	No	No	Fast
QTOF	2,500 –80,000	20–500000	Yes	No	Fast
DFS	~60,000	2–6000	No	No	Fast
Orbitrap	70,000 –1,000,000	40–80,000	Yes	Yes (not all models)	Slow
Fourier transform –ion cyclotron resonance	>200,000	10–10,000	Yes	Yes	Slow
Ion mobility QTOF	2,500 –40,000	20–500000	Yes	No	Fast

Adapted from Beale et al., 2018.

certain frequency (for example, once per hour), since temperature fluctuation can change the time-of-flight of metabolites.

Low-resolution quadrupole and triple quadrupole systems were commonly used, for example, in GC-MS–based applications. Especially triple quadrupole GC-MS/MS (together with ion trap instruments) has good sensitivity and an excellent dynamic range but often slower scan range and much lower mass accuracy than high resolution–based systems. The main disadvantage of low-resolution mass spectrometers is difficulty in the structure elucidation of unknown molecules, without mass spectra available in the libraries.

Generally, the possible applications of quadrupole and ion trap–based systems are rather limited to targeted applications and to some extent to the identification of the compounds based on the library searches.

Recently, comprehensive commercial databases such as National Institute of Standards and Technology (NIST), Wiley, and Metlin (Guijas et al., 2018) have been complemented by extensive open source databases like mzCloud, MassBank (Horai et al., 2010; Stravs et al., 2013), and Human Metabolome Database (Wishart et al., 2018).

Over the years, GC-MS databases were mainly generated with LRMS. Recently, with the greater availability of GC-HR-QTOF and GC-Orbitrap instruments, the number of GC high-resolution mass spectra in different libraries is growing. However, until the date, LC-MS libraries still contain more high-resolution spectra.

It is important to note that due to the complexity of the samples used in metabolomics instruments, GC and LC-MS peaks are not necessarily made up of individual compounds but could represent mixtures of coeluting metabolites. Cleaner spectra of putative metabolites are usually being obtained from overlapping peaks by applying deconvolution methods.

Various deconvolution algorithms are being developed and are supplied by instrument and software vendors (Lu et al., 2008). A number of these tools are also being developed by different research groups and available freely, often as a part of the more complex data processing tools. In the recent review, Spicer and coauthors (Spicer et al., 2017) discuss c. 200 common free tools being used for data processing in metabolomics (inclusive NMR and MS methods), for example, AMDIS (http://www.nist.gov), MS-DIAL (Tsugawa et al., 2015), ADAP-GC (Smirnov et al., 2018), eRah (Domingo-Almenara et al., 2016), Metab (Aggio et al., 2011), Mzmine (Hu et al., 2016; Katajamaa et al., 2006; Myers et al., 2017; Olivon et al., 2017; Pluskal et al., 2010), XCMS (Forsberg et al., 2018; Gowda et al., 2014; Huan et al., 2017; Libiseller et al., 2015; Mahieu et al., 2016; Myers et al., 2017; Siuzdak, 2014; Smith et al., 2006; Tautenhahn et al., 2012), and Metaboanalyst (Chong et al., 2018; Xia et al., 2009, 2012, 2015; Xia and Wishart, 2011).

3.2.1 Gas chromatography coupled to mass spectrometry

GC-MS is one of the most efficient platforms used in the metabolomics field. It is robust and possesses excellent separation capability, sensitivity, and reproducibility (Mastrangelo et al., 2015; Villas-Boas et al., 2005). GC-MS is typically less prone to common LC-MS problems such as matrix effects and ion suppression of the coeluting compounds. It tends to achieve greater chromatographic resolution however; it can only be used to identify volatile compounds with a low molecular weight (Beale et al., 2018; Gowda and Djukovic, 2014; Mastrangelo et al., 2015).

Detection of polar, nonvolatile metabolites requires chemical derivatization prior to analyses. It introduces additional issues especially for the nontargeted metabolomics. Derivatives will be measured instead of an actual compound, making structure elucidation even more complicated task. In addition, an excess of the derivatizing reagent can obscure responses of other chemicals.

Although commonly used capillary GC columns provide high resolution and peak capacity, the complexity of biological samples, being a subject of metabolomics-based experiments, is still substantially exceeding their resolving power.

For complex samples, the use of two-dimensional (2D)-GC (GC × GC) that utilizes two columns having different stationary-phase selectivities and are connected serially can increase peak capacities and resolving power of the separation. In 2D systems usually, a long nonpolar column is focused using a cryogenic modulator. Then the sample is rapidly transferred into a second, typically short polar column, for the second-dimension separation. It results in a fast second-dimension separation and enhancement in the sensitivity. GC × GC is coupled to fast detectors, for example, flame ionization detector or EI fast acquisition time-of-flight mass spectrometer (TOFMS). The latter is one of the preferred detection methods for the GC × GC today because it enables deconvolution of overlapping peaks, quantification, and identification of analytes via mass spectral libraries. In the past, GC × GC−based metabolomics was lacking appropriate data processing, alignment, and analysis tools. Today a variety of algorithms are available. However, the entire procedure of obtaining biological knowledge from raw data is still a time-consuming and challenging task in need of complete automation. In addition, the precision of deconvolution approaches must be further optimized (Almstetter et al., 2012). GC × GC-MS might become a premier metabolomics tool once the challenges of automated and high-throughput data processing have been resolved (Bedair and Sumner, 2008).

GC-MS−based methods are utilizing EI, classical chemical ionization (CI), and other soft ionization approaches such as APCI and atmospheric pressure photo ionization (APPI).

Application of EI where the ionization typically is performed at 70 eV (considered as "hard") leads to the reproducible fragmentation of molecules into well-characterized spectra that could be used for comparison of the data acquired at the instruments manufactured by different vendors and across instruments with different analyzers. This feature and relatively large availability of mass spectral databases and libraries provided by many instrumental suppliers (for example, Agilent and Thermo), NIST, or publicly available, for example, MassBank, MassBank of North America, or GOLM Metabolome Database (Kopka et al., 2005) resulted in most of the metabolomics applications applying EI.

CI has been applied less extensively as EI in the field of metabolomics (Jaeger et al., 2016) as being a soft ionization technique does not produce a very distinctive fragmentation pattern. However, it does provide a distinctive molecular ion and its popularity is increasing especially in targeted metabolomics (Warren, 2013). CI techniques should be considered as complementary to EI and the applicability of the CI techniques for the nontargeted approaches should not be underestimated. Selectivity and sensitivity differences between EI and CI could also be observed (Turner et al., 2011). EI as a hard ionization technique, with a molecular ion often not present, may result in false-positive identifications in library search, causing a need of further data curation (Rostkowski et al., 2019). Besides, the lack of molecular ion makes annotation of putatively identified compounds and unknowns a difficult and often an impossible task.

Other soft ionization sources (APCI and APPI) are typically coupled to high-resolution systems including TOF, Orbitrap, and ion cyclotron resonance systems.

APCI sources on the same mass spectrometer can also be exchangeable with LC and GC with a change of one chromatographic system to another being done in a concise time (minutes).

Like CI, APCI sources generate molecular ion. APCI has been used for many metabolomics applications (Hurtado-Fernandez et al., 2013, 2014, 2015; Jaeger et al., 2014; Olmo-Garcia et al., 2018; Pacchiarotta et al., 2015; Ruttkies et al., 2015; Shackleton et al., 2018).

APPI source for the GC was first introduced in 2007 by Waters (McEwen, 2007), soon followed by Bruker (Carrasco-Pancorbo et al., 2009). In 2016, GC-APPI become available for Thermo Orbitrap (Kersten et al., 2016).

3.2.2 Liquid chromatography coupled to mass spectrometry

The application of LC-MS in metabolomics has recently been increasing (Fig. 2.1) (Miggiels et al., 2019). This platform, typically with ESI, is the most common analytical platform used in nontargeted metabolomics (Chetwynd and David, 2018; Gika et al., 2014; Theodoridis et al., 2012).

LC is a universal technique of separation that can be employed for both targeted and nontargeted approaches with an added value of analyte recovery by fraction collection and/or concentration. All these are far more challenging when GC separations are applied (Bedair and Sumner, 2008).

The application of LC-MS systems offers several advantages over GC-MS. They operate at lower temperatures, which enable the analyses of heat-labile metabolites that are known to degrade in GC analysis. LC analyses typically do not require chemical derivatization, which makes the sample preparation steps and identification of the metabolites an easier task. Furthermore, the use of ESI source also provides a soft ionization process which allows structural elucidation (Naz et al., 2014; Roux et al., 2011; Vuckovic, 2012).

However, the use of LC-ESI-MS—driven metabolomics is not without shortcoming, and the main issue for this technique is ion suppression. This phenomenon and the mechanisms of ion suppression in the case of LC-ESI-MS have been already extensively described in several reviews (Annesley, 2003; Antignac et al., 2005; Trufelli et al., 2011). Basically, the main cause of ion suppression is a change in the spray droplet solution properties (i.e., increase in viscosity and surface tension) caused by the presence of nonvolatile or less volatile solutes (Annesley, 2003; Antignac et al., 2005; Trufelli et al., 2011). These interfering or coeluting nonvolatile chemicals decrease the evaporation efficiency and therefore reduce the number of charged ion in the gas phase that ultimately reaches the detector (Annesley, 2003; Antignac et al., 2005).

This sensitivity issue caused by ion suppression can be overcome to some extent by miniaturization of LC and ESI to nanoflow LC systems (nLC) and nanospray ionization sources (nESI) (Chetwynd and David, 2018). The increased sensitivity of the nLC-nESI can be mainly attributed to nESI since the droplets formed in the ESI plume are 100- to 1000-fold smaller than the typical droplets emitted from conventional ESI emitters (Wilm and Mann, 1996). The generation of significantly smaller droplets with lower volumes considerably increases the rate of desolvation and therefore the ionization efficiency (Karas et al., 2000; Marginean et al., 2008, 2014; Wilm and Mann, 1996). However, sample preparation is an important aspect to address to fully enjoy the benefits of these nLC-nESI platforms. Narrow bore columns, emitters, and connections of nLC-nESI are prone to blockages. Consequently, sample clean-up methods such as SPE are recommended to efficiently remove matrix components such as proteins and salts (David et al., 2014).

Another issue affecting the robustness of the method is the contamination of the MS source and adduct formation. For efficient and successful identification and/or accurate quantification of metabolites, it is essential to detect these artifacts, before normalization of the data.

The identification of metabolites in LC-MS is based on the determination of an accurate mass, MS/MS analysis, and/or coupling to nuclear magnetic resonance spectroscopy (LC-MS-NMR).

Another major disadvantage of the LC-MS in comparison with GC-MS profiling is very limited availability of transferable mass spectral libraries that would facilitate metabolite identification. The mass spectra generated on different LC-MS systems vary a lot in terms of relative abundances, adduct formation, in-source fragmentation, MS/MS fragment ions, and the lack of LC retention indices that could compensate for instrumental and experimental variations. Therefore, usually within research groups, own in house libraries are being created. However, open-source, community-driven initiatives, gathering properly curated spectra, that were already mentioned earlier, are a step forward in automated metabolite identifications.

The development of novel LC technologies with an increased sample throughput and metabolite coverage has been moving at a greater pace since the commercial introduction of the first ultrahigh-pressure LC (2004). This improvement enabled the use of high operating pressures (to date up to 1500 bar), flow rates (up to 5 mL/min), and the use of sub-2-μm columns particle sizes. It resulted in significantly improved chromatographic resolution and efficiency (Wilson et al., 2005), which contributes to increasing the resolution of complex mixtures in nontargeted applications.

The implementation of 2D-LC-MS for metabolomics has lagged behind that of 2D-GC-MS, due to a complicated experimental setup and loss of sensitivity due to a sample dilution effect in the second dimension (Lei et al., 2011).

3.2.3 Ion mobility mass spectrometry

Ion mobility spectroscopy (IMS) is a separation technique utilizing differential mobility of the ions in a buffer gas in the weak uniform electric field (Eiceman et al., 2013). It separates ions based on their shape, size, and charge, which makes it very useful in separations of, for example, isomeric structures. The measured drift time can be converted into the collisional cross section, a unique property of the ion. Ion mobility mass spectrometry (IM-MS) can add another layer to confidence in the identification of unknown metabolites (Zhang et al., 2018). The implementation of ion mobility spectroscopy (IMS) for application in untargeted MS studies has been recently extensively reviewed by Metz et al. (2017).

The authors argued that IMS appears as a very appealing option for improving the coverage, dynamic range, and throughput of measurements in current MS-based analytical methods (Metz et al., 2017). Several IMS-based platforms such as drift tube IMS (DTIMS), traveling wave IMS (TWIMS), trapped IMS, overtone IM, differential IMS, field asymmetric IMS, and transversal modulation IMS are currently available (Metz et al.,

2017). However, among all these techniques, DTIMS offers the advantage of enabling the direct determination of molecular structural information without the external calibration approaches that are required by other IMS-based platforms (Hines et al., 2016). Other strengths of DTIMS lie in its ability to resolve isomers that are difficult to distinguish using LC-MS alone, its high reproducibility of measurements, and its capacity to be used coupled with TOFMS in an ultrahigh-throughput manner (Metz et al., 2017).

Although IM-MS has already its value in metabolomics, it also has a few limitations to overcome; it can reduce the sensitivity of analysis, increase the size of the raw data files, and complicate high-throughput data treatment (Miggiels et al., 2019).

3.2.4 Capillary electrophoresis mass spectrometry

Although CE-MS emerged as a useful analytical technique for the profiling of (highly) polar and charged metabolites (Begou et al., 2017; Hirayama and Soga, 2018; Kuehnbaum and Britz-McKibbin, 2013; Ramautar et al., 2019; Sanchez-Lopez et al., 2019) until now, it has only been used by a limited number of research groups for metabolomics studies.

CE is mainly hyphenated to mass spectrometers via ESI with a sheath gas interface (Miggiels et al., 2019), although such coupling and achieving sufficient sensitivity and robustness can be rather challenging. CE has superior separation efficiency compared to LC; however, migration times are often less reproducible and sensitivity is usually lower as compared to LC-MS (Scalbert et al., 2009). Recent developments and applications were extensively reviewed by Ramautar et al. (2019). The papers showed that the potential of CE-MS has been recognized by multiple research groups over the past 2 years and resulted in over 40 publications. Many of the reported studies were focused on the analysis of volume-limited samples, such as saliva, sweat, dried blood spots, and similar. Due to developments of the platform (for example, MS interfacing), an improvement in sensitivity and stability has been reported (Ramautar et al., 2019). It is anticipated that CE-MS will become a key technique for volume-restricted metabolomics, complementing other platforms in comprehensive metabolomics studies (Ramautar et al., 2019).

4. Summary and future trends

The ongoing and recently very rapid development of sample preparation methods and analytical instrumentation, including GC, HPLC, UHPLC, CE coupled to high and ultrahigh-resolution mass spectrometry, could facilitate separation, detection, characterization, and quantification of known and discovery of new metabolites and related metabolic pathways. Due to the complexity of the metabolome (e.g., the wide range of physicochemical properties of metabolites) and the dynamic ranges of concentrations observed, no single analytical platform can be applied to detect all metabolites in a biological sample. The combined use of modern instrumental analytical approaches is needed to increase the coverage of detected metabolites that cannot be achieved by single-analysis techniques. Continued development of these analytical platforms will accelerate widespread use and integration of not only metabolomics into systems biology but also in exposomics studies since metabolomics is the only omics techniques also providing holistic information on environmental exposures.

Acknowledgments

Arthur David was funded by a research chair of excellence (2016-52/IdeX University of Sorbonne Paris Cité). Pawel Rostkowski was funded by NILU (E-116055).

References

Aggio, R., Villas-Boas, S.G., Ruggiero, K., 2011. Metab: an R package for high-throughput analysis of metabolomics data generated by GC-MS. Bioinformatics 27, 2316–2318.

Al-Salhi, R., Abdul-Sada, A., Lange, A., Tyler, C.R., Hill, E.M., 2012. The xenometabolome and novel contaminant markers in fish exposed to a wastewater treatment works effluent. Environ. Sci. Technol. 46, 9080–9088.

Almstetter, M.F., Oefner, P.J., Dettmer, K., 2012. Comprehensive two-dimensional gas chromatography in metabolomics. Anal. Bioanal. Chem. 402, 1993–2013.

Annesley, T.M., 2003. Ion suppression in mass spectrometry. Clin. Chem. 49, 1041–1044.

Antignac, J.P., de Wasch, K., Monteau, F., De Brabander, H., Andre, F., Le Bizec, B., 2005. The ion suppression phenomenon in liquid chromatography–mass spectrometry and its consequences in the field of residue analysis. Anal. Chim. Acta 529, 129–136.

Balog, J., Sasi-Szabó, L., Kinross, J., Lewis, M.R., Muirhead, L.J., Veselkov, K., Mirnezami, R., Dezső, B., Damjanovich, L., Darzi, A., Nicholson, J.K., Takáts, Z., 2013. Intraoperative tissue identification using rapid evaporative ionisation mass spectrometry. Sci. Transl. Med. 5.

Beale, D.J., Pinu, F.R., Kouremenos, K.A., Poojary, M.M., Narayana, V.K., Boughton, B.A., Kanojia, K., Dayalan, S., Jones, O.A.H., Dias, D.A., 2018. Review of recent developments in GC-MS approaches to metabolomics-based research. Metabolomics 14.

Bedair, M., Sumner, L.W., 2008. Current and emerging mass-spectrometry technologies for metabolomics. Trends Anal. Chem. 27, 238–250.

Begou, O., Gika, H.G., Wilson, I.D., Theodoridis, G., 2017. Hyphenated MS-based targeted approaches in metabolomics. Analyst 142, 3079–3100.

Bojko, B., Reyes-Garcés, N., Bessonneau, V., Goryński, K., Mousavi, F., Souza Silva, E.A., Pawliszyn, J., 2014. Solid-phase microextraction in metabolomics. Trends Anal. Chem. 61, 168–180.

Bonvallot, N., David, A., Chalmel, F., Chevrier, C., Cordier, S., Cravedi, J.-P., Zalko, D., 2018. Metabolomics as a powerful tool to decipher the biological effects of environmental contaminants in humans. Curr. Opin. Toxicol. 8, 48–56.

Bowen, B.P., Northen, T.R., 2010. Dealing with the unknown: metabolomics and metabolite atlases. J. Am. Soc. Mass Spectrom. 21, 1471–1476.

Bruce, S.J., Tavazzi, I., Parisod, V., Rezzi, S., Kochhar, S., Guy, P.A., 2009. Investigation of human blood plasma sample preparation for performing metabolomics using ultrahigh performance liquid chromatography/mass spectrometry. Anal. Chem. 81, 3285–3296.

Carrasco-Pancorbo, A., Nevedomskaya, E., Arthen-Engeland, T., Zey, T., Zurek, G., Baessmann, C., Deelder, A.M., Mayboroda, O.A., 2009. Gas chromatography/atmospheric pressure chemical ionisation-time of flight mass spectrometry: analytical validation and applicability to metabolic profiling. Anal. Chem. 81, 10071–10079.

Chen, Y., Xu, J., Zhang, R., Abliz, Z., 2016. Methods used to increase the comprehensive coverage of urinary and plasma metabolomes by MS. Bioanalysis 8, 981−997.

Chetwynd, A.J., David, A., 2018. A review of nanoscale LC-ESI for metabolomics and its potential to enhance the metabolome coverage. Talanta 182, 380−390.

Chetwynd, A.J., Abdul-Sada, A., Hill, E.M., 2015. Solid-phase extraction and nanoflow liquid chromatography-nanoelectrospray ionisation mass spectrometry for improved global urine metabolomics. Anal. Chem. 87, 1158−1165.

Chong, J., Soufan, O., Li, C., Caraus, I., Li, S.Z., Bourque, G., Wishart, D.S., Xia, J.G., 2018. MetaboAnalyst 4.0: towards more transparent and integrative metabolomics analysis. Nucleic Acids Res. 46, W486−W494.

David, A., Abdul-Sada, A., Lange, A., Tyler, C.R., Hill, E.M., 2014. A new approach for plasma (xeno) metabolomics based on solid-phase extraction and nanoflow liquid chromatography-nanoelectrospray ionisation mass spectrometry. J. Chromatogr. A 1365, 72−85.

David, A., Lange, A., Abdul-Sada, A., Tyler, C.R., Hill, E.M., 2017. Disruption of the prostaglandin metabolome and characterization of the pharmaceutical exposome in fish exposed to wastewater treatment works effluent as revealed by nanoflow-nanospray mass spectrometry-based metabolomics. Environ. Sci. Technol. 51, 616−624.

Dettmer, K., Aronov, P.A., Hammock, B.D., 2007. Mass spectrometry-based metabolomics. Mass Spectrom. Rev. 26, 51−78.

Domingo-Almenara, X., Brezmes, J., Vinaixa, M., Samino, S., Ramirez, N., Ramon-Krauel, M., Lerin, C., Diaz, M., Ibanez, L., Correig, X., Perera-Lluna, A., Yanes, O., 2016. eRah: a computational tool integrating spectral deconvolution and alignment with quantification and identification of metabolites in GC/MS-based metabolomics. Anal. Chem. 88, 9821−9829.

Eiceman, G.A., Karpas, Z., Hill Jr., H.H., 2013. Ion mobility spectrometry, third edition. CRPC Press, Boca Raton, FL.

Fernández-Peralbo, M.A., Luque de Castro, M.D., 2012. Preparation of urine samples prior to targeted or untargeted metabolomics mass-spectrometry analysis. Trends Anal. Chem. 41, 75−85.

Ferreira, A.M.C., Laespada, M.E.F., Pavón, J.L.P., Cordero, B.M., 2013. In situ aqueous derivatisation as sample preparation technique for gas chromatographic determinations. J. Chromatogr. A 1296, 70−83.

Forsberg, E.M., Huan, T., Rinehart, D., Benton, H.P., Warth, B., Hilmers, B., Siuzdak, G., 2018. Data processing, multi-omic pathway mapping, and metabolite activity analysis using XCMS Online. Nat. Protoc. 13, 633−651.

G Marshall, A., T Blakney, G., Chen, T., K Kaiser, N., M McKenna, A., P Rodgers, R., M Ruddy, B., Xian, F., 2013. Mass Resolution and Mass Accuracy: How Much Is Enough? Mass Spectrometry (Tokyo, Japan), vol. 2. S0009−S0009.

Gika, H.G., Theodoridis, G.A., Plumb, R.S., Wilson, I.D., 2014. Current practice of liquid chromatography-mass spectrometry in metabolomics and metabonomics. J. Pharmaceut. Biomed. Anal. 87, 12−25.

Gowda, G.A., Djukovic, D., 2014. Overview of mass spectrometry-based metabolomics: opportunities and challenges. Methods Mol. Biol. 1198, 3−12.

Gowda, H., Ivanisevic, J., Johnson, C.H., Kurczy, M.E., Benton, H.P., Rinehart, D., Nguyen, T., Ray, J., Kuehl, J., Arevalo, B., Westenskow, P.D., Wang, J.H., Arkin, A.P., Deutschbauer, A.M., Patti, G.J., Siuzdak, G., 2014. Interactive XCMS online: simplifying advanced metabolomic data processing and subsequent statistical analyses. Anal. Chem. 86, 6931−6939.

Guijas, C., Montenegro-Burke, J.R., Domingo-Almenara, X., Palermo, A., Warth, B., Hermann, G., Koellensperger, G., Huan, T., Uritboonthai, W., Aisporna, A.E., Wolan, D.W., Spilker, M.E., Benton, H.P., Siuzdak, G., 2018. METLIN: a technology platform for identifying knowns and unknowns. Anal. Chem. 90, 3156−3164.

Hines, K.M., May, J.C., McLean, J.A., Xu, L., 2016. Evaluation of collision cross section calibrants for structural analysis of lipids by traveling wave ion mobility-mass spectrometry. Anal. Chem. 88, 7329–7336.

Hirayama, A., Soga, T., 2018. CHAPTER 7 CE-MS for anionic and cationic metabolic profiling: system optimization and applications, capillary electrophoresis–mass spectrometry for metabolomics. Roy. Soc. Chem. 134–160.

Holmes, E., Loo, R.L., Cloarec, O., Coen, M., Tang, H.R., Maibaum, E., Bruce, S., Chan, Q., Elliott, P., Stamler, J., Wilson, I.D., Lindon, J.C., Nicholson, J.K., 2007. Detection of urinary drug metabolite (Xenometabolome) signatures in molecular epidemiology studies via statistical total correlation (NMR) spectroscopy. Anal. Chem. 79, 2629–2640.

Horai, H., Arita, M., Kanaya, S., Nihei, Y., Ikeda, T., Suwa, K., Ojima, Y., Tanaka, K., Tanaka, S., Aoshima, K., Oda, Y., Kakazu, Y., Kusano, M., Tohge, T., Matsuda, F., Sawada, Y., Hirai, M.Y., Nakanishi, H., Ikeda, K., Akimoto, N., Maoka, T., Takahashi, H., Ara, T., Sakurai, N., Suzuki, H., Shibata, D., Neumann, S., Iida, T., Tanaka, K., Funatsu, K., Matsuura, F., Soga, T., Taguchi, R., Saito, K., Nishioka, T., 2010. MassBank: a public repository for sharing mass spectral data for life sciences. J. Mass Spectrom. 45, 703–714.

Hu, M., Krauss, M., Brack, W., Schulze, T., 2016. Optimization of LC-Orbitrap-HRMS acquisition and MZmine 2 data processing for nontarget screening of environmental samples using design of experiments. Anal. Bioanal. Chem. 408, 7905–7915.

Huan, T., Forsberg, E.M., Rinehart, D., Johnson, C.H., Ivanisevic, J., Benton, H.P., Fang, M.L., Aisporna, A., Hilmers, B., Poole, F.L., Thorgersen, M.P., Adams, M.W.W., Krantz, G., Fields, M.W., Robbins, P.D., Niedernhofer, L.J., Ideker, T., Majumder, E.L., Wall, J.D., Rattray, N.J.W., Goodacre, R., Lairson, L.L., Siuzdak, G., 2017. Systems biology guided by XCMS Online metabolomics. Nat. Methods 14, 461–462.

Hurtado-Fernandez, E., Pacchiarotta, T., Longueira-Suarez, E., Mayboroda, O.A., Fernandez-Gutierrez, A., Carrasco-Pancorbo, A., 2013. Evaluation of gas chromatography-atmospheric pressure chemical ionisation-mass spectrometry as an alternative to gas chromatography-electron ionisation-mass spectrometry: avocado fruit as example. J. Chromatogr. A 1313, 228–244.

Hurtado-Fernandez, E., Pacchiarotta, T., Mayboroda, O.A., Fernandez-Gutierrez, A., Carrasco-Pancorbo, A., 2014. Quantitative characterization of important metabolites of avocado fruit by gas chromatography coupled to different detectors (APCI-TOF MS and FID). Food Res. Int. 62, 801–811.

Hurtado-Fernandez, E., Pacchiarotta, T., Mayboroda, O.A., Fernandez-Gutierrez, A., Carrasco-Pancorbo, A., 2015. Metabolomic analysis of avocado fruits by GC-APCI-TOF MS: effects of ripening degrees and fruit varieties. Anal. Bioanal. Chem. 407, 547–555.

Hyötyläinen, T., 2013. CHAPTER 2 Sample Collection, Storage and Preparation, Chromatographic Methods in Metabolomics. The Royal Society of Chemistry, pp. 11–42.

Jaeger, C., Tellstrom, V., Zurek, G., Konig, S., Eimer, S., Kammerer, B., 2014. Metabolomic changes in *Caenorhabditis elegans* lifespan mutants as evident from GC-EI-MS and GC-APCI-TOF-MS profiling. Metabolomics 10, 859–876.

Jaeger, C., Hoffmann, F., Schmitt, C.A., Lisec, J., 2016. Automated annotation and evaluation of in-source mass spectra in GC/atmospheric pressure chemical ionisation-MS-based metabolomics. Anal. Chem. 88, 9386–9390.

Jézéquel, T., Deborde, C., Maucourt, M., Zhendre, V., Moing, A., Giraudeau, P., 2015. Absolute quantification of metabolites in tomato fruit extracts by fast 2D NMR. Metabolomics 11, 1231–1242.

Johnson, C.H., Patterson, A.D., Idle, J.R., Gonzalez, F.J., 2012a. Xenobiotic metabolomics: major impact on the metabolome. Annu. Rev. Pharmacol. Toxicol. 52, 37–56.

Johnson, C.H., Patterson, A.D., Idle, J.R., Gonzalez, F.J., 2012b. Xenobiotic metabolomics: major impact on the metabolome. Annu. Rev. Pharmacol. Toxicol. 52, 37–56.

Karas, M., Bahr, U., Dulcks, T., 2000. Nano-electrospray ionisation mass spectrometry: addressing analytical problems beyond routine. Fresenius' J. Anal. Chem. 366, 669–676.

Katajamaa, M., Miettinen, J., Oresic, M., 2006. MZmine: toolbox for processing and visualization of mass spectrometry based molecular profile data. Bioinformatics 22, 634−636.

Kaufmann, A., 2014. Combining UHPLC and high-resolution MS: a viable approach for the analysis of complex samples? Trends Anal. Chem. 63, 113−128.

Kersten, H., Kroll, K., Haberer, K., Brockmann, K.J., Benter, T., Peterson, A., Makarov, A., 2016. Design study of an atmospheric pressure photoionisation interface for GC-MS. J. Am. Soc. Mass Spectrom. 27, 607−614.

Kopka, J., Schauer, N., Krueger, S., Birkemeyer, C., Usadel, B., Bergmuller, E., Dormann, P., Weckwerth, W., Gibon, Y., Stitt, M., Willmitzer, L., Fernie, A.R., Steinhauser, D., 2005. the golm metabolome database. Bioinformatics 21, 1635−1638. GMD@CSB.DB:.

Kuehnbaum, N.L., Britz-McKibbin, P., 2013. New advances in separation science for metabolomics: resolving chemical diversity in a post-genomic era. Chem. Rev. 113, 2437−2468.

Lai, Z., Fiehn, O., 2018. Mass spectral fragmentation of trimethylsilylated small molecules. Mass. Spectrom. Rev 37, 245−257.

Lei, Z., Huhman, D.V., Sumner, L.W., 2011. Mass spectrometry strategies in metabolomics. J. Biol. Chem. 286, 25435−25442.

Libiseller, G., Dvorzak, M., Kleb, U., Gander, E., Eisenberg, T., Madeo, F., Neumann, S., Trausinger, G., Sinner, F., Pieber, T., Magnes, C., 2015. IPO: a tool for automated optimization of XCMS parameters. BMC Bioinf. 16.

Lu, H.M., Dunn, W.B., Shen, H.L., Kell, D.B., Liang, Y.Z., 2008. Comparative evaluation of software for deconvolution of metabolomics data based on GC-TOF-MS. Trends Anal. Chem. 27, 215−227.

Mahieu, N.G., Genenbacher, J.L., Patti, G.J., 2016. A roadmap for the XCMS family of software solutions in metabolomics. Curr. Opin. Chem. Biol. 30, 87−93.

Marginean, I., Kelly, R.T., Prior, D.C., LaMarche, B.L., Tang, K., Smith, R.D., 2008. Analytical characterization of the electrospray ion source in the nanoflow regime. Anal. Chem. 80, 6573−6579.

Marginean, I., Tang, K., Smith, R.D., Kelly, R.T., 2014. Picoelectrospray ionisation mass spectrometry using narrow-bore chemically etched emitters. J. Am. Soc. Mass Spectrom. 25, 30−36.

Markley, J.L., Bruschweiler, R., Edison, A.S., Eghbalnia, H.R., Powers, R., Raftery, D., Wishart, D.S., 2017. The future of NMR-based metabolomics. Curr. Opin. Biotechnol. 43, 34−40.

Mastrangelo, A., Ferrarini, A., Rey-Stolle, F., Garcia, A., Barbas, C., 2015. From sample treatment to biomarker discovery: a tutorial for untargeted metabolomics based on GC-(EI)-Q-MS. Anal. Chim. Acta 900, 21−35.

Matyash, V., Liebisch, G., Kurzchalia, T.V., Shevchenko, A., Schwudke, D., 2008. Lipid extraction by methyl-tert-butyl ether for high-throughput lipidomics. J. Lipid Res. 49, 1137−1146.

McEwen, C.N., 2007. GC/MS on an LC/MS instrument using atmospheric pressure photoionisation. Int. J. Mass Spectrom. 259, 57−64.

Metz, T.O., Baker, E.S., Schymanski, E.L., Renslow, R.S., Thomas, D.G., Causon, T.J., Webb, I.K., Hann, S., Smith, R.D., Teeguarden, J.G., 2017. Integrating ion mobility spectrometry into mass spectrometry-based exposome measurements: what can it add and how far can it go? Bioanalysis 9, 81−98.

Michopoulos, F., Lai, L., Gika, H., Theodoridis, G., Wilson, I., 2009. UPLC-MS-based analysis of human plasma for metabonomics using solvent precipitation or solid phase extraction. J. Proteome Res. 8, 2114−2121.

Miggiels, P., Wouters, B., van Westen, G.J.P., Dubbelman, A.-C., Hankemeier, T., 2019. Novel technologies for metabolomics: more for less. Trends Anal. Chem. 120, 115323.

Mitra, S., 2003. Sample Preparation Techniques in Analytical Chemistry. Wiley, Hobo-ken, New Jersey.

Moser, E., Laistler, E., Schmitt, F., Kontaxis, G., 2017. Corrigendum: ultra-high field NMR and MRI—the role of magnet technology to increase sensitivity and specificity. Front. Phys. 5.

Myers, O.D., Sumner, S.J., Li, S.Z., Barnes, S., Du, X.X., 2017. Detailed investigation and comparison of the XCMS and MZmine 2 chromatogram construction and chromatographic peak detection methods for preprocessing mass spectrometry metabolomics data. Anal. Chem. 89, 8689−8695.

Naz, S., Vallejo, M., Garcia, A., Barbas, C., 2014. Method validation strategies involved in non-targeted metabolomics. J. Chromatogr. A 1353, 99–105.

Olivon, F., Grelier, G., Roussi, F., Litaudon, M., Touboul, D., 2017. MZmine 2 data-preprocessing to enhance molecular networking reliability. Anal. Chem. 89, 7836–7840.

Olmo-Garcia, L., Kessler, N., Neuweger, H., Wendt, K., Olmo-Peinado, J.M., Fernandez-Gutierrez, A., Baessmann, C., Carrasco-Pancorbo, A., 2018. Unravelling the distribution of secondary metabolites in *Olea europaea* L.: exhaustive characterization of Eight Olive-Tree derived matrices by complementary platforms (LC-ESI/APCI-MS and GC-APCI-MS). Molecules 23.

Pacchiarotta, T., Derks, R.J., Nevedomskaya, E., van der Starre, W., van Dissel, J., Deelder, A., Mayboroda, O.A., 2015. Exploratory analysis of urinary tract infection using a GC-APCI-MS platform. Analyst 140, 2834–2841.

Patti, G.J., Yanes, O., Siuzdak, G., 2012. Innovation: metabolomics: the apogee of the omics trilogy. Nat. Rev. Mol. Cell Biol. 13, 263–269.

Pluskal, T., Castillo, S., Villar-Briones, A., Oresic, M., 2010. MZmine 2: modular framework for processing, visualizing, and analyzing mass spectrometry-based molecular profile data. BMC Bioinf. 11.

Ramaswamy, V., Hooker, J.W., Withers, R.S., Nast, R.E., Brey, W.W., Edison, A.S., 2013. Development of a 13C-optimized 1.5-mm high temperature superconducting NMR probe. J. Magn. Reson. 235, 58–65.

Ramautar, R., Somsen, G.W., de Jong, G.J., 2019. CE-MS for metabolomics: developments and applications in the period 2016-2018. Electrophoresis 40, 165–179.

Raterink, R.-J., Lindenburg, P.W., Vreeken, R.J., Ramautar, R., Hankemeier, T., 2014. Recent developments in sample-pretreatment techniques for mass spectrometry-based metabolomics. Trends Anal. Chem. 61, 157–167.

Roemisch-Margl, W., Prehn, C., Bogumil, R., Roehring, C., Suhre, K., Adamski, J., 2011. Procedure for tissue sample preparation and metabolite extraction for high-throughput targeted metabolomics. Metabolomics 8, 133–142.

Rostkowski, P., Haglund, P., Aalizadeh, R., Alygizakis, N., Thomaidis, N., Arandes, J.B., Nizzetto, P.B., Booij, P., Budzinski, H., Brunswick, P., Covaci, A., Gallampois, C., Grosse, S., Hindle, R., Ipolyi, I., Jobst, K., Kaserzon, S.L., Leonards, P., Lestremau, F., Letzel, T., Magnér, J., Matsukami, H., Moschet, C., Oswald, P., Plassmann, M., Slobodnik, J., Yang, C., 2019. The strength in numbers: comprehensive characterization of house dust using complementary mass spectrometric techniques. Anal. Bioanal. Chem. 411, 1957–1977.

Roux, A., Lison, D., Junot, C., Heilier, J.F., 2011. Applications of liquid chromatography coupled to mass spectrometry-based metabolomics in clinical chemistry and toxicology: a review. Clin. Biochem. 44, 119–135.

Ruttkies, C., Strehmel, N., Scheel, D., Neumann, S., 2015. Annotation of metabolites from gas chromatography/atmospheric pressure chemical ionisation tandemmass spectrometry data using an in silico generated compound database and MetFrag. Rapid Commun. Mass Spectrom. 29, 1521–1529.

Saggiomo, V., Velders, A.H., 2015. Simple 3D printed scaffold-removal method for the fabrication of intricate microfluidic devices. Adv. Sci. 2, 1500125.

Sanchez-Lopez, E., Kammeijer, G.S.M., Crego, A.L., Marina, M.L., Ramautar, R., Peters, D.J.M., Mayboroda, O.A., 2019. Sheathless CE-MS based metabolic profiling of kidney tissue section samples from a mouse model of Polycystic Kidney Disease. Sci. Rep. 9, 806.

Scalbert, A., Brennan, L., Fiehn, O., Hankemeier, T., Kristal, B.S., van Ommen, B., Pujos-Guillot, E., Verheij, E., Wishart, D., Wopereis, S., 2009. Mass-spectrometry-based metabolomics: limitations and recommendations for future progress with particular focus on nutrition research. Metabolomics 5, 435–458.

Scalbert, A., Brennan, L., Manach, C., Andres-Lacueva, C., Dragsted, L.O., Draper, J., Rappaport, S.M., van der Hooft, J.J., Wishart, D.S., 2014. The food metabolome: a window over dietary exposure. Am. J. Clin. Nutr. 99, 1286–1308.

Shackleton, C., Pozo, O.J., Marcos, J., 2018. GC/MS in recent years has defined the normal and clinically disordered steroidome: will it soon Be surpassed by LC/tandem MS in this role? J. Endocr. Soc. 2, 974–996.

Sitnikov, D.G., Monnin, C.S., Vuckovic, D., 2016. Systematic assessment of seven solvent and solid-phase extraction methods for metabolomics analysis of human plasma by LC-MS. Sci. Rep. 6, 38885.

Siuzdak, G., 2014. Cloud-based XCMS: a metabolomic technology for environmental exposure. Abstr. Pap. Am. Chem. Soc. 248.

Smirnov, A., Jia, W., Walker, D.I., Jones, D.P., Du, X.X., 2018. ADAP-GC 3.2: graphical software tool for efficient spectral deconvolution of gas chromatography-high-resolution mass spectrometry metabolomics data. J. Proteome Res. 17, 470–478.

Smith, C.A., Want, E.J., O'Maille, G., Abagyan, R., Siuzdak, G., 2006. XCMS: processing mass spectrometry data for metabolite profiling using nonlinear peak alignment, matching, and identification. Anal. Chem. 78, 779–787.

Söderholm, S.L., Damm, M., Kappe, C.O., 2010. Microwave-assisted derivatisation procedures for gas chromatography/mass spectrometry analysis. Mol. Divers. 14, 869–888.

Sostare, J., Di Guida, R., Kirwan, J., Chalal, K., Palmer, E., Dunn, W.B., Viant, M.R., 2018. Comparison of modified Matyash method to conventional solvent systems for polar metabolite and lipid extractions. Anal. Chim. Acta 1037, 301–315.

Soule, M.C.K., Longnecker, K., Johnson, W.M., Kujawinski, E.B., 2015. Environmental metabolomics: analytical strategies. Mar. Chem. 177, 374–387.

Spicer, R., Salek, R.M., Moreno, P., Canueto, D., Steinbeck, C., 2017. Navigating freely-available software tools for metabolomics analysis. Metabolomics 13.

Steehler, J.K., 2004. Sample preparation techniques in analytical chemistry (Mitra, Somenath). J. Chem. Educ. 81, 199.

Stravs, M.A., Schymanski, E.L., Singer, H.P., Hollender, J., 2013. Automatic recalibration and processing of tandem mass spectra using formula annotation. J. Mass Spectrom. 48, 89–99.

Stravs, M.A., Mechelke, J., Ferguson, P.L., Singer, H., Hollender, J., 2016. Microvolume trace environmental analysis using peak-focusing online solid-phase extraction-nano-liquid chromatography-high-resolution mass spectrometry. Anal. Bioanal. Chem. 408, 1879–1890.

Sumner, L.W., Mendes, P., Dixon, R.A., 2003. Plant metabolomics: large-scale phytochemistry in the functional genomics era. Phytochemistry 62, 817–836.

Tautenhahn, R., Patti, G.J., Rinehart, D., Siuzdak, G., 2012. XCMS online: a web-based platform to process untargeted metabolomic data. Anal. Chem. 84, 5035–5039.

Theodoridis, G.A., Gika, H.G., Want, E.J., Wilson, I.D., 2012. Liquid chromatography-mass spectrometry based global metabolite profiling: a review. Anal. Chim. Acta 711, 7–16.

Tiziani, S., Emwas, A.H., Lodi, A., Ludwig, C., Bunce, C.M., Viant, M.R., Gunther, U.L., 2008. Optimized metabolite extraction from blood serum for 1H nuclear magnetic resonance spectroscopy. Anal. Biochem. 377, 16–23.

Trufelli, H., Palma, P., Famiglini, G., Cappiello, A., 2011. An overview of matrix effects in liquid chromatography-mass spectrometry. Mass Spectrom. Rev. 30, 491–509.

Tsugawa, H., Cajka, T., Kind, T., Ma, Y., Higgins, B., Ikeda, K., Kanazawa, M., VanderGheynst, J., Fiehn, O., Arita, M., 2015. MS-DIAL: data-independent MS/MS deconvolution for comprehensive metabolome analysis. Nat. Methods 12, 523–526.

Tulipani, S., Llorach, R., Urpi-Sarda, M., Andres-Lacueva, C., 2013. Comparative analysis of sample preparation methods to handle the complexity of the blood fluid metabolome: when less is more. Anal. Chem. 85, 341–348.

Turner, M.A., Guallar-Hoyas, C., Kent, A.L., Wilson, I.D., Thomas, C.L.P., 2011. Comparison of metabolomic profiles obtained using chemical ionisation and electron ionisation MS in exhaled breath. Bioanalysis 3, 2731–2738.

Viant, M.R., 2007. Metabolomics of aquatic organisms the new 'omics' on the block. Mar. Ecol. Prog. Ser. 332, 301−306.

Villas-Boas, S.G., Mas, S., Akesson, M., Smedsgaard, J., Nielsen, J., 2005. Mass spectrometry in metabolome analysis. Mass Spectrom. Rev. 24, 613−646.

Vineis, P., Chadeau-Hyam, M., Gmuender, H., Gulliver, J., Herceg, Z., Kleinjans, J., Kogevinas, M., Kyrtopoulos, S., Nieuwenhuijsen, M., Phillips, D.H., Probst-Hensch, N., Scalbert, A., Vermeulen, R., Wild, C.P., 2017. The exposome in practice: design of the EXPOsOMICS project. Int. J. Hyg Environ. Health 220, 142−151.

Vuckovic, D., 2012. Current trends and challenges in sample preparation for global metabolomics using liquid chromatography-mass spectrometry. Anal. Bioanal. Chem. 403, 1523−1548.

Vuckovic, D., Pawliszyn, J., 2011. Systematic evaluation of solid-phase microextraction coatings for untargeted metabolomic profiling of biological fluids by liquid chromatography-mass spectrometry. Anal. Chem. 83, 1944−1954.

Want, E.J., O'Maille, G., Smith, C.A., Brandon, T.R., Uritboonthai, W., Qin, C., Trauger, S.A., Siuzdak, G., 2006. Solvent-dependent metabolite distribution, clustering, and protein extraction for serum profiling with mass spectrometry. Anal. Chem. 78, 743−752.

Warren, C.R., 2013. Use of chemical ionisation for GC-MS metabolite profiling. Metabolomics 9, S110−S120.

Wilm, M., Mann, M., 1996. Analytical properties of the nanoelectrospray ion source. Anal. Chem. 68, 1−8.

Wilson, I.D., Nicholson, J.K., Castro-Perez, J., Granger, J.H., Johnson, K.A., Smith, B.W., Plumb, R.S., 2005. High resolution "ultra performance" liquid chromatography coupled to oa-TOF mass spectrometry as a tool for differential metabolic pathway profiling in functional genomic studies. J. Proteome Res. 4, 591−598.

Wishart, D.S., Feunang, Y.D., Marcu, A., Guo, A.C., Liang, K., Vazquez-Fresno, R., Sajed, T., Johnson, D., Li, C., Karu, N., Sayeeda, Z., Lo, E., Assempour, N., Berjanskii, M., Singhal, S., Arndt, D., Liang, Y., Badran, H., Grant, J., Serra-Cayuela, A., Liu, Y., Mandal, R., Neveu, V., Pon, A., Knox, C., Wilson, M., Manach, C., Scalbert, A., 2018. HMDB 4.0: the human metabolome database for 2018. Nucleic Acids Res. 46, D608−D617.

Wu, H., Southam, A.D., Hines, A., Viant, M.R., 2008. High-throughput tissue extraction protocol for NMR- and MS-based metabolomics. Anal. Biochem. 372, 204−212.

Xia, J.G., Wishart, D.S., 2011. Web-based inference of biological patterns, functions and pathways from metabolomic data using MetaboAnalyst. Nat. Protoc. 6, 743−760.

Xia, J.G., Psychogios, N., Young, N., Wishart, D.S., 2009. MetaboAnalyst: a web server for metabolomic data analysis and interpretation. Nucleic Acids Res. 37, W652−W660.

Xia, J.G., Mandal, R., Sinelnikov, I.V., Broadhurst, D., Wishart, D.S., 2012. MetaboAnalyst 2.0-a comprehensive server for metabolomic data analysis. Nucleic Acids Res. 40, W127−W133.

Xia, J.G., Sinelnikov, I.V., Han, B., Wishart, D.S., 2015. MetaboAnalyst 3.0-making metabolomics more meaningful. Nucleic Acids Res. 43, W251−W257.

Yanes, O., Tautenhahn, R., Patti, G.J., Siuzdak, G., 2011. Expanding coverage of the metabolome for global metabolite profiling. Anal. Chem. 83, 2152−2161.

Yang, Y., Cruickshank, C., Armstrong, M., Mahaffey, S., Reisdorph, R., Reisdorph, N., 2013. New sample preparation approach for mass spectrometry-based profiling of plasma results in improved coverage of metabolome. J. Chromatogr. A 1300, 217−226.

Zhang, X., Quinn, K., Cruickshank-Quinn, C., Reisdorph, R., Reisdorph, N., 2018. The application of ion mobility mass spectrometry to metabolomics. Curr. Opin. Chem. Biol. 42, 60−66.

CHAPTER 3

Metabolic profiling of biofluids in fish for identifying biomarkers of exposure and effects for assessing aquatic pollution

Pedro Carriquiriborde

Centro de Investigaciones del Medioambiente (CIM), Facultad de Ciencias Exactas, Universidad Nacional de la Plata — CONICET, La Plata, Buenos Aires, Argentina

Chapter Outline

1. Introduction

Biofluids can be defined as the liquid media produced by multicellular organism's cells that circulate through vessels or body cavities (such as blood or cerebrospinal fluid), excreted (such as urine or sweat) or secreted (such as breast milk or bile), or developed as a result of a pathological process (such as blister or cyst fluid). The metabolic composition of biofluids closely reflects the physiological status of the organisms. Today, it is well known in medicine that plasma glucose will increase above basal levels after a meal (Meyer et al., 1971), that plasmatic cortisol levels will decrease below basal levels after repeated traumatic situation (Steudte-Schmiedgen et al., 2016), or that the presence of bilirubin in human urine is consistent with liver injuries (Foley and Wasserman, 2014).

Because of the physiological information they provide, metabolites of different biofluids have been largely used in peripheral biochemical monitoring in human medicine for

Environmental Metabolomics. https://doi.org/10.1016/B978-0-12-818196-6.00003-0

disease diagnosis (Heikenfeld et al., 2019). Historically, target-analysis looking for specific metabolic was used in the clinical diagnosis. However, the advent of OMICs technologies has changed the analytical capabilities, and then a new era of opportunities for exploring metabolic changes in biofluid has opened. OMICs platforms have allowed system biology the comprehensive assessment of thousands of features at different levels of analysis from the genes (genomics), its expression (transcriptomics), associated proteins (proteomics), and finally the metabolites (metabolomics) (Dettmer et al., 2007). In particular, metabolomics is the comprehensive and quantitative study of metabolites in a living system. As it is situated at the end of the OMICs cascade, therefore, it is more predictive of phenotype and can be regarded as the ultimate answer of an organism to genetic alterations, disease, or environmental influences (Dettmer and Hammock, 2004). Due to the different meanings usually given to the terms metabolomics and metabonomics (Lindon et al., 2003), it is important to remark, that in this chapter, the term metabolomics will be used in its widest sense including terms such as metabonomics, metabolite profiling, and metabolic fingerprinting.

In the last years, metabolomics has been extendedly used for assessing the metabolome of different kind of biofluids in almost all the groups of multicellular organisms from plants to diverse groups of animals. *The objective of this chapter we will focus on the use of biofluids metabolomics for identifying biomarkers of exposure and effects for assessing aquatic pollution.*

2. Metabolomics of biofluid

As defined above, biofluids are liquid media produced by multicellular living systems and its metabolic composition has been very useful to assess the physiological status of the organisms. Biofluids are easy to withdraw with a needle usually without the need of killing the organism. Moreover, sample pretreatment used to be relatively simple in comparison with tissue preparations. All these reasons made biofluids a good biological material for metabolomic studies, and several studies on different kinds of organism, including plants, invertebrates, and vertebrates, can be enumerated as examples since the discipline arose (Table 3.1).

The xylem and phloem saps are important biofluids circulating inside the plants; they play key roles in long and short distance transport of signals and nutrients and act as a barrier against local and systemic pathogen infection. Xylem sap is a biofluid that circulates through a hollow tube—conduction system formed by dead cells. Its main function is to transport nutrients and water from the soil interface to stems and leaves and provide mechanical support and storage. The xylem sap composition consists primarily of a watery solution of hormones, mineral elements, and other nutrients (Brodersen et al., 2019). Phloem sap is the biofluid circulating through the sieve elements (specialized plant cells).

Table 3.1: Examples of metabolomics studies on biofluids from different groups of organisms.

Group	Biofluid	Platform	References
		Plants	
	Xylem sap	GC/MS	Rellán-Álvarez et al. (2011)
	Phloem sap	LC/MS; GC/MS	Zhang et al. (2010)
		Invertebrates	
	Hemolymph	FT-ICR/MS; GC/MS	Lu et al. (2014), Poynton et al. (2011);,Schock et al. (2010), Zhang et al. (2013), Zhou et al. (2015)
	Celom	NMR; GC/MS	Bundy et al. (2001), Griffith et al. (2018), Lenz et al. (2005), Yuk et al. (2012)
		Vertebrates	
	Serum	NMR; GG/MS	Carrizo et al. (2017), Szabo et al. (2017)
	Plasma	LC/QTOF/MS	O'Kane et al. (2013), Yang et al. (2013)
	Lymph	IM/QTOF/MS	Kaplan et al. (2009)
	Cerebrospinal	LC/MS; GC/MS	Lestaevel et al. (2016), Noga et al. (2011), Wishart et al. (2008)
	Amniotic fluid	NMR; GC/MS; LC/MS	Palmas et al. (2016), Wan et al. (2017)
	Bile	LC/MS	Al-Salhi et al. (2012), Carriquiriborde et al. (2012)
	Urine	NMR; GC/MS; LC/MS	Lee et al. (2007), Wang et al. (2016b), Xu et al. (2016)
	Milk	GC-TOF/MS; NMR	Sun et al. (2015), Wang et al. (2019)
	Saliva	NMR; GC/MS; LC/MS	Dame et al. (2015), Laiakis et al. (2016)
	Sweat	NMR	Delgado-Povedano et al. (2016), Kutyshenko et al. (2011)
	Tears	GC/MS; LC/MS	Ahamad et al. (2017), Cicalini et al. (2019)

FT-ICR/MS, Fourier-transform ion cyclotron resonance—mass spectrometry; *GC/MS*, gas chromatography—mass spectrometry; *IM/QTOF/MS*, Ion mobility QTOF/MS; *LC/MS*, liquid chromatography—mass spectrometry; *LC/QTOF/MS*, liquid chromatography quadrupole time-of-flight mass spectrometry; *NMR*, nuclear magnetic resonance.

Its main function is to transports the soluble organic compounds made during photosynthesis to various parts of plant where they are needed. The phloem sap is a complex matrix which consists of water, sugars, amino acids, organic acids, secondary metabolites, peptides, and hormones along with ions and a number of macromolecules, including proteins, small RNAs, and mRNAs (Turgeon and Wolf, 2009). Phloem metabolome has been used to know the differences between fascicular and extrafascicular Cucurbitaceous (Zhang et al., 2010) or the response to iron deficiency in tomato, lupines, and sugar beet (Rellán-Álvarez et al., 2011).

In invertebrates, the metabolome of the hemolymph, the fluid filling the open circulatory system of arthropods and molluscs, has been analyzed for studying metabolic differences in developmental stages of insects (Lu et al., 2014; Zhang et al., 2013; Zhou et al., 2015) and the metabolome response of arthropods exposed to pollutants (Poynton et al., 2011; Schock et al., 2010). Also, the celom fluid of annelids has been studied for characterizing endogenous metabolites (Lenz et al., 2005) and assessing the metabolic response to pollutants, both in aquatic and terrestrial ecosystems (Bundy et al., 2007; Griffith et al., 2018).

In vertebrates, several different biofluids were assessed in metabolomics studies. Plasma and serum are two blood biofluids obtained using different sampling and processing methods. These are probably the most studied biofluids of vertebrates, including humans. While plasma is obtained using additives to prevent coagulation (anticoagulants), serum is obtained after the coagulation process has taken place. Therefore, the metabolite profile is quite different among each other (Liu et al., 2018a). Just a few cases are given here as examples of environmental metabolomics using vertebrate plasma or serum. For example, the metabolomics profile of serum was assessed either under laboratory conditions to evaluate responses in mice orally exposed to brominated flame retardant (Szabo et al., 2017), or under field conditions to assess human populations with different degree of exposure to persistent organic pollutants (Carrizo et al., 2017). Plasma metabolome in rats has been used to assess specific compounds, like dichlorvos (Yang et al., 2013), or even PCBs and dioxins in environmental samples, such as incinerator soot (O'Kane et al., 2013).

The lymph metabolome has been much less studied. The lymph is the biofluid that flows through the lymphatic system. It is originated from the drainage of the interstitial fluid (the fluid which is between the cells in all body tissues) to the lymph capillaries. Then it is transported via progressively larger lymphatic vessels, passing through the lymph nodes and ultimately empties into the subclavian vein to return to the blood system. It plays important roles in blood pressure homeostasis, metabolite transport from the bowel to the blood (e.g., proteins, lipids, fat-soluble vitamins), and immunological function. Available studies were focused on basic physiology, for example, related to the response of mice to fasting (Kaplan et al., 2009), but not assessing the effects of environmental pollutants.

Another important circulating biofluid of vertebrates is the cerebrospinal fluid, the secretion product of the central nervous system that fills the ventricles and the subarachnoid space of the brain and spinal column. Most metabolomics studies of the cerebrospinal fluid were focused on the composition and pathological alterations in humans (Wishart et al., 2008) and rats (Noga et al., 2011). One environmental application of the cerebrospinal fluid metabolome was to assess the exposure to natural uranium via lactation and drinking water using rats (Lestaevel et al., 2016).

The amniotic fluid has been also used in metabolomics studies, but mostly used in medicine or veterinary studies (Wan et al., 2017).

Bile is a dark-green-to-yellowish-brown fluid secreted by the hepatocytes that aid in emulsifying and absorption of lipids from the intestine of vertebrates via the action of bile acids or bile salts (Guyton and Hall, 2011). Bile also acts as the medium for excretion of many endogenous and exogenous substances from the blood and liver that are not excreted through the kidneys (Grosell et al., 2000), and therefore bile composition has been traditionally used for assessing exposure to environmental pollutants in fish (Statham et al., 1976). In the last decades, the bile metabolomics studies have been applied to study the response of fish to pesticides (Carriquiriborde et al., 2012) or even environmental samples like wastewaters (Al-Salhi et al., 2012). However, metabolomics of this biofluid has been less studied in other vertebrates.

Urine is the fluid produced by kidneys to rid the body of waste materials that are either ingested or produced by metabolism and to control the volume and composition of the body fluids (Guyton and Hall, 2011). In a similar way than serum and plasma, urine has been largely used in human (Bouatra et al., 2013) and veterinary (Lunn, 2011) medicine, and in toxicology. In addition, urine has long been a "favored" biofluid among metabolomics researchers, since it is sterile, easy to obtain in large volumes, largely free from interfering proteins or lipids, and chemically complex (Bouatra et al., 2013). Therefore, just a couple of examples are mentioned here. The urine metabolome response was assessed either under laboratory conditions in rat exposed to, for example, nonylphenol (Lee et al., 2007) or cadmium (Wang et al., 2016b) or in field studies in woman from small villages exposed to high environmental cadmium levels (Xu et al., 2016).

Milk is the fluid secretion of mammary glands, characteristic of mammals. It has been commonly used in targeting analysis of many environmental pollutants (Duedahl-Olesen et al., 2005; Steinborn et al., 2016). However, recent metabolomics studies have been mostly focused to animal production (Sun et al., 2015) and human health (Slupsky, 2019), and no studies have been found assessing milk metabolome in response to the exposure to environmental pollutants. Only one study can be mentioned assessing the cow milk metabolome, but in response to the exposure to a mycotoxin, the aflatoxin b1 (Wang et al., 2019).

Other less studied biofluids in metabolomics research were the secretion of the salivary, sweat, and lacrimal glands. Saliva has been mostly used in human health basic metabolomics studies (Dame et al., 2015) and to assess the response to gamma irradiation in mice (Laiakis et al., 2016). Sweat metabolomics has been only applied in human health studies, characterizing its composition (Kutyshenko et al., 2011) or for cancer diagnosis (Delgado-Povedano et al., 2016). The metabolome of tears has been assessed in medicine for multiple sclerosis diagnosis (Cicalini et al., 2019) and, curiously, for describing its metabolic composition in camels (Ahamad et al., 2017).

As it is observed, the major biofluids used in environmental metabolomics have been plasma, serum, urine, and bile. The other biofluids have been still relatively underutilized in the environmental field.

3. Methodological approach and platforms

Two main approaches can be followed in metabolomics regarding the biological questions and experimental designs. The **targeted metabolomics** is focused on the quantitative analysis of preselected metabolites, commonly driven by a specific hypothesis on specific pathways. On the other hand, the **untargeted metabolomics** is focused on the comprehensive (qualitative and semi/quantitative) analysis of whole, or a large part, of the metabolites present in a given biological sample. It aims to build a global profile that allows detecting patterns or fingerprints of metabolites under normal or abnormal conditions, to reveal the involvement of metabolic pathways, without the need of a previous hypothesis, following a "top-down" strategy (Alonso et al., 2015; Raterink et al., 2014). Usually, the term metabolomics is coined to the untargeted approach, and the typical methodological pipeline followed in that kind of study is shown in Fig. 3.1. From all steps in the pipeline, sampling and sample pretreatment are usually the ones more diverse among different biofluids.

3.1 Sampling

Blood has been the most studied biofluid in fish, and the most applied method for its extraction has been the following one. Blood was collected from the caudal vessels using heparinized ice-cooled syringes and the plasma separated by centrifugation. Samples were later snap-frozen until processing (Al-Salhi et al., 2012; David et al., 2014, 2017; Kokushi et al., 2012, 2015; Li et al., 2015, 2016; Samuelsson et al., 2006). In another study, 0.5 mol EDTA was used as an anticoagulant (Ziarrusta et al., 2018). In some cases, where fish was too small, blood was collected by directly cutting the caudal peduncle (Hajirezaee et al., 2017). When serum instead plasma was analyzed, the blood sample withdrawn from the caudal vein without anticoagulant was allowed to clog for 30 min, then centrifuge and frozen (Khan et al., 2016; Zhang and Zhao, 2017). Álvarez-Muñoz et al. used a slightly different approach (Álvarez-Muñoz et al., 2014), blood was also collected from the caudal vein, combined with double volume of methanol, and stored at −80°C. Then, frozen blood was thawed, centrifuged, and plasma collected from the supernatant.

Another collection method used in human medicine and promising for using in fish metabolomics is the "dry blood (or biofluid) spot." In this approach, the biofluid was collected on special types of absorbent paper, which was then dried to be ready for

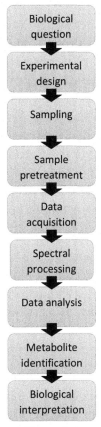

Figure 3.1

Basic workflow for nontarget metabolomics *Adapted from Raterink, R.J., Lindenburg, P.W., Vreeken, R.J., Ramautar, R., Hankemeier, T., 2014. Recent developments in sample-pretreatment techniques for mass spectrometry-based metabolomics. Trac. Trends Anal. Chem. 61, 157—167 and Alonso, A., Marsal, S., Julià, A., 2015. Analytical methods in untargeted metabolomics: state of the art in 2015. Front. Bioeng. Biotechnol. 3, 23.*

subsequent extraction and analysis. This simple, inexpensive method is easy to automate and can store only a few μL of the sample (Wilson, 2011).

Bile has been another biofluid often studied in fish metabolomics. It was obtained by puncturing the gall bladder with a needle and drawing it into a syringe (Carriquiriborde et al., 2012; Ekman et al., 2015). Skin mucus has also been less used. It was collected by placing fish on its side and then laying a small piece of glass fiber filter paper on the side of the fish. The filter paper was then placed into a microfuge tube, frozen in liquid nitrogen, and stored at −80°C. Urine has been used too. It was expelled from fish by applying gentle pressure on the abdomen and collected directly from the cloaca into nonheparinized microcapillary tubes and frozen until analysis (Ekman et al., 2007).

3.2 Sample pretreatment

The sample pretreatment strategy largely depends on the nature of the sample to be analyzed. In addition, diverse analytical platforms have different requirements with regard to sample pretreatment. However, in general terms, the biological sample should preferably be analyzed with minimal pretreatment in order to prevent the potential loss of metabolites. Methods should extract, as straightforward as possible, the low molecular weight compounds from proteins and/or lipids. Sometimes, a fractionated sample obtained after liquid—liquid extraction (LLE) and/or solid-phase extraction (SPE) steps was also analyzed. Evaporation and reconstitution often comprise the final step directed to concentrate the analytes and/or to change to analysis-compatible solvents. In these last steps, attention should be paid to analyte solubility, and potential oxidation of analytes, such as thiol-containing molecules (Raterink et al., 2014).

In particular, for the study of biofluids, nonselective sample pretreatment methods, such as "dilute-and-shoot" and solvent protein precipitation (PPT), are often used, since they enable broad metabolite coverage and high-throughput analysis. The protein content of the biofluid is an important aspect to consider, as most subsequent analysis methods are not compatible with protein-rich samples, such as plasma or serum. In contrast, healthy urine contains relatively low amounts of protein, and centrifugation followed by dilution is often all that is required prior to analysis.

The "dilution and shoot" approach was commonly applied in NMR analysis. Plasma or serum samples were diluted in deuterium water (D_2O), usually with an internal standard, and phosphate buffer (Hajirezaee et al., 2017; Khan et al., 2016; Kokushi et al., 2010, 2013; Samuelsson et al., 2006). It was also used in an LC-MS—based study (Álvarez-Muñoz et al., 2014). Plasma was diluted with one-third of water and filtered before analysis. The "dilution and shoot" technique has been also used for bile analysis by LC-MS. Bile was 50-fold diluted in methanol-water (1:1, v/v) and filtered through 96-well Strata Protein Precipitation Plates (Al-Salhi et al., 2012). Similarly, bile was diluted in the same volume of mobile-phase acetonitrile-water (1:1, v/v) and filtered through 0.22 μm syringe filters (Carriquiriborde et al., 2012). In the case of urine samples, they were lyophilized overnight and subsequently dissolved in phosphate-buffered deuterium oxide containing the internal standard. It was then centrifuged and the resulting supernatants used for NMR analysis (Ekman et al., 2007).

The solvent-based PPT followed by centrifugation has been the most common deproteinization method. Protein denaturation using heat or inorganic acids has been also used but usually with lower metabolite coverage. Other protein removal methods such as surface-functionalized magnetic beads and turbulent flow chromatography, as well as PPT in a 96-well filtration plate, were used also looking for robotic automation solutions (Raterink et al., 2014).

PPT pretreatment for NMR analysis has been used by Li et al. (2015, 2016). Plasma was mixed with a precooled solvent methanol:water (1:1 v/v), vortexed, and centrifuged. The supernatant was lyophilized to dryness, and dried samples were stored at $-80°C$. Samples were then dissolved in D_2O and phosphate buffer containing the internal standard. PPT and centrifugation were employed by Ziarrusta et al. (2018), using chloroform:methanol (20:80 v/v), homogenized and centrifuged, collecting the single-phase supernatant before LC-MS analysis. In the study of Al-Salhi et al. (2012) plasma was mixed with 80% ice-cold methanol and centrifuged. The supernatant was then evaporated to dryness under nitrogen, redissolved in methanol:water (1:1 v/v), and filtered through 96-well Strata Protein Precipitation Plates. A similar approach was also used in the pretreatment of samples for GC-MS analysis. Plasma samples were extracted with methanol:water (1:1 v/v) and then centrifuged. The supernatant was lyophilized to dryness, treated first with methoxamine hydrochloride in pyridine, and later on with bis(trimethylsilyl) trifluoroacetamide and trimethylchlorosilane for derivatization (Zhang and Zhao, 2017). PPT was also applied to the skin mucus pretreatment. Mucus was extracted from glass fiber filter strips with ice-cold methanol:water (8:1, v/v) and the extraction solution was subsequently transferred to an AcroPrep 96-well 0.2 μm PTFE (polytetrafluoroethylene) filter plate and centrifuged previous LC-MS analysis.

The LLE allows the extraction of metabolites into two fractions (aqueous and organic phase) that separately contain polar and apolar compounds and that can be then independently analyzed. In addition, it has also been used for sample clean up, particularly lipid removal. The traditional extraction method used chloroform and methanol (Bligh and Dyer, 1959; Folch et al., 1957) but now methyl tert-butyl ether-methanol (Matyash et al., 2008) is preferred since the organic phase remains at the top and insoluble phase (proteins) on the bottom of the tube, and adapted for whole sample preparation in an LC vial called "in-vial dual extraction" (Whiley et al., 2012). Other LLE-derived methods are available (e.g., supported liquid extraction, liquid-phase microextraction, and ultrasound liquid-phase extraction) but less explored in metabolomics of fluids. LLE method was used by Carriquiriborde et al. (2012), to assess polar and apolar metabolites in bile using diethyl-ether, both with previously acidified and not acidified water-diluted bile samples. Each phase was then separately analyzed by LC-MS.

SPE has been also effectively used for the removal of interfering substances and for the enrichment of analytes. Diverse extraction sorbents are available like phenyl groups or polymeric material, weak and strong ion-exchange materials, and mixed-mode materials. However, due to sorbents selectivity, obtaining high metabolite coverage may be challenging, and therefore this kind of sample pretreatment is usually more suitable for targeted than for untargeted metabolomics (Raterink et al., 2014).

Different combinations of previous pretreatment methods are possible. PPT in a 96-well filtration plate followed by SPE was used for plasma extraction for LC-MS (xeno) metabolomics in trout after sewage effluent exposure (David et al., 2014, 2017).

4. Fish biofluid metabolome profiling in environmental studies

Fish have been historically used in ecotoxicology as testing and sentinel organisms to assess aquatic pollution. In addition, as vertebrates, they have been frequently used as surrogate species to estimate the effect of pollutants on human health. Environmental studies assessing metabolomics responses in fish can be traced back to the beginning of the 21st century, and since that moment, 150 peer-reviewed articles were found in the Scopus database conducting the following search: TITLE-ABS-KEY (fish AND metabolomics) AND (LIMIT-TO (SUBAREA, "ENVI")). Those studies specifically dealing with the assessing of global metabolic responses to environmental pollutants are listed in Table 3.2. Studies were classified in laboratory and field studies. In addition, among laboratory studies were distinguished those assessing single compounds, mixtures, and environmental samples (effluents or surface waters). From the 83 selected studies, only 17 have used some biofluid and most have been done using some tissue (52) or the whole the organisms (31). Despite the poor biological meaning, the use of whole fish, it is an increasing trend in laboratory studies using zebrafish (*Danio rerio*) or medaka (*Oryzias* sp). Among the tissues, the most studied has been the liver (31) followed by the brain (12), kidney, and gill. On the other hand, blood (plasma or serum) has been the most studied biofluid (13), followed by the bile (2), urine (1), and mucus (1).

Plasma and serum have been used to assess metabolomics changes induced by single compounds as a diverse group of pharmaceuticals (17α-ethinylestradiol), sun screens (oxybenzone), pesticides (chlorpyrifos, diazinon, acetamiprid, and halosulfuron-methyl), surfactants (alcohol polyethoxylated), and microplastics (polystyrene nanoparticles). In addition, plasma metabolome has been used to assess the toxic effects of complex mixtures such as heavy oil or pesticide formulations of avermectin and glyphosate. It has been also used to assess environmental samples, such as sewage effluents, under controlled laboratory conditions. Curiously, in the broad list of studies shown in Table 3.2, the metabolome of biofluids has not been yet used to assess environmental pollution under field studies.

It is particularly interesting to note that the species used in plasma or serum metabolomics are usually big or medium size fish, since it is easier for the blood withdraw. Salmonids (*Oncorhynchus* sp.), carps (*Cyprinus carpio* and *Carassius* sp.), the flatfish (*Solea senegalensis*), the roach (*Rutilus rutilus*), the sturgeon (*Acipenser* sp.), and the gilt-head bream (*Sparus aurata*) are among the most used ones. The increasingly frequent use of small species in the laboratory is probably one of the reasons explaining why blood

Table 3.2: Studies assessing the effects of pollutants on the fish metabolome.

Laboratory studies				
(a) Single compounds				
Year	Pollutant	Species	Tissue	Platform	References
2005	Trichloroethylene	*Oryzias latipes*	Embryos	NMR	Viant et al. (2005)
2006	17α-Ethinylestradiol	*Oncorhynchus mykiss*	Plasma	NMR	Samuelsson et al. (2006)
2007	Vinclozolin	*Pimephales promelas*	Urine	NMR	Ekman et al. (2007)
2008	17α-Ethinylestradiol	*P. promelas*	Liver	NMR	Ekman et al. (2008)
2009	17α-Ethinylestradiol	*P. promelas*	Liver	NMR	Ekman et al. (2009)
2010	1,2:5,6-Dibenzanthracene	*Gasterosteus aculeatus*	Liver	NMR	Williams et al. 2009
	Copper	*G. aculeatus*	Liver	NMR	Santos et al. (2010)
	17α-Ethinylestradiol	*G. aculeatus*	Liver	NMR	Katsiadaki et al. (2010)
	17α-Ethinylestradiol	*Rutilus rutilus*	Liver and gonad	UPLC-TOFMS	Flores-Valverde et al. (2010)
2011	17β-Trenbolone	*P. promelas*	Liver	NMR	Ekman et al. (2011)
	Fenitrothion	*R. rutilus*	Liver and testes	NMR	Southam et al. (2011)
2012	Malatión	*O. latipes*	Whole fish	GC-MS	Uno et al. (2012)
	Cypermethrin	*Odontesthes bonariensis*	Bile	HPLC-MS	Carriquiriborde et al. (2012)
2013	Chlorpyrifos	*Cyprinus carpio*	Plasma	NMR	Kokushi et al. (2015)
2014	Alcohol-polyethoxylated	*Solea senegalensis*	Liver and plasma	UPLC-TOF-MS	Álvarez-Muñoz et al. (2014)
	Lambda-cyhalothrin	*Carassius auratus*	Gill, heart, liver, and kidney	NMR	Li et al. (2014)
2015	Butachlor	*C. auratus*	Gill, brain, liver, and kidney	NMR	Xu et al. (2015)
	Polystyrene nanoparticles	*Carassius carassius*	Blood, gill, muscle, liver, and brain	NMR	Mattsson et al. (2015)
	Bisphenol A	*P. promelas*	Mucus	HPLC-MS/MS	Ekman et al. (2015)
	Benzo[a]anthracene and benzo[a]anthracene-7,12-dione	*D. rerio*	Embryos	LC-MS/MS	Elie et al. (2015)
2016	Fipronil	*D. rerio*	Whole larvae	GC-MS	Wang et al. (2016a)
	Polystyrene microplastics	*D. rerio*	Liver	NMR	Lu et al. (2016)
	Sulfamethazine	*D. rerio*	Whole fish	NMR and HPLC-QTOF-MS	de Sotto et al. (2016)

Continued

Table 3.2: Studies assessing the effects of pollutants on the fish metabolome.—cont'd

Laboratory studies					
Year	Pollutant	Species	Tissue	Platform	References
2017	13 Chemicals	*D. rerio*	Whole larvae	HPLC-MS/MS	Huang et al. (2016)
	Fipronil	*D. rerio*	Whole larvae	GC-MS	Yan et al. (2016)
	Methomyl and methidathion	*D. rerio*	Whole larvae	NMR	Yoon et al. (2016)
	Chlorpyrifos	*D. rerio*	Muscle	HPLC-HRMS	Gómez-Canela et al. (2017)
	2,20,4,40-Tetrabromo diphenyl-ether	*Oryzias melastigma*	Brain	HPLC-MS/MS	Lei et al. (2017)
	4 Pharmaceuticals	*D. rerio*	Whole larvae	HPLC-MS/MS	Huang et al. (2017)
2017	Bisphenol A	*D. rerio*	Whole larvae	HPLC-QTOF-MS	Ortiz-Villanueva et al. (2017)
	Epoxiconazole	*D. rerio*	Whole larvae	NMR	Wang et al. (2017a)
	Spironolactone	*P. promelas*	Liver	HPLC-HRMS	Davis et al. (2017)
	Diazinon	*Acipenser persicus*	Plasma	NMR	Hajirezaee et al. (2017)
	Diniconazole	*D. rerio*	Whole larvae	NMR	Wang et al. (2017b)
	Fluoxetine	*D. rerio*	Whole larvae	GC-MS	Mishra et al. (2017)
	Microcystin-LR	*D. rerio*	Liver	NMR	Chen et al. (2017)
	Dithiocarbamate	*Pagrus major, Verasper variegatus, and Pleuronectes yokohamae*	Liver	GC-MS	Hano et al. (2017)
	Bisphenol A	*D. rerio*	Whole larvae	NMR	Yoon et al. (2017)
	Microcystin-LR	*Perca fluviatilis, R. rutilus, and C. carassius*	Liver	NMR	Sotton et al. (2017)
	Imazalil	*D. rerio*	Liver	GC-MS	Jin et al. (2017)
	2,4-Dichlorophenol	*O. latipes*	Whole fish	GC-MS	Kokushi et al. (2017)
2018	Difenoconazole	*D. rerio*	Whole larvae	NMR and HPLC-MS/MS	Teng et al. (2018a)
	Microcystin-LR	*O. latipes*	Liver	HPLC-QTOF-MS	Le Manach et al. (2018)
	Ammonia	*O. melastigma*	Whole larvae	NMR	Zhu et al. (2018)
	Sulfamethazine	*O. melastigma*	Whole larvae	GC × GC-TOF/MS	Liu et al. (2018b)
	Methylmercury	*P. promelas*	Whole larvae	GC-MS	Bridges et al. (2018)
	Bisphenol A, perfluorooctanesulfonate, and tributyltin	*D. rerio*	Whole larvae	HPLC-HRMS	Ortiz-Villanueva et al. (2018)

Continued

Table 3.2: Studies assessing the effects of pollutants on the fish metabolome.—cont'd

		Laboratory studies			
Year	Pollutant	Species	Tissue	Platform	References
2019	Ibuprofen	*D. rerio*	Brain	UPLC-TOF/MS	Song et al. (2018)
	Flutolanil	*D. rerio*	Whole larvae	NMR	Teng et al. (2018b)
	Oxybenzone	*Sparus aurata*	Brain, liver, and plasma	UHPLC qOrbitrap/MS	Ziarrusta et al. (2018)
	Isocarbophos	*D. rerio*	Whole larvae	NMR	Jia et al. (2018)
	Triclosan	*D. rerio*	Whole larvae	GC-MS	Fu et al. (2019)
	Polystyrene microplastics	*D. rerio*	Whole larvae	GC-MS	Wan et al. (2019)
	Fluoxetine	*D. rerio*	Whole larvae	GC-MS	Mishra et al. (2019)
	Polystyrene microplastics	*D. rerio*	Intestine	NMR	Qiao et al. (2019)
	Amitriptyline	*S. aurata*	Brain and liver	UHPLC qOrbitrap/MS	Ziarrusta et al. (2019)
	Hexaconazole and epoxiconazole	*D. rerio*	Whole larvae	NMR	Jia et al. (2019)
	Dydrogesterone	*D. rerio*	Gonads	LC-QTOF and GC-QTOF	Jiang et al. (2019)
	Brevetoxin PbTx-1	*O. melastigma*	Brain	HPLC-QTrap/MS	Yau et al. (2019)
		(b) Mixtures			
2009	Crude and dispersed oil	*Oncorhynchus tshawytscha*	Liver and muscle	NMR	Lin et al. (2009)
2010	Crude and dispersed oil	*O. tshawytscha*	Muscle	NMR	Van Scoy et al. (2010)
2012	Crude and dispersed oil	*Atherinops affinis*	Muscle and embryos	NMR	Van Scoy et al. (2012)
	Heavy oil	*C. carpio*	Plasma	NMR	Kokushi et al. (2012)
2015	Avermectin formulation	*C. auratus*	Brain, kidney, liver, and plasma	NMR	Li et al. (2015)
2016	Glyphosate formulation	*C. auratus*	Brain, kidney, liver, and plasma	NMR	Li et al. (2016)
	Lead and cadmium	*C. auratus*	Serum	NMR	Khan et al. (2016)
2017	Glyphosate formulation	*C. auratus*	Brain, kidney, and liver	NMR	Li et al. (2017)
	Acetamiprid and halosulfuron-methyl	*Brachydanio rerio*	Liver, head, and serum		Zhang and Zhao (2017)
	Clarithromycin, florfenicol, and sulfamethazine	*D. rerio*	Whole fish	HPLC-QTOF-MS	de Sotto et al. (2017)

Continued

Table 3.2: Studies assessing the effects of pollutants on the fish metabolome.—cont'd

		Laboratory studies			
Year	Pollutant	Species	Tissue	Platform	References
	Short-chain chlorinated paraffin	*D. rerio*	Whole larvae	UHPLC/ QTOF-MS	Ren et al. (2018)
	Lead	*D. rerio*	Liver	GC-MS	Xia et al. (2018)
	Triclosan and methyl-triclosan	*D. rerio*	Whole larvae	GC-MS	Fu et al. (2019)
		(c) Environmental samples			
2011	Sewage effluents	*O. mykiss*	Plasma	NMR	Samuelsson et al. (2011)
2012	Sewage effluents	*O. mykiss and R. rutilus*	Plasma and bile	UPLC-TOFMS	Al-Salhi et al. (2012)
2014	Sewage effluents	*P. promelas*	Liver	NMR and CG/MS	Skelton et al. (2014)
2017	Sewage effluents	*R. rutilus*	Plasma	nUHPLC-nESI-TOFMS	David et al. (2017)
2018	Sewage effluents	*P. promelas*		HPLC-HRMS	Mosley et al. (2018)
		Field studies			
2011	Seven estuarine locations	*Platichthys flesus*	Liver	NMR	Williams et al. (2011)
2016	Mercury contaminate site	*Liza aurata*	Gill	NMR	Cappello et al. (2016)
2018	Sewage effluents receiving waters	*P. promelas*	Liver	NMR	Ekman et al. (2018)
	Polychlorinated biphenyl contaminated sites	*Fundulus heteroclitus*	Liver	HPLC-MS/MS	Glazer et al. (2018)
	Metal(loid)-polluted site	*Gambusia holbrooki*	Whole fish	NMR	Melvin et al. (2018)
2019	Microcystin-polluted sites	*P. fluviatilis and Lepomis gibbosus*	Liver	NMR	Sotton et al. (2019)

Data obtained from Scopus (TITLE-ABS-KEY (fish AND metabolomics) AND (LIMIT-TO (SUBAREA, "ENVI")) Accession: June 2019.

metabolomics is not as common as it would be expected. On the other hand, it is difficult to explain why it has not been used more often in field studies.

Although some metabolites and metabolic pathways responding in the plasma and serum differed among single exposure to pollutants, others were common among studies. The metabolic response to EE_2 found by Samuelsson et al. (2006) was characterized by the response of the amino acid alanine, phospholipids, and cholesterol. In the study of Kokushi et al. (2015), the metabolic response to the organophosphorus pesticide chlorpyrifos was related to the metabolites involved in energy production, such as glucose, glycerol, valine, leucine, isoleucine, lactate, alanine, 3-dhydroxybutyrates, and acetoacetate. A similar response was observed in response to another organophosphorus pesticide, diazinon, not only affecting energy metabolites like glucose, lactate, acetate, and acetoacetate but also decreasing levels of creatine, trimethylamine-N-oxide, choline, taurine, betaine, N,N-dimethylglycine, and almost all amino acids during short-term exposure. Under long-term exposure, levels of lipid oxidation metabolites and almost all amino acids increase and creatine, trimethylamine-N-oxide, N,N-dimethylglycine, betaine, choline, glucose, and taurine decrease in response to the organophosphorus (Hajirezaee et al., 2017). The exposure to the sunscreen, oxybenzone, affected the fatty acid elongation, a linolenic acid metabolism, biosynthesis of unsaturated fatty acids and fatty acid metabolism, as well as the amino acid serine, explained as possible energy metabolism modification and oxidative stress (Ziarrusta et al., 2018). The exposure to the surfactant alcohol polyethoxylated induced the increase of the circulating levels of C24 bile acids and C27 bile alcohols, disturbance of glucocorticoid and lipid metabolism, and the decrease in levels of the fatty acid transport molecule palmitoylcarnitine (Álvarez-Muñoz et al., 2014). From these studies, it is possible to identify in the serum and plasma metabolic responses to different pollutants some common patterns linked to energy regulation and oxidative stress, and other more specific of each compound.

Two kinds of studies can be differentiated working with mixtures: those assessing the impact of complex mixtures and those evaluating binary mixtures in comparison with the single compounds. Among the first group, it was observed that the metabolic response to a complex mixture of hydrocarbons named heavy oil was characterized by the increase of amino acids, 3-D-hydroxybutyrate, and glycerol and the reduction of formate, indicating a disturbance of the TCA (tricarboxylic acid) cycle. Also, isobutyrate and creatinine were increased, markers of anoxia and kidney dysfunction, respectively (Kokushi et al., 2012). Within this group, it is also possible to include studies that have evaluated commercial formulations of pesticides. Amino acids and energy metabolism were affected by both, avermectin and glyphosate formulations, while avermectin exposure also affected osmotic equilibrium and redox equilibrium (Li et al., 2015, 2016). Two other studies were found assessing the serum metabolic response of binary mixtures in comparison with the single compounds. The serum metabolic response against the exposure to a single or binary mixture of Cd and Pb showed additive effects on the amino acids, neurotransmitters, energy, osmoregulatory, and aerobic/anaerobic metabolisms (Khan et al., 2016). A similar response was observed to the combined or single exposure to acetamiprid and halosulfuron-methyl, that also disturbed amino acid metabolism, the TCA cycle, and the balance of neurotransmitters (Zhang and Zhao, 2017).

Finally, studies assessing global metabolic response in plasma of fish exposed to environmental samples have been focused mainly on sewage effluents. Perturbations in the plasma concentrations of bile acids and lipids were found in trout and the increase of cyprinol sulfate and taurocholic acid, lysophospholipids, and a decrease in sphingosine levels were reported for roach (Al-Salhi et al., 2012). Disruptions in bile acid and lipid metabolism were also reported for roach in another study, but together with a widespread reduction in prostaglandin and disruption of tryptophan/serotonin metabolism (David et al., 2017). Response at HDL, LDL, VLDL (high-, low-, or very low–density lipoprotein, respectively), and glycerol containing lipids, cholesterol, glucose, phosphatidylcholine, glutamine, and alanine were the main metabolic changes observed in the plasma of trout exposed to six differently treated sewage effluents (Samuelsson et al., 2011).

Fewer studies have evaluated the bile metabolomics response to single pollutants, and all performed under laboratory conditions. The global metabolic profiling of this biofluid has helped to identify new biotransformation routes in fish for the pesticide cypermethrin (Carriquiriborde et al., 2012). In addition, bile xenometabolome was useful to identify the fish exposome to sewage effluents but was less sensitive than plasma to identify endogenous metabolome responses (Al-Salhi et al., 2012).

Only one publication has focused on the urine metabolome response in fish exposed the fungicide vinclozolin. That study was able to identify potential hepatotoxicity effects indicated by a large increase in taurine, lactate, acetate, and formate with a marked decrease in hippurate (Ekman et al., 2007). Similarly, only one study has used the mucus to evaluate the metabolomics response in fish exposed to bisphenol A (BPA). Skin mucus metabolome demonstrated to be sexually dimorphic as well as its response to BPA. The main response showed that BPA was able to stimulate the purine degradation pathway in males, but any particular pathway could be ascribed to the metabolic response in female (Ekman et al., 2015).

All the mentioned studies demonstrate the usefulness of biofluid metabolomics to assess exposure and effects induced by pollutants on fish not only under controlled laboratory conditions but also in field studies.

5. Advantages, drawbacks, and future perspectives

The major fish biofluids that have been used in environmental metabolomics were blood plasma or serum, bile, urine, and mucus. Others like cerebrospinal fluid or lymph have not been assessed yet in fish. As the metabolome of biofluids closely reflect the physiological status of the organisms, it has demonstrated to be suitable to assess responses induced by environmental pollutants and to identify biomarkers of exposure and effect. Those kinds of studies have been conducted either under laboratory or field conditions. In addition, biofluid samples are easy to obtain and usually are less invasive if compared with tissues samples. Despite the mentioned advantages, in the last years, biofluids have been less studied than expected, and the whole organism metabolomics using small model fish species has gained more attention.

From the methodological point of view, biofluids not only are easy to obtain but also need less complex sample pretreatments methods. This could facilitate robotic automation for taking large sample number in environmental monitoring assessment. In particular, low protein biofluids, such as urine, are prone to direct analysis through "dilute-and-shoot" methods. Mucus is another interesting and nondestructive biofluid that has been underused. Bile required the sacrifice of the fish, but metabolite extraction is simple and was useful to study the exposome, analyzing the exogenous metabolome. Blood plasma and serum were by far the most used biofluids, but due to the high protein and lipid content, they usually required more intensive pretreatment, and therefore, several strategies for deproteinization and delipidation have been explored. Despite the methodological advances achieved in the last year, the development of standardized methods involving sample collection and pretreatment, data acquisition, processing, and analysis is still lacking, and it is probably one of the future challenges. This is particularly important for the metabolome of biofluids because it presents higher biological variability in comparison to tissue metabolome due to a lesser homeostatic control.

Metabolite identification has shown a huge progress but it still remains as a general challenge in metabolomics, and biofluids metabolic profiling is not excluded. Standardization of instrumental conditions, as for GC-MS, would be useful for LC-MS platforms due to the diversity of available instruments and ionization conditions. The incorporation of extra parameters in metabolomic databases, complementing the mass spectra and ionization source information, could help to facilitate the metabolite identification. For example, the column type (i.e., RFC, HILIC, etc.) and the relative retention time could help to decide if a given feature showing a particular *m/z* could be, or not, one of the metabolites listed by the database with similar probabilities. Other useful information included in the databases could be the fragments and collision energy if spectra are available.

Biological significance and biomarker identification are also key issues in metabolomics of biofluids. Most of the here-listed studies have appealed to the metabolic pathway analysis to explain observed responses and have reaching interesting conclusions. However, sometimes, too general metabolic pathways have been proposed, and it was not clear if they were general stress responses or specifically related to the toxic mode of action. Discussions like the one in Ekman et al. (2015) or the classification models to identify and validate metabolites as potential biomarkers employed by Samuelsson et al. (2006) encourage the further use of fish biofluids in environmental studies.

Acknowledgment

Pedro Carriquiriborde is Senior Researcher of the National Research Council of Argentina (CONICET) and was granted by the CONICET (PIP2014-2016-0090) and the ANPCyT (PICT2014-1690).

References

Ahamad, S.R., Raish, M., Yaqoob, S.H., Khan, A., Shakeel, F., 2017. Metabolomics and trace element analysis of camel tear by GC-MS and ICP-MS. Biol. Trace Elem. Res. 177, 251−257.

Al-Salhi, R., Abdul-Sada, A., Lange, A., Tyler, C.R., Hill, E.M., 2012. The xenometabolome and novel contaminant markers in fish exposed to a wastewater treatment works effluent. Environ. Sci. Technol. 46, 9080−9088.

Alonso, A., Marsal, S., Julià, A., 2015. Analytical methods in untargeted metabolomics: state of the art in 2015. Front. Bioeng. Biotechnol. 3, 23.

Álvarez-Muñoz et al., 2014Álvarez-Muñoz, D., Al-Salhi, R., Abdul-Sada, A., González-Mazo, E., Hill, E.M., 2014. Global metabolite profiling reveals transformation pathways and novel metabolomic responses in *Solea Senegalensis* after exposure to a non-ionic surfactant. Environ. Sci. Technol. 48, 5203−5210.

Bligh, E.G., Dyer, W.J., 1959. A rapid method of total lipid extraction and purification. Can. J. Biochem. Physiol. 37, 911−917.

Bouatra, S., Aziat, F., Mandal, R., Guo, A.C., Wilson, M.R., Knox, C., Bjorndahl, T.C., Krishnamurthy, R., Saleem, F., Liu, P., Dame, Z.T., Poelzer, J., Huynh, J., Yallou, F.S., Psychogios, N., Dong, E., Bogumil, R., Roehring, C., Wishart, D.S., 2013. The human urine metabolome. PLoS One 8.

Bridges, K.N., Zhang, Y., Curran, T.E., Magnuson, J.T., Venables, B.J., Durrer, K.E., Allen, M.S., Roberts, A.P., 2018. Alterations to the intestinal microbiome and metabolome of *Pimephales promelas* and *Mus musculus* following exposure to dietary methylmercury. Environ. Sci. Technol. 52, 8774−8784.

Brodersen, C.R., Roddy, A.B., Wason, J.W., McElrone, A.J., 2019. Functional status of xylem through time. Annu. Rev. Plant Biol. 70, 407−433.

Bundy, J.G., Keun, H.C., Sidhu, J.K., Spurgeon, D.J., Svendsen, C., Kille, P., Morgan, A.J., 2007. Metabolic profile biomarkers of metal contamination in a sentinel terrestrial species are applicable across multiple sites. Environ. Sci. Technol. 41, 4458−4464.

Bundy, J.G., Osborn, D., Weeks, J.M., Lindon, J.C., Nicholson, J.K., 2001. An NMR-based metabonomic approach to the investigation of coelomic fluid biochemistry in earthworms under toxic stress. Fed. Eur. Biochem. Soc. Lett. 500, 31−35.

Cappello, T., Brandão, F., Guilherme, S., Santos, M.A., Maisano, M., Mauceri, A., Canário, J., Pacheco, M., Pereira, P., 2016. Insights into the mechanisms underlying mercury-induced oxidative stress in gills of wild fish (*Liza aurata*) combining 1H NMR metabolomics and conventional biochemical assays. Sci. Total Environ. 548−549, 13−24.

Carriquiriborde, P., Marino, D.J., Giachero, G., Castro, E.A., Ronco, A.E., 2012. Global metabolic response in the bile of pejerrey (*Odontesthes bonariensis*, Pisces) sublethally exposed to the pyrethroid cypermethrin. Ecotoxicol. Environ. Saf. 76, 46−54.

Carrizo, D., Chevallier, O.P., Woodside, J.V., Brennan, S.F., Cantwell, M.M., Cuskelly, G., Elliott, C.T., 2017. Untargeted metabolomic analysis of human serum samples associated with exposure levels of Persistent organic pollutants indicate important perturbations in Sphingolipids and Glycerophospholipids levels. Chemosphere 168, 731−738.

Chen, L., Hu, Y., He, J., Chen, J., Giesy, J.P., Xie, P., 2017. Responses of the proteome and metabolome in livers of zebrafish exposed chronically to environmentally relevant concentrations of microcystin-LR. Environ. Sci. Technol. 51, 596−607.

Cicalini, I., Rossi, C., Pieragostino, D., Agnifili, L., Mastropasqua, L., Di Ioia, M., De Luca, G., Onofrj, M., Federici, L., Del Boccio, P., 2019. Integrated lipidomics and metabolomics analysis of tears in multiple sclerosis: an insight into diagnostic potential of lacrimal fluid. Int. J. Mol. Sci. 20.

Dame, Z.T., Aziat, F., Mandal, R., Krishnamurthy, R., Bouatra, S., Borzouie, S., Guo, A.C., Sajed, T., Deng, L., Lin, H., Liu, P., Dong, E., Wishart, D.S., 2015. The human saliva metabolome. Metabolomics 11, 1864−1883.

David, A., Abdul-Sada, A., Lange, A., Tyler, C.R., Hill, E.M., 2014. A new approach for plasma (xeno) metabolomics based on solid-phase extraction and nanoflow liquid chromatography-nanoelectrospray ionisation mass spectrometry. J. Chromatogr. A 1365, 72−85.

David, A., Lange, A., Abdul-Sada, A., Tyler, C.R., Hill, E.M., 2017. Disruption of the prostaglandin metabolome and characterization of the pharmaceutical exposome in fish exposed to wastewater treatment works effluent as revealed by nanoflow-Nanospray mass spectrometry-based metabolomics. Environ. Sci. Technol. 51, 616−624.

Davis, J.M., Ekman, D.R., Skelton, D.M., LaLone, C.A., Ankley, G.T., Cavallin, J.E., Villeneuve, D.L., Collette, T.W., 2017. Metabolomics for informing adverse outcome pathways: androgen receptor activation and the pharmaceutical spironolactone. Aquat. Toxicol. 184, 103−115.

de Sotto, R.B., Medriano, C., Cho, Y., Seok, K.S., Park, Y., Kim, S., 2016. Significance of metabolite extraction method for evaluating sulfamethazine toxicity in adult zebrafish using metabolomics. Ecotoxicol. Environ. Saf. 127, 127−134.

de Sotto, R.B., Medriano, C.D., Cho, Y., Kim, H., Chung, I.Y., Seok, K.S., Song, K.G., Hong, S.W., Park, Y., Kim, S., 2017. Sub-lethal pharmaceutical hazard tracking in adult zebrafish using untargeted LC−MS environmental metabolomics. J. Hazard Mater. 339, 63−72.

Delgado-Povedano, M.M., Calderón-Santiago, M., Priego-Capote, F., Jurado-Gámez, B., Luque de Castro, M.D., 2016. Recent advances in human sweat metabolomics for lung cancer screening. Metabolomics 12.

Dettmer, K., Aronov, P.A., Hammock, B.D., 2007. Mass spectrometry-based metabolomics. Mass Spectrom. Rev. 26, 51−78.

Dettmer, K., Hammock, B.D., 2004. Metabolomics - a new exciting field within the "omics" sciences. Environ. Health Perspect. 112, A396−A397.

Duedahl-Olesen, L., Cederberg, T., Pedersen, K.H., Højgård, A., 2005. Synthetic musk fragrances in trout from Danish fish farms and human milk. Chemosphere 61, 422−431.

Ekman, D.R., Keteles, K., Beihoffer, J., Cavallin, J.E., Dahlin, K., Davis, J.M., Jastrow, A., Lazorchak, J.M., Mills, M.A., Murphy, M., Nguyen, D., Vajda, A.M., Villeneuve, D.L., Winkelman, D.L., Collette, T.W., 2018. Evaluation of targeted and untargeted effects-based monitoring tools to assess impacts of contaminants of emerging concern on fish in the South Platte River. CO. Environ. Pollut. 706−713.

Ekman, D.R., Skelton, D.M., Davis, J.M., Villeneuve, D.L., Cavallin, J.E., Schroeder, A., Jensen, K.M., Ankley, G.T., Collette, T.W., 2015. Metabolite profiling of fish skin mucus: a novel approach for minimally-invasive environmental exposure monitoring and surveillance. Environ. Sci. Technol. 49, 3091−3100.

Ekman, D.R., Teng, Q., Jensen, K.M., Martinovic, D., Villeneuve, D.L., Ankley, G.T., Collette, T.W., 2007. NMR analysis of male fathead minnow urinary metabolites: a potential approach for studying impacts of chemical exposures. Aquat. Toxicol. 85, 104−112.

Ekman, D.R., Teng, Q., Villeneuve, D.L., Kahl, M.D., Jensen, K.M., Durhan, E.J., Ankley, G.T., Collette, T.W., 2008. Investigating compensation and recovery of fathead minnow (*Pimephales promelas*) exposed to 17a-ethynylestradiol with metabolite profiling. Environ. Sci. Technol. 42, 4188−4194.

Ekman, D.R., Teng, Q., Villeneuve, D.L., Kahl, M.D., Jensen, K.M., Durhan, E.J., Ankley, G.T., Collette, T.W., 2009. Profiling lipid metabolites yields unique information on sex- and time-dependent responses of fathead minnows (*Pimephales promelas*) exposed to 17α-ethynylestradiol. Metabolomics 5, 22−32.

Ekman, D.R., Villeneuve, D.L., Teng, Q., Ralston-Hooper, K.J., Martinović-Weigelt, D., Kahl, M.D., Jensen, K.M., Durhan, E.J., Makynen, E.A., Ankley, G.T., Collette, T.W., 2011. Use of gene expression, biochemical and metabolite profiles to enhance exposure and effects assessment of the model androgen 17β-trenbolone in fish. Environ. Toxicol. Chem. 30, 319−329.

Elie, M.R., Choi, J., Nkrumah-Elie, Y.M., Gonnerman, G.D., Stevens, J.F., Tanguay, R.L., 2015. Metabolomic analysis to define and compare the effects of PAHs and oxygenated PAHs in developing zebrafish. Environ. Res. 140, 502−510.

Flores-Valverde, A.M., Horwood, J., Hill, E.M., 2010. Disruption of the steroid metabolome in fish caused by exposure to the environmental estrogen 17α-ethinylestradiol. Environ. Sci. Technol. 44, 3552–3558.

Folch, J., Lees, M., Sloane Stanley, G.H., 1957. A simple method for the isolation and purification of total lipides from animal tissues. J. Biol. Chem. 226, 497–509.

Foley, K.F., Wasserman, J., 2014. Are unexpected positive dipstick urine bilirubin results clinically significant? A retrospective review. Lab. Med. 45, 59–61.

Fu, J., Gong, Z., Bae, S., 2019. Assessment of the effect of methyl-triclosan and its mixture with triclosan on developing zebrafish (*Danio rerio*) embryos using mass spectrometry-based metabolomics. J. Hazard Mater. 186–196.

Glazer, L., Kido Soule, M.C., Longnecker, K., Kujawinski, E.B., Aluru, N., 2018. Hepatic metabolite profiling of polychlorinated biphenyl (PCB)-resistant and sensitive populations of Atlantic killifish (*Fundulus heteroclitus*). Aquat. Toxicol. 205, 114–122.

Gómez-Canela, C., Prats, E., Piña, B., Tauler, R., 2017. Assessment of chlorpyrifos toxic effects in zebrafish (*Danio rerio*) metabolism. Environ. Pollut. 220, 1231–1243.

Griffith, C.M., Morgan, M.A., Dinges, M.M., Mathon, C., Larive, C.K., 2018. Metabolic profiling of chloroacetanilide herbicides in earthworm coelomic fluid using 1 H NMR and GC-MS. J. Proteome Res. 17, 2611–2622.

Grosell, M., O'Donnell, M.J., Wood, C.M., 2000. Hepatic versus gallbladder bile composition: in vivo transport physiology of the gallbladder in rainbow trout. Am. J. Physiol. Regul. Integr. Comp. Physiol. 278, R1674–R1684.

Guyton, A.C., Hall, J.E., 2011. Textbook of Medical Physiology, twelfth ed. Saunders Elsevier, Philadelphia.

Hajirezaee, S., Mirvaghefi, A., Farahmand, H., Agh, N., 2017. NMR-based metabolomic study on the toxicological effects of pesticide, diazinon on adaptation to sea water by endangered Persian sturgeon, *Acipenser persicus* fingerlings. Chemosphere 185, 213–226.

Hano, T., Ohkubo, N., Mochida, K., 2017. A hepatic metabolomics-based diagnostic approach to assess lethal toxicity of dithiocarbamate fungicide polycarbamate in three marine fish species. Ecotoxicol. Environ. Saf. 138, 64–70.

Heikenfeld, J., Jajack, A., Feldman, B., Granger, S.W., Gaitonde, S., Begtrup, G., Katchman, B.A., 2019. Accessing analytes in biofluids for peripheral biochemical monitoring. Nat. Biotechnol. 37, 407–419.

Huang, S.S.Y., Benskin, J.P., Chandramouli, B., Butler, H., Helbing, C.C., Cosgrove, J.R., 2016. Xenobiotics produce distinct metabolomic responses in zebrafish larvae (*Danio rerio*). Environ. Sci. Technol. 50, 6526–6535.

Huang, S.S.Y., Benskin, J.P., Veldhoen, N., Chandramouli, B., Butler, H., Helbing, C.C., Cosgrove, J.R., 2017. A multi-omic approach to elucidate low-dose effects of xenobiotics in zebrafish (*Danio rerio*) larvae. Aquat. Toxicol. 182, 102–112.

Jia, M., Wang, Y., Teng, M., Wang, D., Yan, J., Miao, J., Zhou, Z., Zhu, W., 2018. Toxicity and metabolomics study of isocarbophos in adult zebrafish (*Danio rerio*). Ecotoxicol. Environ. Saf. 163, 1–6.

Jia, M., Wang, Y., Wang, D., Teng, M., Yan, J., Yan, S., Meng, Z., Li, R., Zhou, Z., Zhu, W., 2019. The effects of hexaconazole and epoxiconazole enantiomers on metabolic profile following exposure to zebrafish (*Danio rerio*) as well as the histopathological changes. Chemosphere 520–533.

Jiang, Y.X., Shi, W.J., Ma, D.D., Zhang, J.N., Ying, G.G., Zhang, H., Ong, C.N., 2019. Dydrogesterone exposure induces zebrafish ovulation but leads to oocytes over-ripening: an integrated histological and metabolomics study. Environ. Int. 390–398.

Jin, C., Luo, T., Zhu, Z., Pan, Z., Yang, J., Wang, W., Fu, Z., Jin, Y., 2017. Imazalil exposure induces gut microbiota dysbiosis and hepatic metabolism disorder in zebrafish. Comp. Biochem. Physiol. C Toxicol. Pharmacol. 202, 85–93.

Kaplan, K., Dwivedi, P., Davidson, S., Yang, Q., Tso, P., Siems, W., Hill, H.H., 2009. Monitoring dynamic changes in lymph metabolome of fasting and fed rats by electrospray ionization-ion mobility mass spectrometry (ESI-IMMS). Anal. Chem. 81, 7944–7953.

Katsiadaki, I., Williams, T.D., Ball, J.S., Bean, T.P., Sanders, M.B., Wu, H., Santos, E.M., Brown, M.M., Baker, P., Ortega, F., Falciani, F., Craft, J.A., Tyler, C.R., Viant, M.R., Chipman, J.K., 2010. Hepatic transcriptomic and metabolomic responses in the stickleback (*Gasterosteus aculeatus*) exposed to ethinyl-estradiol. Aquat. Toxicol. 97, 174–187.

Khan, S.A., Liu, X., Li, H., Zhu, Y., Fan, W., Zhou, P., Rehman, Z., 2016. 1 H NMR-based serum metabolic profiling of *Carassius auratus gibelio* under the toxicity of Pb 2^+ and Cd 2^+. Int. J. Environ. Sci. Technol. 13, 2597–2608.

Kokushi, E., Shintoyo, A., Koyama, J., Uno, S., 2017. Evaluation of 2,4-dichlorophenol exposure of Japanese medaka, *Oryzias latipes*, using a metabolomics approach. Environ. Sci. Poll. Res. 24, 27678–27686.

Kokushi, E., Uno, S., Harada, T., Koyama, J., 2010. 1H NMR-based metabolomics approach to assess toxicity of bunker a heavy oil to freshwater carp, *Cyprinus carpio*. Environ. Toxicol. 27, 404–414.

Kokushi, E., Uno, S., Pal, S., Koyama, J., 2013. Effects of chlorpyrifos on the metabolome of the freshwater carp, *Cyprinus Carpio*. Environ. Toxicol. 30 (3), 253–260.

Kutyshenko, V.P., Molchanov, M., Beskaravayny, P., Uversky, V.N., Timchenko, M.A., 2011. Analyzing and mapping sweat metabolomics by high-resolution NMR spectroscopy. PLoS One 6.

Laiakis, E.C., Strawn, S.J., Brenner, D.J., Fornace Jr., A.J., 2016. Assessment of saliva as a potential biofluid for biodosimetry: a pilot metabolomics study in mice. Radiat. Res. 186, 92–97.

Le Manach, S., Sotton, B., Huet, H., Duval, C., Paris, A., Marie, A., Yépremian, C., Catherine, A., Mathéron, L., Vinh, J., Edery, M., Marie, B., 2018. Physiological effects caused by microcystin-producing and non-microcystin producing *Microcystis aeruginosa* on medaka fish: a proteomic and metabolomic study on liver. Environ. Pollut. 234, 523–537.

Lee, S.H., Woo, H.M., Jung, B.H., Lee, J., Kwon, O.S., Pyo, H.S., Choi, M.H., Chung, B.C., 2007. Metabolomic approach to evaluate the toxicological effects of nonylphenol with rat urine. Anal. Chem. 79, 6102–6110.

Lei, E.N.Y., Yau, M.S., Yeung, C.C., Murphy, M.B., Wong, K.L., Lam, M.H.W., 2017. Profiling of selected functional metabolites in the central nervous system of marine medaka (*Oryzias melastigma*) for environmental neurotoxicological assessments. Arch. Environ. Contam. Toxicol. 72, 269–280.

Lenz, E.M., Weeks, J.M., Lindon, J.C., Osborn, D., Nicholson, J.K., 2005. Qualitative high field 1H-NMR spectroscopy for the characterization of endogenous metabolites in earthworms with biochemical biomarker potential. Metabolomics 1, 123–136.

Lestaevel, P., Grison, S., Favé, G., Elie, C., Dhieux, B., Martin, J.C., Tack, K., Souidi, M., 2016. Assessment of the central effects of natural uranium via behavioural performances and the cerebrospinal fluid metabolome. Neural Plast. 1–11. https://doi.org/10.1155/2016/9740353.

Li, M., Wang, J., Lu, Z., Wei, D., Yang, M., Kong, L., 2014. NMR-based metabolomics approach to study the toxicity of lambda-cyhalothrin to goldfish (*Carassius auratus*). Aquat. Toxicol. 146, 82–92.

Li, M.H., Ruan, L.Y., Liu, Y., Xu, H.D., Chen, T., Fu, Y.H., Jiang, L., Wang, J.S., 2015. Insight into biological system responses in goldfish (*Carassius auratus*) to multiple doses of avermectin exposure by integrated 1H NMR-based metabolomics. Toxicol. Res. 4, 1374–1388.

Li, M.H., Ruan, L.Y., Zhou, J.W., Fu, Y.H., Jiang, L., Zhao, H., Wang, J.S., 2017. Metabolic profiling of goldfish (*Carassius auratus*) after long-term glyphosate-based herbicide exposure. Aquat. Toxicol. 188, 159–169.

Li, M.H., Xu, H.D., Liu, Y., Chen, T., Jiang, L., Fu, Y.H., Wang, J.S., 2016. Multi-tissue metabolic responses of goldfish (*Carassius auratus*) exposed to glyphosate-based herbicide. Toxicol. Res. 5, 1039–1052.

Lin, C.Y., Anderson, B.S., Phillips, B.M., Peng, A.C., Clark, S., Voorhees, J., Wu, H.D.I., Martin, M.J., McCall, J., Todd, C.R., Hsieh, F., Crane, D., Viant, M.R., Sowby, M.L., Tjeerdema, R.S., 2009. Characterization of the metabolic actions of crude versus dispersed oil in salmon smolts via NMR-based metabolomics. Aquat. Toxicol. 95, 230–238.

Lindon, J.C., Holmes, E., Nicholson, J.K., 2003. Peer reviewed: so what's the deal with metabonomics? Anal. Chem. 75, 384 A–391 A.

Liu, X., Hoene, M., Wang, X., Yin, P., Häring, H.U., Xu, G., Lehmann, R., 2018a. Serum or plasma, what is the difference? Investigations to facilitate the sample material selection decision making process for metabolomics studies and beyond. Anal. Chim. Acta 1037, 293–300.

Liu, Y., Wang, X., Li, Y., Chen, X., 2018b. Metabolomic analysis of short-term sulfamethazine exposure on marine medaka (*Oryzias melastigma*) by comprehensive two-dimensional gas chromatography-time-of-flight mass spectrometry. Aquat. Toxicol. 198, 269−275.

Lu, Y., Zhang, Y., Deng, Y., Jiang, W., Zhao, Y., Geng, J., Ding, L., Ren, H., 2016. Uptake and accumulation of polystyrene microplastics in zebrafish (*Danio rerio*) and toxic effects in liver. Environ. Sci. Technol. 50, 4054−4060.

Lu, Y.X., Zhang, Q., Xu, W.H., 2014. Global metabolomic analyses of the hemolymph and brain during the initiation, maintenance, and termination of pupal diapause in the cotton bollworm, *Helicoverpa armigera*. PLoS One 9.

Lunn, K.F., 2011. The kidney in critically Ill small animals. Vet. Clin. Small Anim. Pract. 41, 727−744.

Mattsson, K., Ekvall, M.T., Hansson, L.A., Linse, S., Malmendal, A., Cedervall, T., 2015. Altered behavior, physiology, and metabolism in fish exposed to polystyrene nanoparticles. Environ. Sci. Technol. 49, 553−561.

Matyash, V., Liebisch, G., Kurzchalia, T.V., Shevchenko, A., Schwudke, D., 2008. Lipid extraction by methyl-tert-butyl ether for high-throughput lipidomics. J. Lipid Res. 49, 1137−1146.

Melvin, S.D., Lanctôt, C.M., Doriean, N.J.C., Carroll, A.R., Bennett, W.W., 2018. Untargeted NMR-based metabolomics for field-scale monitoring: temporal reproducibility and biomarker discovery in mosquitofish (*Gambusia holbrooki*) from a metal(loid)-contaminated wetland. Environ. Pollut. 243, 1096−1105.

Meyer, F.L., Mattox, H., Bolick, M., MacDonald, C., 1971. Metabolic changes after test meals with different carbohydrates: blood levels of pyruvic acid, glucose, and lactic dehydrogenase. Am. J. Clin. Nutr. 24, 615−621.

Mishra, P., Gong, Z., Kelly, B.C., 2017. Assessing biological effects of fluoxetine in developing zebrafish embryos using gas chromatography-mass spectrometry based metabolomics. Chemosphere 188, 157−167.

Mishra, P., Gong, Z., Kelly, B.C., 2019. Assessing pH-dependent toxicity of fluoxetine in embryonic zebrafish using mass spectrometry-based metabolomics. Sci. Total Environ. 650, 2731−2741.

Mosley, J.D., Ekman, D.R., Cavallin, J.E., Villeneuve, D.L., Ankley, G.T., Collette, T.W., 2018. High-resolution mass spectrometry of skin mucus for monitoring physiological impacts and contaminant biotransformation products in fathead minnows exposed to wastewater effluent. Environ. Toxicol. Chem. 37, 788−796.

Noga, M.J., Dane, A., Shi, S., Attali, A., van Aken, H., Suidgeest, E., Tuinstra, T., Muilwijk, B., Coulier, L., Luider, T., Reijmers, T.H., Vreeken, R.J., Hankemeier, T., 2011. Metabolomics of cerebrospinal fluid reveals changes in the central nervous system metabolism in a rat model of multiple sclerosis. Metabolomics 1−11.

O'Kane, A.A., Chevallier, O.P., Graham, S.F., Elliott, C.T., Mooney, M.H., 2013. Metabolomic profiling of in vivo plasma responses to dioxin-associated dietary contaminant exposure in rats: implications for identification of sources of animal and human exposure. Environ. Sci. Technol. 47, 5409−5418.

Ortiz-Villanueva, E., Jaumot, J., Martínez, R., Navarro-Martín, L., Piña, B., Tauler, R., 2018. Assessment of endocrine disruptors effects on zebrafish (*Danio rerio*) embryos by untargeted LC-HRMS metabolomic analysis. Sci. Total Environ. 635, 156−166.

Ortiz-Villanueva, E., Navarro-Martín, L., Jaumot, J., Benavente, F., Sanz-Nebot, V., Piña, B., Tauler, R., 2017. Metabolic disruption of zebrafish (*Danio rerio*) embryos by bisphenol A. An integrated metabolomic and transcriptomic approach. Environ. Pollut. 231, 22−36.

Palmas, F., Fattuoni, C., Noto, A., Barberini, L., Dessì, A., Fanos, V., 2016. The choice of amniotic fluid in metabolomics for the monitoring of fetus health. Expert Rev. Mol. Diagn. 16, 473−486.

Poynton, H.C., Taylor, N.S., Hicks, J., Colson, K., Chan, S., Clark, C., Scanlan, L., Loguinov, A.V., Vulpe, C., Viant, M.R., 2011. Metabolomics of microliter hemolymph samples enables an Improved understanding of the combined metabolic and transcriptional responses of Daphnia magna to cadmium. Environ. Sci. Technol. 45, 3710−3717.

Qiao, R., Sheng, C., Lu, Y., Zhang, Y., Ren, H., Lemos, B., 2019. Microplastics induce intestinal inflammation, oxidative stress, and disorders of metabolome and microbiome in zebrafish. Sci. Total Environ. 662, 246−253.

Raterink, R.J., Lindenburg, P.W., Vreeken, R.J., Ramautar, R., Hankemeier, T., 2014. Recent developments in sample-pretreatment techniques for mass spectrometry-based metabolomics. Trac. Trends Anal. Chem. 61, 157−167.

Rellán-Álvarez, R., El-Jendoubi, H., Wohlgemuth, G., Abadía, A., Fiehn, O., Abadía, J., Alvarez-Fernández, A., 2011. Metabolite profile changes in xylem sap and leaf extracts of strategy I plants in response to iron deficiency and resupply. Front. Plant Sci. 2, 66-66.

Ren, X., Zhang, H., Geng, N., Xing, L., Zhao, Y., Wang, F., Chen, J., 2018. Developmental and metabolic responses of zebrafish (*Danio rerio*) embryos and larvae to short-chain chlorinated paraffins (SCCPs) exposure. Sci. Total Environ. 622−623, 214−221.

Samuelsson, L.M., Björlenius, B., Förlin, L., Larsson, D.G.J., 2011. Reproducible 1H NMR-based metabolomic responses in fish exposed to different sewage effluents in two separate studies. Environ. Sci. Technol. 45, 1703−1710.

Samuelsson, L.M., Förlin, L., Karlsson, G., Adolfsson-Erici, M., Larsson, D.G.J., 2006. Using NMR metabolomics to identify responses of an environmental estrogen in blood plasma of fish. Aquat. Toxicol. 78, 341−349.

Santos, E.M., Ball, J.S., Williams, T.D., Wu, H., Ortega, F., Van Aerle, R., Katsiadaki, I., Falciani, F., Viant, M.R., Chipman, J.K., Tyler, C.R., 2010. Identifying health impacts of exposure to copper using transcriptomics and metabolomics in a fish model. Environ. Sci. Technol. 44, 820−826.

Schock, T.B., Stancyk, D.A., Thibodeaux, L., Burnett, K.G., Burnett, L.E., Boroujerdi, A.F.B., Bearden, D.W., 2010. Metabolomic analysis of Atlantic blue crab, *Callinectes sapidus*, hemolymph following oxidative stress. Metabolomics 6, 250−262.

Skelton, D.M., Ekman, D.R., Martinović-Weigelt, D., Ankley, G.T., Villeneuve, D.L., Teng, Q., Collette, T.W., 2014. Metabolomics for in situ environmental monitoring of surface waters impacted by contaminants from both point and nonpoint sources. Environ. Sci. Technol. 48, 2395−2403.

Slupsky, C.M., 2019. Metabolomics in Human Milk Research. Nestle Nutrition Institute Workshop Series, pp. 179−190.

Song, Y., Chai, T., Yin, Z., Zhang, X., Zhang, W., Qian, Y., Qiu, J., 2018. Stereoselective effects of ibuprofen in adult zebrafish (*Danio rerio*) using UPLC-TOF/MS-based metabolomics. Environ. Pollut. 241, 730−739.

Sotton, B., Paris, A., Le Manach, S., Blond, A., Duval, C., Qiao, Q., Catherine, A., Combes, A., Pichon, V., Bernard, C., Marie, B., 2019. Specificity of the metabolic signatures of fish from cyanobacteria rich lakes. Chemosphere 183−191.

Sotton, B., Paris, A., Le Manach, S., Blond, A., Lacroix, G., Millot, A., Duval, C., Qiao, Q., Catherine, A., Marie, B., 2017. Global metabolome changes induced by cyanobacterial blooms in three representative fish species. Sci. Total Environ. 590−591, 333−342.

Southam, A.D., Lange, A., Hines, A., Hill, E.M., Katsu, Y., Iguchi, T., Tyler, C.R., Viant, M.R., 2011. Metabolomics reveals target and off-target toxicities of a model organophosphate pesticide to roach (*Rutilus rutilus*): Implications for biomonitoring. Environ. Sci. Technol. 45, 3759−3767.

Statham, C.N., Melancon Jr., M.J., Lech, J.J., 1976. Bioconcentration of xenobiotics in trout bile: a proposed monitoring aid for some waterborne chemicals. Science 193, 680−681.

Steinborn, A., Alder, L., Michalski, B., Zomer, P., Bendig, P., Martinez, S.A., Mol, H.G.J., Class, T.J., Costa Pinheiro, N., 2016. Determination of glyphosate levels in breast milk samples from Germany by LC-MS/MS and GC-MS/MS. J. Agric. Food Chem. 64, 1414−1421.

Steudte-Schmiedgen, S., Kirschbaum, C., Alexander, N., Stalder, T., 2016. An integrative model linking traumatization, cortisol dysregulation and posttraumatic stress disorder: insight from recent hair cortisol findings. Neurosci. Biobehav. Rev. 69, 124−135.

Sun, H.Z., Wang, D.M., Wang, B., Wang, J.K., Liu, H.Y., Guan, L.L., Liu, J.X., 2015. Metabolomics of four biofluids from dairy cows: potential biomarkers for milk production and quality. J. Proteome Res. 14, 1287−1298.

Szabo, D.T., Pathmasiri, W., Sumner, S., Birnbaum, L.S., 2017. Serum metabolomic profiles in neonatal mice following oral brominated flame retardant exposures to hexabromocyclododecane (HBCD) alpha, gamma, and commercial mixture. Environ. Health Perspect. 125, 651−659.

Teng, M., Zhu, W., Wang, D., Qi, S., Wang, Y., Yan, J., Dong, K., Zheng, M., Wang, C., 2018a. Metabolomics and transcriptomics reveal the toxicity of difenoconazole to the early life stages of zebrafish (*Danio rerio*). Aquat. Toxicol. 194, 112−120.

Teng, M., Zhu, W., Wang, D., Yan, J., Qi, S., Song, M., Wang, C., 2018b. Acute exposure of zebrafish embryo (*Danio rerio*) to flutolanil reveals its developmental mechanism of toxicity via disrupting the thyroid system and metabolism. Environ. Pollut. 242, 1157−1165.

Turgeon, R., Wolf, S., 2009. Phloem transport: cellular pathways and molecular trafficking. Annu. Rev. Plant Biol. 60, 207−221.

Uno, S., Shintoyo, A., Kokushi, E., Yamamoto, M., Nakayama, K., Koyama, J., 2012. Gas chromatography-mass spectrometry for metabolite profiling of Japanese medaka (*Oryzias latipes*) juveniles exposed to malathion. Environ. Sci. Pollut. Res. 19, 2595−2605.

Van Scoy, A.R., Anderson, B.S., Philips, B.M., Voorhees, J., McCann, M., De Haro, H., Martin, M.J., McCall, J., Todd, C.R., Crane, D., Sowby, M.L., Tjeerdema, R.S., 2012. NMR-based characterization of the acute metabolic effects of weathered crude and dispersed oil in spawning topsmelt and their embryos. Ecotoxicol. Environ. Saf. 78, 99−109.

Van Scoy, A.R., Yu Lin, C., Anderson, B.S., Philips, B.M., Martin, M.J., McCall, J., Todd, C.R., Crane, D., Sowby, M.L., Viant, M.R., Tjeerdema, R.S., 2010. Metabolic responses produced by crude versus dispersed oil in Chinook salmon pre-smolts via NMR-based metabolomics. Ecotoxicol. Environ. Saf. 73, 710−717.

Viant, M.R., Bundy, J.G., Pincetich, C.A., de Ropp, J.S., Tjeerdema, R.S., 2005. NMR-derived developmental metabolic trajectories: an approach for visualizing the toxic actions of trichloroethylene during embryogenesis. Metabolomics 1, 149−158.

Wan, J., Jiang, F., Zhang, J., Xu, Q., Chen, D., Yu, B., Mao, X., Yu, J., Luo, Y., He, J., 2017. Amniotic fluid metabolomics and biochemistry analysis provides novel insights into the diet-regulated foetal growth in a pig model. Sci. Rep. 7.

Wan, Z., Wang, C., Zhou, J., Shen, M., Wang, X., Fu, Z., Jin, Y., 2019. Effects of polystyrene microplastics on the composition of the microbiome and metabolism in larval zebrafish. Chemosphere 646−658.

Wang, C., Qian, Y., Zhang, X., Chen, F., Zhang, Q., Li, Z., Zhao, M., 2016a. A metabolomic study of fipronil for the anxiety-like behavior in zebrafish larvae at environmentally relevant levels. Environ. Pollut. 211, 252−258.

Wang, D., Zhang, P., Wang, X., Wang, Y., Zhou, Z., Zhu, W., 2016b. NMR- and LC−MS/MS-based urine metabolomic investigation of the subacute effects of hexabromocyclododecane in mice. Environ. Sci. Pollut. Res. 23, 8500−8507.

Wang, Q., Zhang, Y., Zheng, N., Guo, L., Song, X., Zhao, S., Wang, J., 2019. Biological system responses of dairy cows to aflatoxin b1 exposure revealed with metabolomic changes in multiple biofluids. Toxins 11.

Wang, Y., Teng, M., Wang, D., Yan, J., Miao, J., Zhou, Z., Zhu, W., 2017a. Enantioselective bioaccumulation following exposure of adult zebrafish (*Danio rerio*) to epoxiconazole and its effects on metabolomic profile as well as genes expression. Environ. Pollut. 229, 264−271.

Wang, Y., Zhu, W., Wang, D., Teng, M., Yan, J., Miao, J., Zhou, Z., 2017b. 1H NMR-based metabolomics analysis of adult zebrafish (*Danio rerio*) after exposure to diniconazole as well as its bioaccumulation behavior. Chemosphere 168, 1571−1577.

Whiley, L., Godzien, J., Ruperez, F.J., Legido-Quigley, C., Barbas, C., 2012. In-vial dual extraction for direct LC-MS analysis of plasma for comprehensive and highly reproducible metabolic fingerprinting. Anal. Chem. 84, 5992−5999.

Williams, T.D., Wu, H., Santos, E.M., Ball, J., Katsiadaki, I., Brown, M.M., Baker, P., Ortega, F., Falciani, F., Craft, J.A., Tyler, C.R., Chipman, J.K., Viant, M.R., 2009. Hepatic transcriptomic and metabolomic responses in the stickleback (Gasterosteus aculeatus) exposed to environmentally relevant concentrations of dibenzanthracene. Environ. Sci. Technol. 43, 6341−6348.

Williams, T.D., Turan, N., Diab, A.M., Wu, H., Mackenzie, C., Bartie, K.L., Hrydziuszko, O., Lyons, B.P., Stentiford, G.D., Herbert, J.M., Abraham, J.K., Katsiadaki, I., Leaver, M.J., Taggart, J.B., George, S.G., Viant, M.R., Chipman, K.J., Falciani, F., 2011. Towards a system level understanding of non-model organisms sampled from the environment: a network biology approach. PLoS Comput. Biol. 7.

Wilson, I., 2011. Global metabolic profiling (metabonomics/metabolomics) using dried blood spots: advantages and pitfalls. Bioanalysis 3, 2255−2257.

Wishart, D.S., Lewis, M.J., Morrissey, J.A., Flegel, M.D., Jeroncic, K., Xiong, Y., Cheng, D., Eisner, R., Gautam, B., Tzur, D., Sawhney, S., Bamforth, F., Greiner, R., Li, L., 2008. The human cerebrospinal fluid metabolome. J. Chromatogr. B 871, 164−173.

Xia, J., Lu, L., Jin, C., Wang, S., Zhou, J., Ni, Y., Fu, Z., Jin, Y., 2018. Effects of short term lead exposure on gut microbiota and hepatic metabolism in adult zebrafish. Comp. Biochem. Physiol. C Toxicol. Pharmacol. 209, 1−8.

Xu, H.D., Wang, J.S., Li, M.H., Liu, Y., Chen, T., Jia, A.Q., 2015. 1H NMR based metabolomics approach to study the toxic effects of herbicide butachlor on goldfish (*Carassius auratus*). Aquat. Toxicol. 159, 69−80.

Xu, Y., Wang, J., Liang, X., Gao, Y., Chen, W., Huang, Q., Liang, C., Tang, L., Ouyang, G., Yang, X., 2016. Urine metabolomics of women from small villages exposed to high environmental cadmium levels. Environ. Toxicol. Chem. 35, 1268−1275.

Yan, L., Gong, C., Zhang, X., Zhang, Q., Zhao, M., Wang, C., 2016. Perturbation of metabonome of embryo/larvae zebrafish after exposure to fipronil. Environ. Toxicol. Pharmacol. 48, 39−45.

Yang, J., Wang, H., Xu, W., Hao, D., Du, L., Zhao, X., Sun, C., 2013. Metabolomic analysis of rat plasma following chronic low-dose exposure to dichlorvos. Hum. Exp. Toxicol. 32, 196−205.

Yau, M.S., Lei, E.N.Y., Ng, I.H.M., Yuen, C.K.K., Lam, J.C.W., Lam, M.H.W., 2019. Changes in the neurotransmitter profile in the central nervous system of marine medaka (*Oryzias melastigma*) after exposure to brevetoxin PbTx-1 − a multivariate approach to establish exposure biomarkers. Sci. Total Environ. 673, 327−336.

Yoon, C., Yoon, D., Cho, J., Kim, S., Lee, H., Choi, H., Kim, S., 2017. 1H-NMR-based metabolomic studies of bisphenol A in zebrafish (*Danio rerio*). J. Environ. Sci. Health Part B Pestic. Food Contam. Agric. Wastes 52, 282−289.

Yoon, D., Kim, S., Lee, M., Yoon, C., Kim, S., 2016. 1H-NMR-based metabolomic study on toxicity of methomyl and methidathion in fish. J. Environ. Sci. Health Part B Pestic. Food Contam. Agric. Wastes 51, 824−831.

Yuk, J., Simpson, M.J., Simpson, A.J., 2012. Coelomic fluid: a complimentary biological medium to assess sub-lethal endosulfan exposure using 1H NMR-based earthworm metabolomics. Ecotoxicology 1−13.

Zhang, B., Tolstikov, V., Turnbull, C., Hicks, L.M., Fiehn, O., 2010. Divergent metabolome and proteome suggest functional independence of dual phloem transport systems in cucurbits. Proc. Natl. Acad. Sci. U.S.A. 107, 13532.

Zhang, H., Zhao, L., 2017. Influence of sublethal doses of acetamiprid and halosulfuron-methyl on metabolites of zebra fish (*Brachydanio rerio*). Aquat. Toxicol. 191, 85−94.

Zhang, Q., Lu, Y.X., Xu, W.H., 2013. Proteomic and metabolomic profiles of larval hemolymph associated with diapause in the cotton bollworm, *Helicoverpa armigera*. BMC Genom. 14.

Zhou, L., Li, H., Hao, F., Li, N., Liu, X., Wang, G., Wang, Y., Tang, H., 2015. Developmental changes for the hemolymph metabolome of silkworm (*Bombyx mori* L.). J. Proteome Res. 14, 2331−2347.

Zhu, L., Gao, N., Wang, R., Zhang, L., 2018. Proteomic and metabolomic analysis of marine medaka (*Oryzias melastigma*) after acute ammonia exposure. Ecotoxicology 27, 267−277.

Ziarrusta, H., Mijangos, L., Picart-Armada, S., Irazola, M., Perera-Lluna, A., Usobiaga, A., Prieto, A., Etxebarria, N., Olivares, M., Zuloaga, O., 2018. Non-targeted metabolomics reveals alterations in liver and plasma of gilt-head bream exposed to oxybenzone. Chemosphere 211, 624−631.

Ziarrusta, H., Ribbenstedt, A., Mijangos, L., Picart-Armada, S., Perera-Lluna, A., Prieto, A., Izagirre, U., Benskin, J.P., Olivares, M., Zuloaga, O., Etxebarria, N., 2019. Amitriptyline at an environmentally relevant concentration alters the profile of metabolites beyond monoamines in gilt-head bream. Environ. Toxicol. Chem. 38, 965−977.

CHAPTER 4

Environmental metallomics and metabolomics in free-living and model organisms: an approach for unraveling metal exposure mechanisms

Julián Blasco[1], Gema Rodríguez-Moro[2,3], Belén Callejón-Leblic[2,3], Sara Ramírez-Acosta[2,3], Francisca Arellano-Beltrán[2,3], Ana Arias-Borrego[2,3], Tamara García-Barrera[2,3], José Luis Gómez-Ariza[2,3]

[1]Institute of Marine Sciences of Andalusia (CSIC), Campus Rio San Pedro, Cádiz, Spain; [2]Department of Chemistry, Faculty of Experimental Sciences, University of Huelva, Campus El Carmen, Huelva, Spain; [3]Research Center of Natural Resources, Health and the Environment (RENSMA), University of Huelva, Campus El Carmen, Huelva, Spain

Chapter Outline

Environmental Metabolomics. https://doi.org/10.1016/B978-0-12-818196-6.00004-2

1. Introduction

Elements have a great importance for living organisms, in some cases because they act as cofactors in many enzymatic reactions, in others due to their undefined role as essential or toxic entities, as is the case of selenium, and finally by the well-established toxicological behavior of some of them (e.g., As, Hg, Cd, and Pb) (Mounicou et al., 2009). In all the cases, it is very important to know the element speciation or chemical form in which it is present. It is well known that about one-third of proteins need a metal cofactor to fulfill their function (Tainer et al., 1991), and many metabolites include any metal or metalloid in their structure (Garcia-Sevillano et al., 2014a). These facts have given rise to the concept of metallomics focused on the nontargeted study of metal-tagged biomolecules, in opposition to metabolomics that considers all the metabolites (regardless of metal) that are overexpressed or inhibited by some contaminant. In addition, metallobiomolecules should not be considered in isolation because their spatial, temporal, and physiological environment has to be taken into account since it conditions the role of an active metallobiomolecule in respect to health or disease. The knowledge about metals traffic and interactions in cells and tissues is critical because they determine theirs toxicological effects, being necessary to consider their transport by biological fluids and the pass through biological barriers (Gómez-Ariza et al., 2011; Maret, 2004).

In the cell, the concentration and distribution of metals among cellular compartments and their incorporation to metalloproteins are carefully controlled (Outten and O'Halloran, 2001), which means that metals have to be sensed and incorporated into other biomolecules (Monicou et al., 2009), but even more, the spatial, temporal, and physiological environments of metallobiomolecules have also to be considered due to their influence in the molecule behavior and action. This means that it is necessary to study and characterize the molecules in the complexity of their environment, in which many other biomolecules of different size and chemical properties coexist, affecting each other's functions (Gierasch and Gershenson, 2009).

In environmental studies, all these remarks have to be considered since there are many factors that can influence the action of potential contaminants on organism's well-being under the genetic and epigenetic framework (Wrobel et al., 2009). In this context, powerful analytical methodologies are needed to conveniently address this complex problem, which requires the combined application of multidimensional analytical platforms involving elemental and molecular mass spectrometry, generally coupled to liquid and/or gas chromatography (Mounicou et al., 2009; García-Sevillano et al., 2014b, 2014f and 2015a).

In this sense, the combination of metallomics and metabolomics approaches has become very successful in deciphering the response of living organisms exposed to environmental contaminants.

Several main points have to be considered for a suitable methodological approach to research complex environmental systems, in order to decipher the multiple metal—biomolecule interactions developed in living organisms under the action of pollutant flows caused by natural or anthropogenic action. A first question considers that the environment can joint multiple contaminants simultaneously, so that both methods and exposure experiment in the laboratory have to include both the presence of individual contaminants (Garcia-Sevillano et al., 2013a) and mixtures of them (Garcia-Sevillano et al., 2014c). In addition, is important the appropriate selection of bioindicators in connection to the environmental area under study, such as mice in terrestrial ecosystems (Garcia-Sevillano et al., 2013a) or crabs (Fernández-Cisnal et al., 2018; Gago-Tinoco et al., 2014) and anemones (Contreras-Acuña et al., 2013) in aquatic ones. Samples used in these studies are very diverse; generally, they include tissues—liver, kidney, brain—and biological fluids—blood, serum, plasma (Garcia-Sevillano et al., 2014h)—as well as cell organelles from the cells of these tissues (Garcia-Sevillano et al., 2015b). Another point is to assure the integrity of organometallic analytes and metabolites to be studied avoiding changes in the chemical species and interactions among them that can provide a blurred vision of the issue. Finally, several questions arise in relation to methods suitability to deal with this complex problem, the multidimensional nature of the approaches required to address the multiplicity of molecules, and their interaction in the complexity of living organisms, being necessary the use of omics (Garcia-Sevillano et al., 2014g), which have to consider the input sources of contaminants into organisms—oral, skin, respiratory (Garcia-Sevillano et al., 2014b)—sampling and sample treatment, metabolic changes associated with the presence of contaminants (Garcia-Sevillano et al., 2015a) and statistic interpretation of results (Garcia-Sevillano et al., 2015a).

In this chapter, several analytical platforms are considered for metabolomics studies focused on environmental problems of metal contamination, taking into account not only the methodological aspects but also the way of approaching the problem. Metallomics issues related to metal exposure of living organisms, especially mice and metabolic response to the presence of metallic species, are the main purposes of this study.

2. Typical analytical approaches for environmental metallomics/ metabolomics

2.1 Metallomics/metabolomics workflows for nontargeted analysis

The application of dual metallomics/metabolomics methodologies in environmental studies requires the use of analytical platforms based on elemental (inductively coupled plasma mass spectrometry, ICP-MS) and molecular (quadrupole time-of-flight mass spectrometry, QTOF-MS) mass spectrometry coupled to different chromatographic devices. However, direct sample introduction into the mass spectrometer can also be performed, using ICP-MS for total trace metal evaluation in tissue distribution or homeostasis studies and direct infusion electrospray quadrupole time-of-flight mass spectrometry (DI-ESI-QTOF-MS) for metabolite identification and metabolomics approaches. These instrumental units provide a multiplicity of possible configurations and analytical dimensions that can be adapted to the problem under consideration. Fig. 4.1 shows a typical metallomics workflow for nontargeted profiling and the characterization of metallobiomolecules in living organisms, especially metalloproteins. This instrumental arrangement is based on a two-dimensional chromatographic device with ICP-MS detection to trace metal-tagged biomolecules. Generally, in the first dimension metal-biomolecules are separated by the size using size-exclusion chromatography (SEC-ICP-MS) followed by the purification of the selected fraction by anion- or cation-exchange chromatography with ICP-MS detection (AEC/CEC-ICP-MS). Finally, the metal—proteinous fraction is collected from the AEC or CEC column and latterly desalted, lyophilized, and tryptic digested to identify the metalloproteins by analysis with nano-ESI-triple quadrupole-TOF-MS and MASCOT database (García-Sevillano et al., 2012a).

This up/down regulation process of metalloproteins and metallobiomolecules caused by environmental pollution usually occurs in conjunction with important metabolic changes, which can be characterized with high-performance liquid chromatography (HPLC) and ultra-performance liquid chromatography using ESI (Viant and Sommer, 2013), as well as gas chromatography—mass spectrometry (GC-MS) (Viant and Sommer, 2013) and two-dimensional gas chromatography with time-of-flight mass spectrometry (GCxGC-TOF MS) (Marney et al., 2014). However, nontargeted direct infusion mass spectrometry (DIMS) is a good alternative in metabolomics analysis, especially for high sample throughput due to short time needed for the analysis of each sample (Viant and Sommer, 2013; Garcia-Sevillano et al., 2014g, 2015a). Generally, samples are introduced as liquid extracts in methanol/water solution which contains either formic acid or ammonium acetate for positive or negative ionization, respectively. However, since separation of the metabolites in DIMS is only based upon m/z values, mass spectrometers with high or ultrahigh mass resolution are necessary, specifically QTOF (Garcia-Sevillano et al., 2014g, 2015a), Orbitrap (Hu et al., 2005), and Fourier-transform ion cyclotron resonance mass spectrometry (FT-ICR-MS) (Brown et al., 2005). In addition, DIMS avoid sample carryover and eliminate changes caused by the spoilage of chromatographic units, especially by columns degradation, which represent an evident benefit of reproducibility in studies with a large number of samples. However, DIMS has two important limitations because of ion suppression caused by complex biological matrices and the overlapping of isobaric metabolites. Despite this, DIMS is very suitable for initial screening of

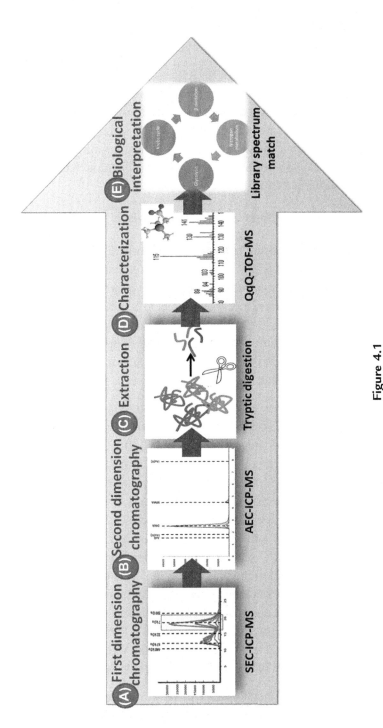

Figure 4.1

Typical workflow in environmental metallomics experiment to study metalloproteins in liver extract of mice from metal-contaminated area: (A) separation of metallobiomolecules by SEC-ICP-MS; (B) purification of 32 kDa fraction by AEC-ICP-MS; (C) and (D) identification of Cu, Zn-peak (32 kDa Cu, Zn-SOD) by tryptic digestion and analysis by nano-ESI-Qq-TOF-MS using MASCOT database; and (E) biological interpretation.

environmental metabolomics studies (Fernández-Ciscal et al., 2018; Garcia-Sevillano et al., 2012a, 2014h) and contaminant exposure experiments (Garcia-Sevillano et al., 2013a, 2014g, 2014h, 2015b), thus allowing the discovery of putative biomarkers of contamination.

Fig. 4.2 shows a typical metabolomics workflow for nontargeted fingerprinting and annotation of metabolites suffering changes as consequence of environmental contamination. Several steps are performed for this purpose: (a) sample treatment and metabolites extraction, (b) metabolites data acquisition by molecular mass spectrometry, (c) statistical data analysis, (d) metabolites annotation and identification, and (e) biological interpretation (Garcia-Sevillano et al., 2014h).

2.2 Metallomics/metabolomics workflows for targeted analysis

Both environmental and laboratory exposure studies of living organisms to toxic metal species, such as Hg^{2+}, $MeHg^+$, As(III) and (V), Cd^{2+} and metal antagonists as Se(IV) and (VI), and Zn^{2+}, trigger multiple interactions that involve organs, tissues, and biological fluids and a large amount of metal species, metallobiomolecules, and metabolites that configure the metal homeostatic equilibrium of these organisms (Gómez-Ariza et al., 2011). The study of these processes requires the complementary application of targeted methods of analysis for metal species next to the overall information provided by the omics methodologies previously described.

Most of the first-generation metal species analyses are based in the coupling of liquid chromatography (with anion- or cation-exchange columns) with ICP-MS detection using commercially available or synthesized standards for methods optimization and quantification (speciation). This is the case of arsenic speciation in the anemone *Anemonia sulcata*, a seafood type very appreciated in the south of Spain with high levels of arsenic. The application of HPLC-(AEC/CEC)−ICP-MS after As-species extraction from the sample with methanol/water (1:1, v/v) assisted by focused microwave revealed the presence of dimethylarsinic acid (DMA^V) followed in concentration by arsenobetaine (AB), As^V, monomethylarsonic acid (MA^V), tetramethylarsonium ion (TETRA), and trimethylarsine oxide (TMAO). In addition, nontargeted arsenic metabolites, such as arsenocholine (AsC), glyceryl phosphorylarsenocholine (GPAsC), and dimethylarsinothioic acid (DMAS), were identified by liquid chromatography coupled to triple quadrupole mass spectrometry (HPLC-QQQ-MS) (Contreras-Acuña et al., 2013). The approach has been applied to study the bioaccessibility of arsenic species in human after ingestion of the anemone analyzing the serum and urine along the time (Contreras-Acuña et al., 2014).

However, complex targeted analysis of metal species using environmental metallomics requires the use of column switching devices and multidimensional approaches involving several chromatographic columns: column switching−based speciation. These approaches

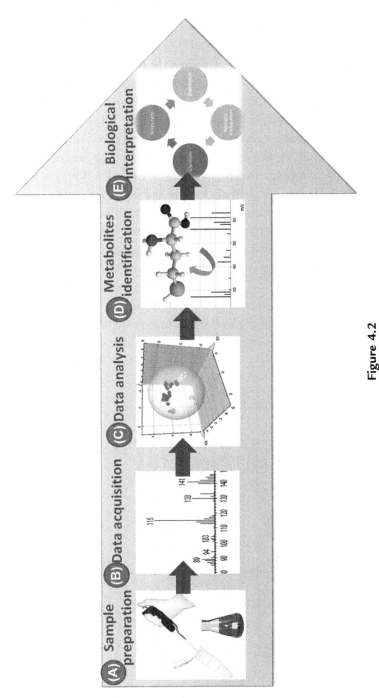

Figure 4.2

Typical workflow for environmental metabolomics: (A) sample treatment and metabolites extraction, (B) metabolites data acquisition by molecular mass spectrometry, (C) statistical data analysis, (D) metabolites annotation and identification, and (E) biological interpretation.

have been proposed for multispeciation analysis as it is the case of different selenium and mercury species including two chiral species of selenium: Se-methylselenocysteine, Se-MeSeCys; selenite, Se(IV); selenate, Se(VI); L-selenomethionine, L-SeMet; D-selenomethionine, D-SeMet; methylmercury, $MeHg^+$; and inorganic mercury, Hg^{2+}, which play an important role in the interactions of Se—Hg in living organisms. The method has been applied to the simultaneous speciation of mercury and selenium in urine and serum (after 10-fold dilution and filtration by 0.45 µm) involving two different mobile phases as well as two chromatographic columns: reverse phase (C18) and chiral columns. The columns were connected using a column-switching system consisting of two Rheodyne six-way valves (Fig. 4.3). The operating conditions of the arrangement and retention times of the chemical species are summarized in Table 4.1. First, in the **Step 1** (0—5 min) the sample is loaded into the 100 µL loop of the first valve that allows, in inject position, 100% mobile phase *A* flows through the C18 column, eluting Se-MeSeCys and Se(IV), while the other species are retained in the C18 column. **Step 2** (5—6.7 min), chiral column is in series connected to C18 column and both chiral species of SeMet are retained in this column. **Step 3** (6.7—9 min), the valves are moved again flowing the phase *A* only through the C18 column, whereby Se(VI) elutes. **Step 4** (9—13 min), valves' positions are again changed to flows phase *A* through both C18+chiral columns eluting L- and D-SeMet. Finally, in the **Step 5** (13—25 min), chiral column is again deactivated and the mobile phase *B* is introduced into the system, which produces the elution of mercury species (Moreno et al., 2010).

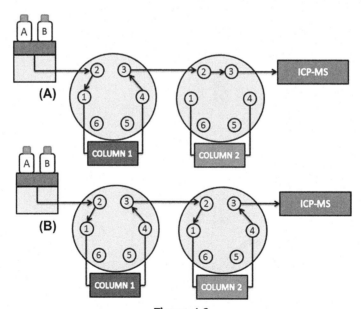

Figure 4.3
Column switching system for Se speciation, selenomethionine chiral speciation and Hg speciation. COLUMN 1 (C18 column), COLUMN 2 (Chiral column), following the description in Table 4.1.

Table 4.1: Steps and chromatographic conditions in the targeted speciation analysis of selenium and mercury species in serum and urine.

Step	Time (min)	Column arrangements	Mobile phase	Species	Retention time (min)
1	0−5	C18	A	SeMeSeCys	3.8
				Se(IV)	4.6
2	5−6.7	C18+chiral	A	SeMe species	
3	6.7−9	C18	A	Se(VI)	8.0
4	9−13	C18+chiral	A	L-SeMet,	12.7
				D-SeMet	13.0
5	13−25	C18	B	$MeHg^+$	23.8
				Hg^{2+}	25.6

Mobile phase A = 0.075% tetraethylammonium chloride, pH 4.5; Mobile phase B = 0.1% 2-mercaptoethanol, 0.06 M ammonium acetate, 5% methanol.

Another example is the simultaneous analysis of selenium species in biological matrices including selenoamino acids and selenoproteins, due to the antioxidant action of selenoprotein P (SeP) and glutathione peroxidase (GPx) complemented with the transporter activity of selenoalbumin (SeAlb) (Suzuki and Ogra, 2002). Inorganic selenium and selenometabolites have also important roles in this scenery contributing to the synthesis of selenoproteins. The methods proposed for the targeted speciation of selenoproteins (SeAlb, SelP, and GPx) in serum are based on the combination of affinity chromatography-HPLC (AF-HPLC) with on-line selenium detection by ICP-MS and quantification by postcolumn isotope dilution (ID) (Hinojosa Reyes et al., 2003; Shigeta et al., 2007; Jitaru et al., 2008). However, this approach suffers from some drawbacks, such as the interferences of Br^- and Cl^- present in the serum and the potential presence of selenoamino acids (selenomethionine and selenocysteine) and inorganic selenium (Se(IV) and Se(VI)), all of which interfere GPx analysis. To solve these problems, two-dimensional chromatographic separation has been proposed. It is based on the use of a double size exclusion column prior to a double affinity chromatographic system arranged on a six-port column-switching valve (Fig. 4.4)—2D/SE-AF-HPLC-SUID−ICP-ORS-MS (Garcia-Sevillano et al., 2013b). The use of two stacked size exclusion columns (5 mL HiTrap Desalting column) in series connected with the dual affinity column arrangement (1 mL heparin-sepharose column (HEP-HP) and 1 mL blue-sepharose column (BLU-HP)) make possible the complete on-line separation of targeted selenoproteins: GPx, SelP, and SeAlb, without the interference of Cl^-, Br^-, selenoamino acids, and inorganic selenium on GPx (Garcia-Sevillano et al., 2013b), following the chromatographic conditions described in Table 4.2. Additionally, the joint analysis of selenium-containing proteins and inorganic selenium can be performed by three-dimensional chromatographic separation based on double size exclusion chromatography prior to double affinity chromatography and AEC. For this purpose, the 2D/SE-AF-HPLC-SUID−ICP-ORS-MS arrangement, previously

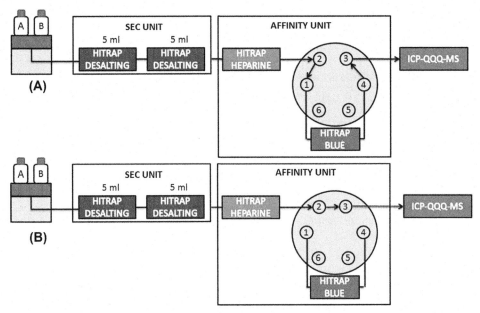

Figure 4.4

Schematic diagram of 2D/SE-AF-HPLC-SUID-ICP-ORS-MS arrangement for selenium-containing protein speciation in plasma and serum samples. HITRAP DESALTING (mini size exclusion column), HIPTRAP HEPARINE (heparine affinity column), HIPTRAP BLUE (blue sepharose affinity column), following the description in Table 4.2.

Table 4.2: Steps and chromatographic conditions in the targeted speciation analysis of selenoproteins in serum and plasma.

Step	Time (min)	Column arrangements	Mobile phase	Species	Retention time (min)
1	1–10	2xSEC	A	GP x	4.7
		HEP-HP (SelP is retained)		Se(IV); Se(VI)	6.1
		BLU-HP (SeAlb is retained)			
2	10–16	HEP-HP	B	SelP	14.8
3	16–20	BLU-HP	A	SeAlb	17.6

Mobile phase A = 0.05 M ammonium acetate pH 7.4; Mobile phase B = 1.5 M ammonium acetate pH 7.4.

described, is complemented with by an 8-port valve interposed in the connection path from the 6-port switching valve to the ICP-MS, previously mentioned (Fig. 4.4). This 8-port valve connects the system to an AEC column (PRP-X100) for the separation of inorganic selenium species preceded by a 3 mL loop (Fig. 4.5). When the system starts to operate, *Step 1*, the flux passes through the two in series SEC mini-columns followed by the two affinity columns, which retain SelP and SeAlb, and fills the 3 mL loop with low molecular selenium species, reaching GPx the ICP-MS detector. In the *Step 2*, the position of the 8-port valve changes and the inorganic selenium species are driven to the

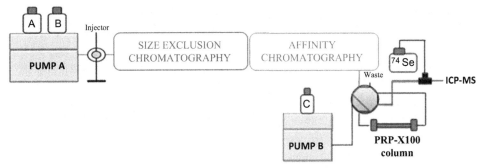

Figure 4.5
Additional arrangement based on a 3-mL loop and 8-port valve with anion-exchange chromatography for inorganic selenium speciation optionally coupled to the 2D/SE-AF-HPLC-SUID—ICP-ORS-MS device.

anion-exchange column by a second auxiliary pump (Pump B), while the flow driven by the main pump (Pump A) is derived to the waste. During the **Steps 3** and **4**, the anion-exchange column and Pump B are deactivated for the elution of SelP and SeAlb (Garcia-Sevillano et al., 2013b), as described in Fig. 4.4. All the steps and chromatographic conditions of this approach are described in Table 4.3.

2.3 Sample treatment in environmental metallomics/metabolomics

Generally, these studies use tissues or fluids from free-living organisms such as mice, clams, anemones, crustaceans, and others, suffering the action of contaminants in the field (Damek-Poprawa and Sawicka-Kapusta, 2004; Sanchez-Chardi et al., 2007, 2009;

Table 4.3: Steps and chromatographic conditions in the targeted speciation analysis of selenoproteins and inorganic selenium in serum and plasma.

Step	Time (min)	Column and device arrangements pump	Mobile phase	Species	Retention time (min)
1	1—9	• 2xSEC • HEP-HP (SelP is retained) • BLU-HP (SeAlb is retained) • 3-mL loop (Se(IV) and Se(VI)) loading Pump A	A	GPx	7.6
2	8—17	• AEC Pump B	B	Se(IV) Se(VI)	13.1 15.2
3	17—26	• HEP-HP Pump A (flow for HEP-HP)	C	SelP	23.1
4	26—35	• BLU-HP Pump A	A	SeAlb	31.2

Mobile phase A = 0.05 M ammonium acetate pH 7.4; Mobile phase B = 0.1 M sodium phosphate pH 8; Mobile phase C = 1.5 M ammonium acetate pH 7.4.

Sawicka-Kapusta et al., 1990; Garcia-Sevillano et al., 2012a; Vioque-Fernandez et al., 2007; Gago-Tinoco et al., 2014). However, others' research considers in vivo specimen exposed to pollutants in laboratory experiments (García-Sevillano et al., 2013a, 2014c,d,h), which allows to control the dose of contaminants and administration. Similar exposure experiments with cell lines also provide valuable knowledge about metal toxicity (Wolters et al., 2012; Nischwitz et al., 2013; Wang et al., 2015; Ogra et al., 2016), but the information about potential changes and interactions of metal species and metabolites through the biological interfaces (digestive tract, tissues, blood–brain barrier) and homeostasis in organisms may be missing.

Sample treatment and storage are very critical points in environmental studies. In metallomics studies, the cytosolic extracts from the different organs (liver, kidneys, brain, and others) can be obtained after tissue disruption by cryogenic homogenization by extraction with a suitable solution. This extraction is performed in a glass/Teflon homogenizer at a constant temperature of 4°C followed by centrifugation at 120,000 g for 1 h at 4°C. The resulting extract is stored at −80°C until analysis (García-Sevillano et al., 2013c, 2012a, 2014d). In the case of blood samples, plasma is obtained for metallomics studies by centrifugation (4000 g, 30 min, 4°C), after addition of heparin (ANTICLOT) as anticoagulant, and the addition of 10 mg of 100 mM PMSF and 100 mM of TCEP mixture (Garcia-Sevillano et al., 2014b,g,h).

In metabolomics studies based on DI-ESI(±)-QTOF-MS, cryohomogenized organs and solid tissues are extracted in a two-step procedure. Firstly, polar metabolites are extracted with a mixture of (1:1, v/v) methanol/water by adding 500 μL to 50 mg of tissue in an Eppendorf tube followed by vigorous vortex shaking during 5 min. Then, the cells are disrupted using a pellet mixer (2 min) under low temperature, followed by centrifugation at 4000 g for 10 min at 4°C. The supernatant is carefully collected and transferred to another Eppendorf tube. The pellet is homogenized again, using a pellet mixer for 2 min with 100 μL of (1:1, v/v) methanol/water mixture for a second time, then it is centrifuged at the same conditions described above and the pellet is kept for further treatment. After that, both supernatants are combined (600 μL), and the resulting polar extract is taken to dryness under nitrogen stream and stored at −80°C until analysis. The remaining pellet is latterly extracted with 500 μL of a mixture (2:1, v/v) chloroform/methanol, using a pellet mixer (2 min), to extract lipophilic metabolites and centrifuged at the same conditions described above. Finally, the resulting supernatant is taken to dryness under nitrogen stream and stored at −80°C until analysis (Garcia-Sevillano et al., 2014d,g).

For metabolomics analysis of plasma samples, proteins are removed by adding 400 μL of methanol/ethanol mixture (1:1, v/v) to 100 μL of plasma with vigorous shaking for 5 min in vortex at room temperature followed by centrifugation at 4000 g for 10 min at 4°C. The supernatant is carefully collected avoiding contamination with the precipitated proteins, and the resulting supernatant is taken to dryness under nitrogen stream for storage at −80°C until analysis. The pellet is homogenized again, with 200 μL of a mixture of (2:1, v/v) chloroform/methanol, using a pellet mixer (2 min), to extract lipophilic metabolites and centrifuged (10,000 g at 4°C for 10 min). Finally, the resulting supernatant is taken to dryness under nitrogen stream and stored at −80°C until analysis (Garcia-Sevillano et al., 2014b).

Reconstitution of polar extracts is performed with 200 μL of a mixture of methanol/water (4:1, v/v) and lipophilic extracts with 200 μL of (1:1) chloroform/water mixture before the analysis by ESI-MS. For data acquisitions from positive ionization, 0.1% (v/v) formic acid is added to polar extract and 30 mM of ammonium acetate in the case of lipophilic extract. In the case of negative ionization, intact extracts are directly infused to the mass spectrometer (Garcia-Sevillano et al., 2014e,g).

When the metabolomics analysis is performed by GC-MS, sample preparation requires a derivatization step to convert metabolites in volatile species. In the case of plasma, 100 μL of sample are mixed with 400 μL of 1:1 methanol/ethanol mixture and vortexed for 5 min at room temperature, followed by centrifugation at 4000 g for 10 min at 4°C. The supernatant is dried under nitrogen stream and metabolites are derivatized with 50 μL of methoxylamine hydrochloride solution (20 mg/mL in pyridine) at 70°C for 40 min for protection of carbonyl groups by methoximation. This is followed by treatment with N-methyl-N-(trimethylsilyl) trifluoroacetamide (MSTFA) containing 1% (w/v) trimethylchlorosilane (TMCS) at 50°C for 40 min, which silanizes primary amines and primary and secondary hydroxy groups in the case of MTSFA. In addition, TMCS aids in the derivatization of amides and secondary amines and hindered hydroxy groups (Jiye et al., 2005). Then, the derivatized samples is vortex-mixed for 2 min before GC analysis, centrifuged at 4000 g for 5 min and the supernatant collected is for its analysis (García-Sevillano et al., 2014c).

2.4 Data processing and analysis in environmental metabolomics

Statistical analysis is performed on mass spectra that is submitted to peak picking and matching of peaks across samples in order to reduce the results into a two-dimensional data matrix of spectral peaks and peak intensities, by using Markerview software (Applied Biosystems). Then, SIMCA-P software (version 11.5, published y UMetrics AB, Umeå, Sweden) is usually employed for statistical processing. Partial least squares discriminant analysis (PLS-DA) is performed to build predictive models in order to find differences between the groups of study and further study of potential biomarkers. Quality of the model is assessed by the R2 and Q2 values, provided by the software (indicative of class separation and predictive power of the model, respectively).

Compound identification of significant compounds is made matching the experimental accurate mass and tandem mass spectra with those available in metabolomics databases (HMDB, METLIN, KEGG, and LIPIDMAPS), using as a general search criteria a mass accuracy of 20 ppm, which is then adjusted based on the response of the database. Moreover, different classes of lipids and metabolites are confirmed based on characteristic fragmentation patterns reported in literature.

3. Application of metallomics/metabolomics approaches to laboratory exposure experiments to toxic metals

3.1 Arsenic exposure

Arsenic is an element with well-known toxicological effects, such as carcinogenic activity, liver damage, and cardiovascular disease. The toxic action of arsenic depends on the chemical form of this element, with a decreasing order of toxicity as followed As(III) > As(V) > monomethyl arsenic (MMA) > dimethylarsenic (DMA) > arsenobetaine (AB) = arsenocholine = arsenosugars. Actually the last three arsenic species are considered as innocuous (Geiszinger et al., 2002; Fattorini et al., 2006). The combination of nontargeted metallomics analysis based on SEC-HPLC-ICP-MS and targeted speciation analysis (AEC-HPLC-ICP-MS) of the fraction lower than 1.35 kDa from liver cytosolic extracts of mice exposed to As_2O_3 in the laboratory (García-Sevillano et al., 2013a) shows that methylated arsenicals (MA^V and DMA^V) are the predominant species in this organ, which is in agreement with previous studies from Suzuki (Suzuki, 2005; Naranmandura and Suzuki, 2008; Suzuki et al., 2002). During the progressive exposure of mice (*Mus musculus*) to arsenic (García-Sevillano et al., 2013a), the concentration of DMA^V decreases, increasing correlatively the presence of MA^V and iAs^{III} (Fig. 4.6), which can be related to the exceeded methylation capacity of liver for inorganic arsenic. This fact has been previously reported in mice for other authors (Vahter and Norin, 1980; Vahter, 1981). In addition, SEC-HPLC-ICP-MS analysis reveals the increasing intensity of a peak traced by As, although Cu and Zn are also present in the molecule. The peak matchs with the 32 kDa standard used in the experiment, and can be related to biomolecules containing thiols groups that bond to As, such as superoxide dismutase (SOD) o carbonic anhydrase, with a molecular mass in the range cited above, which include cysteine residues able to form arsenic-monothiol of arsenic-dithiol bonds (Kitchin and Wallace, 2005). This fact can produce the inhibition of the activity of these enzymes and correlatively the carcinogenic action of As (Kitchin and Wallace, 2008). As a consequence, the activity of SOD decreases under the action of As exposure (García-Sevillano et al., 2013a).

The changes of metal-containing proteins and metallometabolites produced by exposure to arsenic are accompanied by general metabolic alterations that affect the basic cycles of metabolic equilibrium, which will be considered jointly in Section 3.5. More specific metabolic alterations associated with mice arsenic exposure are related to arsenic

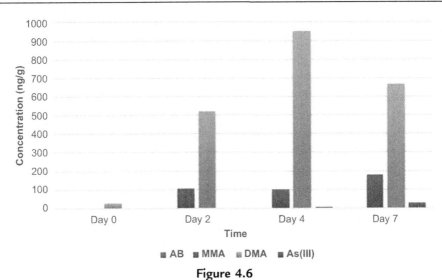

Figure 4.6

Concentration of arsenic metabolites (µg/g) in liver of mice *Mus musculus* after 7 days exposure to this element (As_2O_3).

methylation process as detoxification mechanism process (Garcia-Sevillano et al., 2013a; Garcia-Sevillano et al., 2014h,c). In this way, it has been observed the reduction of reduced glutathione levels in liver and plasma under exposure, increasing correlatively the concentration of L-cysteine. The glutathione is a metabolite that actively participates in arsenic methylation by the formation of arsenic glutathione complexes that are substrates for the action of arsenic methyltransferase (Hayakawa et al., 2005). Glutathione is synthesized from the amino acids L-cysteine, L-glutamic acid, and glycine, and in the reduced form (GSH) it protects living organisms against free radicals produced by arsenic, contributing additionally to arsenic methylation (Jin et al., 2010). Consequently, glutathione levels decrease by the oxidative stress caused by the presence of arsenic and correlatively increases L-cysteine and L-glutamic acid involved in the production of this metabolite. In addition, methionine increases since it is involved in the methylation of this element by the conversion in S-adenosyl-methionine (SAM) under the action of methionine adenosyltransferase. Later, SAM donates methyl groups to arsenic forming in turn methyl and dimethylarsenic (Thomas et al., 2001), although, as previously mentioned, in the case of an acute exposure to arsenic the formation of DMA decreases, increasing the biosynthesis of MMA (Fig. 4.6).

3.2 Cadmium exposure

The toxicity of cadmium is also related with its carcinogenic activity (Elinder et al., 1985), the adverse effect of cadmium mainly affects kidney and bone (Åkesson et al., 2005, 2006), and endocrine system (Piasek et al., 2001; Åkesson et al., 2008). It is also well known the antagonist action of selenium against the toxicity of cadmium (Messaoudi et al., 2009), which can be studied in depth using metallomics/metabolomics approaches.

The biological response of mice *M. musculus* exposed to cadmium has been studied by the coupling SEC-ICP-MS, which provides the profile of metallobiomolecules and their changes during exposure. The interactions of Cd with Se, this later acting as protector element, normally in form of selenoproteins can be also considered using the targeted multidimensional arrangement 2D/SE-AF-HPLC-SUID−ICP-ORS-MS, previously described (García-Sevillano et al., 2014h). The SEC-ICP-MS profiles traced by Cd in tissues and fluids reveal the affinity of this element for thiol groups of proteins, mainly in metallothioneins (MTs), in competence with other elements such as Cu and Zn, also linked to metallothioneins. Transport proteins such as albumin (BSA, 67 kDa) and transferrin (Tf, 79 kDa), traced by Cu, also show an increasing intensity in the SEC-ICP-MS profile under Cd exposure, correlatively with the increases of 32 kDa Cu-peak, corresponding to the antioxidative stress protein superoxide dismutase (SOD). These facts reflect the mechanisms to alleviate Cd toxicity based on the transport of the element to the kidneys (BSA and Tf) for excretion by the urine in the form of Cd-MT (Suzuki and Yoshikawa, 1981). Simultaneously, SOD increases to reduce the oxidative stress caused by Cd. On the other hand, multidimensional chromatography, 2D/SE-AF-HPLC-SUID−ICP-ORS-MS, of plasma samples shows peaks matching with SeP standard that increases with Cd exposure, this fact concurs with the increase of the SEC-ICP-MS peak traced by Cd (with molecular mass 55 kDa corresponding to SeP). These results suggest the bind of Cd to SeP (Srivastava et al., 1988).

3.3 Arsenic + cadmium exposure

The toxicological effects of environmental harmful elements in experimental animals are usually considered in isolation. Independent exposure to toxic elements have been previously considered in experiments with As (García-Sevillano et al., 2013a), Cd (García-Sevillano et al., 2014h), and Hg (García-Sevillano et al., 2014g). However, in the field living organisms suffer the simultaneous action of all the pollutants present in the environment. To the date, the biological response of bioindicators to the simultaneous action of multiple toxic elements has been scarcely studied.

New studies based on the joint exposure to several contaminants can contribute to clarify the potential interactions between them, as well as their influence on the metabolic cycles. This is the case of the interaction of the pair As/Cd, which results more harmful than the action of the elements separately (Yáñez et al., 1991), such as lipid peroxidation and overexpression of both glutathione and metallothioneins. For this purpose, a combination of two metallomics approaches, SEC-ICP-MS and 2D-SEC-AF-SUID-ICP-MS, has been used to deepen the mechanisms occurring in mice under the presence of arsenic plus cadmium in an experiment performed during 12 days (García-Sevillano et al., 2014c). It was observed the appearance of low molecular mass arsenic species in liver with a peak of higher intensity in the As/Cd exposure during 6 days (Fig. 4.7). In addition, the SEC-

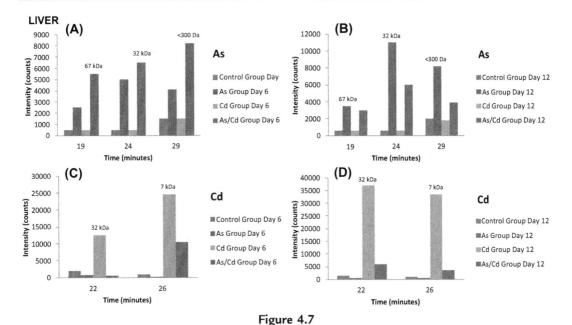

Figure 4.7

Profiles of metallobiomolecules in liver of *Mus musculus* under arsenic/cadmium exposure. (A) and (B) profiles traced by As after 6 and 12 days of exposure, respectively. (C) and (D) profiles traced by Cd after 6 and 12 days of exposure, respectively.

ICP-MS profile traced by As shows a peak that matches with the standard of superoxide dismutase at 32 kDa, which can be associated with the interaction of As carbonic anhydrase (35 kDa) and superoxide dismutase (32 kDa). The intensity of this peak is higher under the exposure to As alone respect to the mixture As/Cd that suggests potential interactions between both elements. At about 70 kDa appears other peak traced by As increasing more markedly with the exposure to As/Cd that may be associated with the affinity of arsenite toward albumin (67 kDa) and hemoglobin (68 kDa) (Bogdan et al., 1994; Lu et al., 2007). It can be also mentioned the intense peak traced by Cd (Fig. 4.7C) that can be related with the presence of Cd-metallothionein (Cd-MT) at 7 kDa, which is more marked when Cd is administered alone. However, the intensity of the peak decreases notably with the simultaneous exposure of As/Cd, possibly due to the antagonist interaction of these elements.

The interaction of these toxic elements with protective and antioxidants selenoproteins can be evaluated applying the approach 2D-SEC-AF-SUID-ICP-MS to serum of bioindicators exposed to these contaminants. The changes in Se-proteins expression depend on the type of contaminant (García-Sevillano et al., 2014c). Fig. 4.8 shows how the levels of Se-proteins decrease under the presence of the toxic elements considered: As and Cd. However, the joint exposure to both elements increases the levels of Se-metabolites and Se-proteins species due to a decrease in toxicity (Saïd et al., 2010).

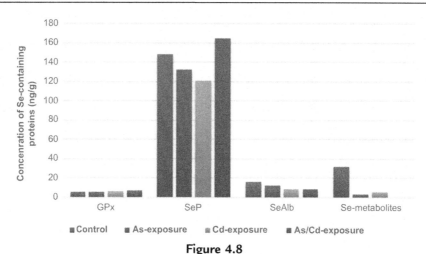

Figure 4.8

Se-protein concentrations in plasma of *Mus musculus* under As and Cd exposure.

3.4 Mercury exposure

Mercury is a well-known contaminant with very adverse consequences on health of human and living organisms. This is due to the affinity of the two main mercury species (divalent inorganic mercury, Hg^{2+}, and methylmercury, CH_3Hg^+) for thiol groups of proteins and metabolites (Clarkson, 1997), such as reduced glutathione (GSH) and cysteine (Cys), in cells, tissues, and biological fluids (Pei et al., 2011; Percy et al., 2007). Besides, mercury exposure has important consequence on metabolic cycles and gut microflora (Wei et al., 2008; Nicholson et al., 1985).

To explore the metallomics mechanisms and interactions of mercury in living organisms, the mouse *M. musculus* has been exposed to inorganic mercury by subcutaneous injection (García-Sevillano et al., 2014g), and the coupling SEC-ICP-ORS-MS has been applied, using a size exclusion column from 3 to 70 kDa. This instrumental approach provides metalloproteins profiles (Fig. 4.9), in which can be remarked the presence of a peak traced by Cu that matches well with the Cu, Zn-SOD standard (32 kDa) whose intensity decreases with mercury exposure (Fig. 4.9A), possibly because it is consumed by its action against the oxidative stress caused by the presence of Hg. However, the intensity of the peak-related mercury bound to this biomolecule (Hg-SOD) increased (Fig. 4.9B), possibly because Hg competes with Cu in Cu-containing proteins. Nevertheless, the concentration of mercury in liver is low (Fig. 4.9B), possibly because it is transported to kidney for excretion (Zalups, 2000). This suggestion is confirmed by the increase of concentration of Cu and Hg in kidney with exposure, especially the later (Fig. 4.9C,D) that is mainly in the form of Hg-metallothionein (Hg-MT) analogously to Cu.

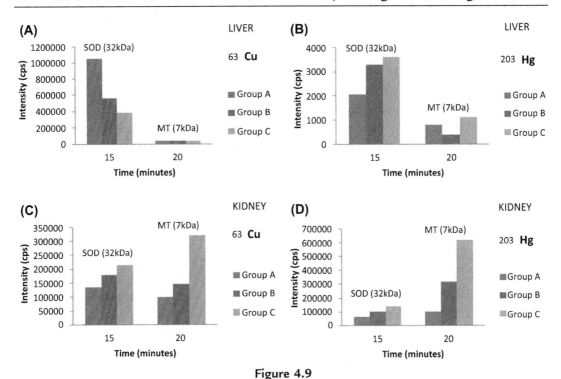

Figure 4.9

Relative intensity of peaks traced by Cu (A) and (C), and Hg (B) and (D) in cytosolic extracts of liver and kidney of mice (*Mus musculus*) exposed to Hg using the coupling SEC-ICP-MS. Time of exposure: Group A, 2 days; Group B, 7 days; Group C, 14 days.

These exposure experiments also allow to deep insight into the antagonist action of selenium on mercury (Chen et al., 2006), which produces changes in the selenium concentration in plasma during exposure (Su et al., 2008). In this way, low molecular mass selenium species drastically decrease during mouse mercury exposure, in the order of 92%, because they are required for the synthesis of SelP (García-Sevillano et al., 2014g). According to this, the synthesis of SelP after mercury administration increases and correlatively the concentration of this protein in plasma is higher. Later on the concentration of SelP gradually decreases during exposure, which can be related to the formation of an Hg–Se complex in plasma (Gailer et al., 2000) that binds to the SelP (Yoneda and Suzuki, 1997).

3.5 General metabolic alterations associated with toxic metal exposure

The metabolic response of living organisms to toxic metals (García-Sevillano et al., 2013a, 2014d,c, 2015b,d) presents a common behavior pattern, when different tissues and fluids, such as liver, kidney, serum red blood cells, and liver mitochondria from mice *M. musculus*, are considered.

3.5.1 Energy metabolism depletion

In general, the levels of metabolites involved in the glycolysis such as glucose, glyceraldehyde-3-phosphate, and pyruvate decrease under toxic metal exposures, such as Hg (García-Sevillano et al., 2014g, 2015d), As (García-Sevillano et al., 2013a), Cd (García-Sevillano et al., 2015c), and As + Cd (García-Sevillano et al., 2014c). The decrease of pyruvate, end product of glycolysis, triggers the increases of acetyl-CoA, which transports the carbon atoms within the acetyl group to the citric acid cycle (Krebs cycle) that are oxidized for energy production. In addition, high levels of citric acid, isocitric acid, α-ketoglutarate, and glutamic acid (intermediate metabolites in the Krebs cycle) have been observed during the exposure.

3.5.2 Cell membrane impairment

Another effect of mice exposure to toxic elements is cell membrane deterioration, which is a consequence of phospholipid degradation (Griffin et al., 2001), in particular phosphatidylcholine (PC) to phosphocholine and choline, as well as the release of free fatty acids from these PCs, including pipecolic, arachidonic, linoleic, and docosahexaenoic acids. This membrane impairment has been observed in mercury exposure (García-Sevillano et al., 2014g), but also with arsenic (García-Sevillano et al., 2013a).

3.5.3 Other metabolic disorders

Several interconnected metabolic cycles also suffer alterations because of the input of toxic metals into living organisms. Taurine involved in both antioxidant activity and osmoregulation presents an opposite trend to reduced glutathione (GSH), increasing its presence with exposure to As (García-Sevillano et al., 2013a). However, arginine, other metabolite with protective effect on oxidative stress (Dasgupta et al., 2006), decreases in mice plasma upon the exposure to As and Hg (García-Sevillano et al., 2014b). In addition, amino acids metabolism is also altered (García-Sevillano et al., 2015d) increasing the concentration of valine and isoleucine due to the depletion of branched amino acids, and As exposure changes several amino acids in the mitochondrial fraction of liver (García-Sevillano et al., 2015b). Finally, the oxidative degradation of amino acids supplements metabolic energy cycle by their conversion in intermediates of TCA cycle, such as pyruvate, α-ketoglutarate, succinyl-CoA, fumarate, oxaloacetate, acetyl-CoA, or acetoacetate.

4. Application of metallomics/metabolomics to environmental issues caused by metal exposure

The binomial approach based on the combined use of metallomics and metabolomics, previously described, provides powerful information about the toxicological mechanisms triggered by the inputs of contaminants, particularly metals, on living organisms, based on exposure experiments in the laboratory using suitable bioindicators. However, a critical point in this research is the ability to translate the methodology and the experience from the laboratory to the field.

In this way, many studies have been performed in order to assess environmental issues in natural areas that can be affected by contamination episodes. A paradigmatic example is Doñana Natural Park located in the southwest of the Iberian Peninsula, which is a wildlife reserve declared as UNESCO World Heritage Site because of its ecological value. However, the Park suffers threats of contamination due to the agricultural, industrial, and mining activities that take place in its surroundings, specially the potential inputs of metals (Grimalt et al., 1999; Bonilla-Valverde et al., 2004; Vioque-Fernández et al., 2009; Gonzalez-Fernández et al., 2011).

Several free-living animals have been used as bioindicators to evaluate the environmental status of the Park, such as spoonbills, storks, and flamingos from which feathers, blood, and lost eggs were sampled (Baos et al., 2006; Barón et al., 2015), as well as mice (*Mus spretus*) (Gonzalez-Fernández et al., 2011) and crabs (*Procambarus clarkii*) (Gago-Tinoco et al., 2014). A singular characteristic of *M. spretus* is its marked genetic homology with the laboratory mouse *M. musculus* already sequenced (Ruiz-laguna et al., 2006; Montes-Nieto et al., 2007), which allows interesting comparisons between the exposure experiments to contaminants made in the laboratory and the results obtained with free-living organisms, applying the combined metabolic/metabolomics approach previously described.

Most of these studies compare potentially contaminated areas, such as "El Matochal" (MAT) affected by agricultural and mining contamination by metals, with sites located at the center of the Park, "Lucio del Palacio" (LDP), that can be considered as control area with low contamination. A primer study based on the use of SEC-ICP-MS coupling combining analytical and preparative columns showed interesting Cu, Zn, and Mn profiles (Gonzalez-Fernández et al., 2011). The higher presence of copper, zinc, and manganese in liver extract is remarkable; however, toxic elements, such as cadmium and especially arsenic, show very low concentrations in both liver and kidney, especially from mice captured in MAT. Using the concentration of metals in these organs classification of contaminated and noncontaminated area can be performed, especially considering Mn that can be used as key element for this purpose taking into account organs, soils, and sediments.

The concentration of manganese is markedly high in liver extracts, since this organ is the major reservoir of this element, which is transported to other organs by Mn-transferrin and albumin (Quintana et al., 2005; Quintana et Al., 2006). The significant presence of Mn-traced biomolecules in liver and kidney from *M. spretus* mice can be studied in the cytosolic fraction of these organs with the SEC-ICP-MS arrangement previously described (Gonzalez-Fernández et al., 2011). The intensity of the peaks is higher in liver than in kidney, showing a peak at the void volume that can be related to Mn-containing enzymes such as catalase (Mr = 345 kDa) and arginase (Mr = 75 kDa). In addition, other two Mn-traced peaks at retention time matching with BSA (66.7 kDa) that can be associated with Mn-albumin, and Cu, Zn-SOD (32 kDa) that could be related to Mn-superoxide dismutase. However, this protein is only present in mitochondria and not in the cytosolic extracts used in this experiment, although the substitution of active sites of Zn-carbonic anhydrase by Mn could explain the presence of this biomolecule (Okrasa and Kazlauskas, 2006), since the intensity of this peak increases with the concentration of Mn in the liver extract LDP > MAT.

On the other hand, the coupling SEC-ICP-MS can be used to get information about the interactions of Cd, Zn, Cu, and Pb in mice *M. musculus* exposed to metal pollution (Teodorova et al., 2003). In the area considered (DNP), the Cu, Zn, and Cd-traced peaks in the cytosolic fractions of liver and kidney of *M. spretus* show similar behavior in metallothioneins expression. The intensity of Cu-MT peak is higher in the contaminated area MAT than in LDP. Another peak traced by Cu at 32 kDa shows higher intensity for samples from LDP and MAT, in connection to this, a peak traced by Zn is observed at 32 kDa that suggests the presence of SOD induced by the contamination, although in this case the highest response corresponds to LDP. To confirm that this Cu, Zn-peak is superoxide dismutase, the 32 kDa fraction from the liver extract is collected using a preparative SEC. The fraction is desalted, lyophilized, and analyzed by AEC, and the corresponding fraction tryptic digested for latter peptides identification by nano-ESI-triple quadrupole-TOF. The mass spectrum of peptides is used for protein identification, using doubly charged peptide ions of m/z 584.31, 684.39, and 756.84 for MS/MS analysis and protein identification in MASCOT database, which confirms the presence of Cu, Zn-superoxide dismutase associated with the peak of 32 kDa (Gonzalez-Fernández et al., 2011; Garcia Sevillano et al., 2012a).

The contamination associated with the presence of arsenic especially in MAT caused the marked presence of inorganic arsenic and particularly dimethylarsenic (DMA) in kidney of *M. spretus* (García-Sevillano et al., 2012b), similarly to the results observed in arsenic exposure experiments with *M. musculus* performed in the laboratory (Fig. 4.5). In order to obtain suitable resolved chromatograms of low molecular mass of As molecules in cytosolic kidney extracts from *M. spretus*, it is necessary to apply a two-dimensional separation based in SEC and AEC both with ICP-MS detection (Garcia-Sevillano et al., 2012a).

The influence of metal pollution on antioxidant mechanisms in *M. spretus* is based on the presence of selenoproteins in plasma estimated using the two-dimensional chromatography-column switching valve coupled to ICP-MS, previously described (2D/SEC-AF-HPLC-SUID-ICP-ORS-qMS). It has been observed that concentrations of SeP, SeAlb, and Se metabolites decrease in mice from MAT as consequence of contamination, in comparison with LDP; however, the concentration of eGPx increases (García-Sevillano et al., 2014b).

Metabolomics changes in *M. spretus* can help to understand metal toxicity mechanisms. As a consequence the levels of reduced glutathione decrease, due to the antioxidant properties of this metabolite that contributes to defend against reactive oxygen species keeping reduced the thiols groups of important proteins. In addition, GSH links to reactive xenobiotics and endogenous metabolites for excretion. As a consequence, the levels of GSH decrease in the contaminated area of DNP surrounding as MAT, and to compensate this reduction the synthesis of L-glutamate, L-cysteine, and glycine is induced, which increases the levels of glutamate observed in mice from MAT. Another consequence of oxidative stress caused by the presence of metals is the damage of cell membranes; this oxidative process affects unsaturated fatty acids of membrane phospholipids, provoking the corresponding degradation. This process has been checked in mice kidney from MAT, which shows the increase of degradation products of phosphatidylcholine, especially lysophosphatidylcholine, and fatty acids, such as arachidonic, oleic, and pipecolic acids, as well as choline. Finally, L-carnitine, which transports fatty acids to the mitochondria contributing to energy production through the β-oxidation pathway, decreases with the presence of metals in MAT and consequently the energy status of exposed mice also decreases (García-Sevillano et al., 2014i).

5. Final remarks

The complexity of living organisms requires powerful tools for a suitable analysis of the molecular world inside them. Metallobiomolecules and metabolites interact with each other, being transported from one organ to another, crossing membranes and passing through biological interfaces in order to assure the organism well-being. The situation is even more complex when organisms suffer the action of pollutants from the environment. Many of these contaminants are toxic metals and for this reason, the application of instrumental analytical approaches based on the use of ICP-MS allows tracing the presence of these metals in complex molecules assisted by separation techniques, mainly chromatography, generally in multidimensional arrangements. In some cases, the problem under study can be approached with targeted analysis, in others nontarget approaches are necessary, usually complementarily supported with organic mass spectrometry to annotate or identify the metallobiomolecules. Besides, simpler instrumental arrangements based on the direct coupling of the chromatographic unit with the ICP-MS are used. This allows obtaining profiles on the presence and evolution of Cu, Zn-superoxide dismutase, ferritin, or metallothioneins, as well as the speciation of some elements such as As. Other couplings based on switching of columns have been proposed before the detection of metallobiomolecules by ICP-MS, generally in a targeted analysis, as is the case of selenium proteins and metabolites. However, due to the critical role of many other nonmetallic metabolites in the detoxification processes associated with the input of metals in the organisms, metabolomics is crucial to decipher the metabolic changes triggered in living organisms under metal pollution. Therefore, this binomial methodology (metallomics/metabolomics) offers great possibilities to deep insight into detoxification processes in living organisms as consequence of contamination and is a valuable tool in exposure experiments in the laboratory, which can be translated to contamination episodes in the field.

Conflict of interest

The authors declare that the research was conducted in the absence of any commercial or financial relationships that could be construed as a potential conflict of interest.

Acknowledgments

This work was supported by the project CTM2015-67902-C2-1-P and CTM2016-75908-R from the Spanish Ministry of Economy and Competitiveness (MINECO) and by the project P12-FQM-0442 from the Regional Ministry of Economy, Innovation, Science and Employment (Andalusian Government, Spain). In addition, authors are grateful to FEDER (European Community) for financial support, Grants UNHU13-1E-1611 and UNHU15-CE-3140. Gema Rodríguez-Moro and Sara Ramírez-Acosta thank to Ministry of Economy and Competitiveness for PhD scholarships BES-2013-064501 and BES-2016-076364, respectively. Francisca Arellano-Beltrán thanks to the Regional Ministry of Economy and Knowledge (Andalusian Government, Spain) and the European Social Fund by the Grant SNGJ-JPI-009.

References

Åkesson, A., Lundh, T., Vahter, M., Bjellerup, P., Lidfeldt, J., Nerbrand, C., Samsioe, G., Strömberg, U., Skerfving, S., 2005. Tubular and glomerular kidney effects in Swedish women with low environmental cadmium exposure. Environ. Health Perspect. 113, 1627−1631.

Åkesson, A., Bjellerup, P., Lundh, T., Lidfeldt, J., Nerbrand, C., Samsioe, G., Skerfving, S., Marie Vahter, M., 2006. Cadmium-induced effects on bone in a population-based study of women. Environ. Health Perspect. 114, 830−834.

Åkesson, A., Julin, B., Wolk, A., 2008. Long-term dietary cadmium intake and postmenopausal endometrial cancer incidence: a population-based prospective cohort study. Cancer Res. 68, 6435−6441.

Baos, R., Jovani, R., Pastor, N., Tella, J.L., Jiménez, B.c, Gómez, G.c, González, M.J., Hiraldo, F., 2006. Evaluation of genotoxic effects of heavy metals and arsenic in wild nestling white storks(*Ciconia ciconia*) and black kites (*Milvus migrans*) from Southwestern Spain after a mining accident. Environ. Toxicol. Chem. 25, 2794−2803.

Barón, E., Bosch, C., Máñez, M., Andreu, A., Sergio, F., Hiraldo, F., Eljarrat, E., Barceló, D., 2015. Temporal Trends in Classical and Alternative Flame Retardants in Bird Eggs from Doñana Natural Space and Surrounding Areas (South-western Spain) between 1999 and 2013. Chemosphere128, pp. 316−323.

Bogdan, G.M., Sampayo-Reyes, A., Vasken Aposhian, H., 1994. Arsenic binding proteins of mammalian systems: I. Isolation of three arsenite-binding proteins of rabbit liver. Toxicology 93, 175−193.

Bonilla-Valverde, D., Ruiz-Laguna, J., Muñoz, A., Ballesteros, J., Lorenzo, F., Gómez-Ariza, J.L., López-Barea, J., 2004. Evolution of biological effects of Aznalcóllar mining spill in the Algerian mouse (*Mus spretus*) using biochemical biomarkers. Toxicology 197, 123−138.

Brown, S.C., Kruppa, G., Dasseux, J.L., 2005. Metabolomics applications of FT-ICR mass spectrometry. Mass Spectrom. Rev 24, 223−231.

Chen, C., Yu, H., Zhao, J., Li, B., Qu, L., Liu, S., Zhang, P., Chai, Z., 2006. The roles of serum selenium and selenoproteins on Mercury toxicity in environmental and occupational exposure. Environ. Health Perspect. 114, 297−301.

Clarkson, T.W., 1997. The toxicology of mercury. Crit. Rev. Clin. Lab Sci. 34, 369−403.

Contreras-Acuña, M., García-Barrera, T., García-Sevillano, M.A., Gómez-Ariza, J.L., 2013. Speciation of arsenic in marine food (*Anemonia sulcata*) by liquid chromatography coupled to inductively coupled plasma mass spectrometry and organic mass spectrometry. J. Chromatogr. A 1282, 133−141.

Contreras-Acuña, M., García-Barrera, T., García-Sevillano, M.A., Gómez-Ariza, J.L., 2014. Arsenic metabolites in human serum and urine after seafood (*Anemonia sulcata*) consumption and bioaccessibility assessment using liquid chromatography coupled to inorganic and organic mass spectrometry. Microchem. J. 112, 56−64.

Damek-Poprawa, M., Sawicka-Kapusta, K., 2004. Histopathological changes in the liver, kidneys, and testes of bank voles environmentally exposed to heavy metal emissions from the steelworks and zinc smelter in Poland. Environ. Res. 96, 72−78.

Dasgupta, T., Hebbel, R.P., Kaul, D.K., 2006. Protective effect of arginine on oxidative stress in transgenic sickle mouse models. Free Radic. Biol. Med. 41, 1771−1780.

Elinder, C.-G., Kjellstrom, T., Hogstedt, C., Andersson, K., Spång, G., 1985. Cancer mortality of cadmium workers. Br. J. Ind. Med. 42, 651−655.

Fattorini, D., Notti, A., Regoli, F., 2006. Characterization of arsenic content in marine organisms from temperate, tropical, and polar environments. Chem. Ecol. 22, 405−414.

Fernández-Cisnal, R., García-Sevillano, M.A., García-Barrera, T., Gómez-Ariza, J.L., Abril, N., 2018. Metabolomic alterations and oxidative stress are associated with environmental pollution in *Procambarus clarkii*. Aquat. Toxicol. 205, 76−88.

Gago-Tinoco, A., González-Domínguez, R., García-Barrera, T., Blasco-Moreno, J., Bebianno, M.J., Gómez-Ariza, J.L., 2014. Metabolic signatures associated with environmental pollution by metals in Doñana National Park using *P. clarkii* as bioindicator. Environ. Sci. Pollut. Res. 21, 13315−13323.

Gailer, J., George, G.N., Pickering, I.J., Madden, S., Prince, R.C., Yu, E.Y., Denton, M.B., Younis, H.S., Aposhian, H.V., 2000. Structural basis of the antagonism between inorganic mercury and selenium in mammals. Chem. Res. Toxicol. 13, 1135−1142.

García-Sevillano, M.A., González-Fernández, M., Jara-Biedma, R., García-Barrera, T., López-Barea, J., Pueyo, C., Gómez-Ariza, J.L., 2012a. Biological response of free-living mouse *Mus spretus* from Doñana National Park under environmental stress based on assessment of metal-binding biomolecules by SEC-ICP-MS. Anal. Bioanal. Chem. 404, 1967−1981.

García-Sevillano, M.A., González-Fernández, M., Jara-Biedma, R., García-Barrera, T., Vioque-Fernández, A., López-Barea, J., Pueyo, C., Gómez-Ariza, J.L., 2012b. Speciation of arsenic metabolites in the free-living mouse *Mus spretus* from Doñana National Park used as a bio-indicator for environmental pollution monitoring. Chem. Pap. 66, 914.

García-Sevillano, M.A., García-Barrera, T., Navarro, F., Gómez-Ariza, J.L., 2013a. Analysis of the biological response of mouse liver (*Mus musculus*) exposed to As_2O_3 based on integrated-omics approaches. Metallomics 5, 1644−1655.

García-Sevillano, M.A., García-Barrera, T., Gómez-Ariza, J.L., 2013b. Development of a new column switching method for simultaneous speciation of selenometabolites and selenoproteins in human serum. J. Chromatogr. A 1318, 171−179.

Garcia-Sevillano, M.A., Jara-Biedma, R., Gonzalez-Fernandez, M., Garcia-Barrera, T., Gomez-Ariza, J.L., 2013. Metal interactions in mice under environmental stress. Biometals 26, 651−666. https://doi.org/10.1007/s10534-013-9642-2.

García-Sevillano, M.A., García-Barrera, T., Navarro, F., Gómez-Ariza, J.L., 2014a. Absolute quantification of superoxide dismutase in cytosol and mitochondria of mice hepatic cells exposed to mercury by a novel metallomic approach. Anal. Chim. Acta 842, 42−50.

García-Sevillano, M.A., García-Barrera, T., Gómez-Ariza, J.L., 2014b. Application of metallomic and metabolomic approaches in exposure experiments on laboratory mice for environmental metal toxicity assessment. Metallomics 6, 237−248.

García-Sevillano, M.A., García-Barrera, T., Navarro-Roldán, F., Montero-Lobato, Z., Gómez-Ariza, J.L., 2014c. A combination of metallomics and metabolomics studies to evaluate the effects of metal interactions in mammals. Application to *Mus musculus* mice under arsenic/cadmium exposure. J. Proteom. 104, 66−79.

García-Sevillano, M.A., Contreras-Acuña, M., García-Barrera, T., Navarro, F., Gómez-Ariza, J.L., 2014d. Metabolomic study in plasma, liver and kidney of mice exposed to inorganic arsenic based on mass spectrometry. Anal. Bioanal. Chem. 406, 1455−1469.

García-Sevillano, M.A., García-Barrera, T., Abril, N., Pueyo, C., López-Barea, J., Gómez-Ariza, J.L., 2014e. Omics technologies and their applications to evaluate metal toxicity in mice *M. spretus* as a bioindicator. J. Proteom. 104, 4−26.

García-Sevillano, M.A., Garcia-Barrera, T., Navarro, F., Abril, N., Pueyo, C., López-Barea, J., Gómez-Ariza, J.L., 2014f. Use of metallomics and metabolomics to assess metal pollution in Doñana National Park (SW Spain). Environ. Sci. Technol. 48, 7747−7755.

García-Sevillano, M.A., García-Barrera, T., Navarro, F., Gailer, J., Gómez-Ariza, J.L., 2014g. Use of elemental and molecular-mass spectrometry to assess the toxicological effects of inorganic mercury in the mouse *Mus musculus*. Anal. Bioanal. Chem. 406, 5853−5865.

García-Sevillano, M.A., García-Barrera, T., Navarro, F., Gómez-Ariza, J.L., 2014h. Cadmium toxicity in *Mus musculus* mice based on a metallomic study. Antagonistic interaction between Se and Cd in the bloodstream. Metallomics 6, 672−681.

García-Sevillano, M.A., García-Barrera, T., Navarro, F., Abril, N., Pueyo, C., Lopez-Barea, J., Gómez-Ariza, J.L., 2014i. Use of metallomics and metabolomics to assess metal pollution in Doñana national Park (SW Spain). Environ. Sci. Technol. 48, 7747−7755.

García-Sevillano, M.A., García-Barrera, T., Gómez-Ariza, J.L., 2015a. Environmental metabolomics: biological markers for metal toxicity. Electrophoresis 36, 2348−2365.

García-Sevillano, M.A., García-Barrera, T., Navarro, F., Montero-Lobato, Z., Gómez-Ariza, J.L., 2015b. Shotgun metabolomic approach based on mass spectrometry for hepatic mitochondria of mice under arsenic exposure. Biometals 28, 341−351.

García-Sevillano, M.A., Abril, N., Fernández-Cisnal, R., Garcia-Barrera, T., Pueyo, C., López-Barea, J., Gómez-Ariza, J.L., 2015c. Functional genomics and metabolomics reveal the toxicological effects of cadmium in *Mus musculus* mice. Metabolomics 11, 1432−1450.

García-Sevillano, M.A., García-Barrera, T., Navarro, F., Abril, N., Pueyo, C., López-Barea, J., Gómez-Ariza, J.L., 2015d. Combination of direct infusion mass spectrometry and gas chromatography mass spectrometry for toxicometabolomic study of red blood cells and serum of mice *Mus musculus* after mercury exposure. J. Chromatogr. B 985, 75−84.

Geiszinger, A.E., Goessler, W., Francesconi, K.A., 2002. The marine polychaete Arenicola marina: its unusual arsenic compound pattern and its uptake of arsenate from seawater. Mar. Environ. Res. 53, 37−50.

Gierasch, L.M., Gershenson, A., 2009. Post-reductionist protein science, or putting Humpty Dumpty back together again. Nat. Chem. Biol. 5, 774−777.

Gómez-Ariza, J.L., Jahromi, E.Z., González-Fernández, M., García-Barrera, T., Gailer, J., 2011. Liquid chromatography-inductively coupled plasma-based metallomic approaches to probe health-relevant interactions between xenobiotics and mammalian organisms. Metallomics 3, 566−577.

Gonzalez-Fernández, M., García-Sevillano, M.A., Jara-Biedma, R., García-Barrera, T., Vioque, A., López-Barea, J., Pueyo, C., Gómez-Ariza, J.L., 2011. Size characterization of metal species in liver and brain from free-living (*Mus spretus*) and laboratory (Mus Musculus) mice by SEC-ICP-MS: application to environmental contamination assessment. J. Anal. Atomic Spectrom. 26, 141–149.

Griffin, J.L., Mann, C.J., Scott, J., Shoulders, C.C., Nicholson, J.K., 2001. Choline containing metabolites during cell transfection: an insight into magnetic resonance spectroscopy detectable changes. Fed. Eur. Biochem. Soc. Lett. 509, 263–266.

Grimalt, J.O., Ferrer, M., MacPherson, E., 1999. The mine tailing accident in Aznalcollar. Sci. Total Environ. 242, 3–11.

Hayakawa, T., Kobayashi, Y., Cui, Y., Hirano, S., 2005. A new metabolic pathway of arsenite: arsenic-glutathione complexes are substrates for human arsenic methyltransferase Cyt19. Arch. Toxicol. 79, 183–191.

Hinojosa Reyes, L., Marchante-Gayón, J.M., García Alonso, J.I., Sanz-Medel, A., 2003. Quantitative speciation of selenium in human serum by affinity chromatography coupled to post-column isotope dilution analysis ICP-MS. J. Anal. Atomic Spectrom. 18, 1210–1216.

Hu, Q.Z., Noll, R.J., Li, H.Y., Makarov, A., Hardman, M., Cooks, R.G., 2005. The Orbitrap: a new mass spectrometer. J. Mass Spectrom. 40, 430–443.

Jin, Y., Zhao, F., Zhong, Y., Yu, X., Sun, D., Liao, Y., Lv, X., Li, G., Sun, G., 2010. Effects of exogenous GSH and methionine on methylation of inorganic arsenic in mice exposed to arsenite through drinking water. Environ. Toxicol. 25, 361–366.

Jitaru, P., Prete, M., Cozzi, G., Turetta, C., Cairns, W., Seraglia, R., Traldi, P., Cescon, P., Barbante, C., 2008. Speciation analysis of selenoproteins in human serum by solid-phase extraction and affinity HPLC hyphenated to ICP-quadrupole MS. J. Anal. Atomic Spectrom. 23, 402–406.

Jiye, A., Trygg, J., Gullberg, J., Johansson, A.I., Jonsson, P., Antti, H., Marklund, S.L., Moritz, T., 2005. Extraction and GC/MS analysis of the human blood. plasma metabolome. Anal. Chem. 77, 8086–8094.

Kitchin, K.T., Wallace, K., 2005. Arsenite binding to synthetic peptides based on the Zn finger region and the estrogen binding region of the human estrogen receptor-α. Toxicol. Appl. Pharmacol. 206, 66–72.

Kitchin, K.T., Wallace, K., 2008. The role of protein binding of trivalent arsenicals in arsenic carcinogenesis and toxicity. J. Inorg. Biochem. 102, 532–539.

Lu, M., Wang, H., Li, X.-F., Arnold, L.L., Cohen, S.M., Le, X.C., 2007. Binding of dimethylarsinous acid to Cys-13α of rat hemoglobin is responsible for the retention of arsenic in rat blood. Chem. Res. Toxicol. 20, 27–37.

Maret, W., 2004. Exploring the zinc proteome. J. Anal. Atomic Spectrom. 19, 15–19.

Marney, L.C., Hoggard, J.C., Skogerboe, K.J., Synovec, R.E., 2014. Methods in discovery-based and targeted metabolite analysis by comprehensive two-dimensional gas chromatography with time-of-flight mass spectrometry detection. In: Laftery, D. (Ed.), Mass Spectrometry in Metabolomics. Springer Science+Business Media), New York, pp. 83–97. https://doi.org/10.1007/978-1-1258-2_6.

Messaoudi, I., El Heni, J., Hammouda, F., Saïd, K., Kerkeni, A., 2009. Protective effects of selenium, zinc, or their combination on cadmium-induced oxidative stress in rat kidney. Biol. Trace Elem. Res. 130, 152–161.

Montes-Nieto, R., Fuentes-Almagro, C.A., Bonilla-Valverde, D., Prieto-Alamo, M.J., Jurado, J., Carrascal, M., Gómez-Ariza, J.L., López-Barea, J., Pueyo, C., 2007. Proteomics in free-living *Mus spretus* to monitor terrestrial ecosystems. Proteomics 7, 4376–4387.

Mounicou, S., Szpunar, J., Lobinski, R., 2009. Metallomics: the concept and methodology. Chem. Soc. Rev. 38, 1119–1138.

Moreno, F., García-Barrera, T., Gómez-Ariza, J.L., 2010. Simultaneous analysis of mercury and selenium species including chiral forms of selenomethionine in human urine and serum by HPLC column-switching coupled to ICP-MS. Analyst 135, 2700–2705. https://doi.org/10.1039/c0an00090f.

Nicholson, J.K., Timbrell, J.A., Sadler, P.J., 1985. Proton NMR spectra of urine as indicators of renal damage. Mercury-induced nephrotoxicity in rats. Mol. Pharmacol. 27, 644–651.

Naranmandura, H., Suzuki, K.T., 2008. Formation of dimethylthioarsenicals in red blood cells. Toxicol. Appl. Pharmacol. 227, 390−399.

Nischwitz, V., Davies, J.T., Marshall, D., González, M., Gómez Ariza, J.L., Goenaga-Infante, H., 2013. Speciation studies of vanadium in human liver (HepG2) cells after in vitro exposure to bis(maltolato)oxovanadium(iv) using HPLC online with elemental and molecular mass spectrometry. Metallomics 5, 1685−1697.

Ogra, Y., Nagasaki, S., Yawata, A., Anan, Y., Hamada, K., Mizutani, A., 2016. Metallomics approach to changes in element concentration during differentiation from fibroblasts into adipocytes by element array analysis. J. Toxicol. Sci. 41, 241−244.

Okrasa, K., Kazlauskas, R.J., 2006. Manganese-substituted carbonic anhydrase as a new peroxidase. Chem. Eur J. 12, 1587−1596.

Outten, C.E., O'Halloran, T.V., 2001. Femtomolar sensitivity of metalloregulatory proteins controlling zinc homeostasis. Science 292, 2488−2492.

Pei, K.L., Sooriyaarachchi, M., Sherrell, D.A., George, G.N., Gailer, J., 2011. Probing the coordination behavior of Hg2+, CH3Hg+, and Cd2+ towards mixtures of two biological thiols by HPLC-ICP-AES. J. Inorg. Biochem. 105, 375−381.

Percy, A.J., Korbas, M., George, G.N., Gailer, J., 2007. Reversed-phase high-performance liquid chromatographic separation of inorganic mercury and methylmercury driven by their different coordination chemistry towards thiols. J. Chromatogr. A 1156, 331−339.

Piasek, M., Blanuša, M., Kostial, K., Laskey, J.W., 2001. Placental cadmium and progesterone concentrations in cigarette smokers. Reprod. Toxicol. 15, 673−681.

Quintana, M., Klouda, A.D., Ochsenkühn-Petropoulou, M., 2005. Size characterization of manganese species from liver extracts using size exclusion chromatography inductively coupled plasma mass spectrometry. Anal. Chim. Acta 554, 130−135.

Quintana, M., Klouda, A.D., Gondikas, A., Ochsenkühn-Petropoulou, M., Michalke, B., 2006. Analysis of size characterized manganese species from liver extracts using capillary zone electrophoresis coupled to inductively coupled plasma mass spectrometry (CZE-ICP-MS). Anal. Chim. Acta 573−574, 172−180.

Ruiz-Laguna, J., Abril, N., García-Barrera, T., Gómez-Ariza, J.L., López-Barea, J., Pueyo, C., 2006. Absolute transcript expression signatures of Cyp and Gst Genes in *Mus spretus* to detect environmental contamination. Environ. Sci. Technol. 40, 3646−3652.

Saïd, L., Banni, M., Kerkeni, A., Saïd, K., Messaoudi, I., 2010. Influence of combined treatment with zinc and selenium on cadmium induced testicular pathophysiology in rat. Food Chem. Toxicol. 48, 2759−2765.

Sanchez-Chardi, A., Penarroja-Matutano, C., Ribeiro, C.A., Nadal, J., 2007. Bioaccumulation of metals and effects of a landfill in small mammals. Part II. The wood mouse, *Apodemus sylvaticus*. Chemosphere 70, 101−109.

Sanchez-Chardi, A., Ribeiro, C.A., Nadal, J., 2009. Metals in liver and kidneys and the effects of chronic exposure to pyrite mine pollution in the shrew *Crocidura russula* inhabiting the protected wetland of Donana. Chemosphere 76, 387−394.

Sawicka-Kapusta, K., Swiergosz, R., Zakrzewska, M., 1990. Bank voles as monitors of environmental contamination by heavy metals. A remote wilderness area in Poland imperilled. Environ. Pollut. 67 (4), 315−324.

Shigeta, K., Sato, K., Furuta, N., 2007. Determination of selenoprotein P in submicrolitre samples of human plasma using micro-affinity chromatography coupled with low flow ICP-MS. J. Anal. Atomic Spectrom. 22, 911−916.

Suzuki, Y., Yoshikawa, H., 1981. Cadmium, copper, and zinc excretion and their binding to metallothionein in urine of cadmium exposed rats. J. Toxicol. Environ. Health 8, 479−487.

Suzuki, K.T., Ogra, Y., 2002. Metabolic pathway for selenium in the body: speciation by HPLC-ICP MS with enriched Se. Food Addit. Contam. 19, 974−983.

Suzuki, K.T., Mandal, B.K., Ogra, Y., 2002. Speciation of arsenic in body fluids. Talanta 58, 111−119.

Suzuki, K.T., 2005. Metabolomics of arsenic based on speciation studies. Anal. Chim. Acta 540, 71−76.

Srivastava, R.C., Ahmad, I., Kaur, G., Hasan, S.K., 1988. Alterations in the metabolism of endogenous trace metals due to cadmium, manganese and nickel-effect of partial hepatectomy. J. Environ. Sci. Health Part A Environ. Sci. Eng. 23, 95–101.

Su, L., Wang, M., Yin, S.T., Wang, H.L., Chen, L., Sun, L.G., Ruan, D.Y., 2008. The interaction of selenium and mercury in the accumulations and oxidative stress of rat tissues. Ecotoxicol. Environ. Saf. 70, 483–489.

Tainer, J.A., Roberts, V.A., Getzoff, E.D., 1991. Metal-binding sites in proteins. Curr. Opin. Biotechnol. 2, 582–591.

Teodorova, S., Metcheva, R., Topashka-Anchevab, M., 2003. Bioaccumulation and damaging action of polymetal industrial dust on laboratory mice *Mus musculus* alba: I. Analysis of Zn, Cu, Pb, and Cd disposition and mathematical model for Zn and Cd bioaccumulations. Environ. Res. 91, 85–94.

Thomas, D.J., Styblo, M., Lin, S., 2001. The cellular metabolism and systemic toxicity of arsenic. Toxicol. Appl. Pharmacol. 176, 127–144.

Vahter, M., Norin, H., 1980. Metabolism of [74]As-labeled trivalent and pentavalent inorganic arsenic in mice. Environ. Res. 21, 446–457.

Vahter, M., 1981. Biotransformation of trivalent and pentavalent inorganic arsenic in mice and rats. Environ. Res. 25, 286–293.

Viant, M.R., Sommer, U., 2013. Mass spectrometry based environmental metabolomics: a primer and review. Metabolomics 9, S144–S158.

Vioque-Fernandez, A., de Almeida, E.A., Ballesteros, J., Garcia-Barrera, T., Gomez-Ariza, J.L., Lopez-Barea, J., 2007. Donana National Park survey using crayfish (*Procambarus clarkii*) as bioindicator: esterase inhibition and pollutant levels. Toxicol. Lett. 168, 260–268.

Vioque-Fernández, A., Alves de Almeida, E., López-Barea, J., 2009. Assessment of Doñana National Park contamination in *Procambarus clarkii*: integration of conventional biomarkers and proteomic approaches. Sci. Total Environ. 407, 1784–1797.

Wang, H., Wu, Z., Chen, B., He, M., Hu, B., 2015. Chip-based array magnetic solid phase microextraction on-line coupled with inductively coupled plasma mass spectrometry for the determination of trace heavy metals in cells. Analyst 140, 5619–5626.

Wei, L., Liao, P., Wu, H., Li, X., Pei, F., Li, W., Wu, Y., 2008. Toxicological effects of cinnabar in rats by NMR-based metabolic profiling of urine and serum. Toxicol. Appl. Pharmacol. 227, 417–429.

Wolters, D.A., Stefanopoulou, M., Dyson, P.J., Groessl, M., 2012. Combination of metallomics and proteomics to study the effects of the metallodrug RAPTA-T on human cancer cells. Metallomics 4, 1185–1196.

Wrobel, K., Wrobel, K., Caruso, J.A., 2009. Epigenetics: an important challenge for ICP-MS in metallomics studies. Anal. Bioanal. Chem. 393, 481–486.

Yáñez, L., Carrizales, L., Zanatta, M.T., de Jesús Mejía, J., Batres, L., Díaz-Barriga, F., 1991. Arsenic-cadmium interaction in rats: toxic effects in the heart and tissue metal shifts. Toxicology 67, 227–234.

Yoneda, S., Suzuki, K.T., 1997. Equimolar Hg-Se complex binds to selenoprotein P. Biochem. Biophys. Res. Commun. 231, 7–11.

Zalups, R.K., 2000. Molecular interactions with mercury in the kidney. Pharmacol. Rev. 52, 113–143.

CHAPTER 5

The metabolic responses of aquatic animal exposed to POPs

Hailong Zhou[1,2], Chien-Min Chen[3], Xiaoping Diao[1,4]
[1]*State Key Laboratory of Marine Resource Utilization in South China Sea, Hainan University, Haikou, Hainan Province, China;* [2]*School of Life and Pharmaceutical Sciences, Hainan University, Haikou, Hainan Province, China;* [3]*Department of Environmental Resources Management, Chia Nan University of Pharmacy & Science, Tainan, Taiwan;* [4]*Ministry of Education Key Laboratory of Tropical Island Ecology, Hainan Normal University, Haikou, Hainan Province, China*

Chapter Outline

1. Introduction

Nature has its own way to deal with chemical substances through different natural processes, namely chemical, physical, and biological; elements and their associated substances have specific spatial and temporal characteristics in terms of occurrence as well as distribution in the environment. Since the beginning of life on earth, different life forms

Environmental Metabolomics. https://doi.org/10.1016/B978-0-12-818196-6.00005-4

used whatever they could acquire in their ambient environment as building or supporting materials for their survival. Meanwhile, they rejected or eliminated whatever was harmful or useless out of their bodies and into their surroundings; however, some substances, even not essential to an organism, sometimes would be absorbed passively and subsequently accumulated during uptake of other useful materials. These biological processes involving constant transportation of substances and transformation of the parent compounds, whether organics or inorganics and whether into or out of organisms, not only created material flows within the abiotic components, i.e., hydrosphere, geosphere, and atmosphere, but also became a vital characteristic of the biosphere. These natural cycles for elements or substances are described by the term biogeochemical cycles.

Most of the scientists, especially in the fields of natural or environmental research, have acknowledged different biogeochemical cycles, such as carbon, nitrogen, sulfur, and phosphorus, for their roles in supporting various human activities and daily life. Some may also be aware of the disruption of such cycles by humans in recent years, but only a few are appreciated in that since we have started to synthesize and massively use certain unnatural and environmentally persistent organic chemicals for different purposes, we have also created new artificial biogeochemical cycles for such substances. The physical and chemical properties of these chemicals render them to be long-ranged transported and widely distributed in different environmental compartments, as well as in different trophic levels of organisms, with limited degradation by various natural processes. Persistent organic pollutants (POPs) have first drawn public attention due to the publication of "*Silent Spring*" by Rachel Carson in 1962 (Carson and Darling, 1962). DDT (dichlorodiphenyltrichloroethane) is the most notorious one, on ecosystems and human health on the local scale. Not until scientists have found various levels of DDT and other chlorinated organic compounds present at almost everywhere throughout the world in 1980s, including naturally formed dioxins, have the term "POPs" been scientifically defined and recognized by general public. The POP substances were then considered global contaminants and became a globally major concern with respect to environmental pollution, especially in the Polar Regions, and for their endocrine-disrupting effects on some animal species including human.

In 1995, United Nations Environmental Programme (UNEP) proposed global actions for controlling POPs, for which they were defined as "chemical substances that persist in the environment, bio-accumulate through the food web, and pose a risk of causing adverse effects to human health and the environment." This call for action to eliminate or restrict the production and use of POPs resulted in the "Stockholm Convention on Persistent Organic Pollutants" global treaty in 2001, which has been effective since May 2004.

Major characteristics of POPs are summarized as follows:

- Resistance to environmental breakdown via biological, chemical, and photolytic processes, with low environmental decay rates and long biological half-lives for as long as several years.
- Limited biodegradation of POPs by organisms leads to bioaccumulation/bio-concentration. This property not only allows POPs to be accumulated but also to move up through the food chain and store in the fatty tissues of organisms at higher trophic levels due to their high lipid solubility, and because of biomagnification, marine mammals as well as humans around the world, particularly in the Arctic, carry high burdens of POPs.
- Can travel long distances from their origin via wind or ocean currents. Some POPs with relatively higher volatility can transport from lower (warmer) to higher (colder) regions through the "grasshopper effect," also referred to the "global distillation," which described the process of POPs vaporized at a relatively high temperature, and then the vapor travels to an area of lower temperature where it condenses. It also explained why relatively high concentrations of POPs had been found in the Arctic (or the Antarctic) and the bodies of local animals and inhabitants, even though most of the chemicals have not been used in that region in appreciable amounts. Human exposures in certain arctic areas are among the highest worldwide.
- Some POPs exhibit endocrine-disruptive effects resulting in a wide spectrum of adverse manifestations, such as in development, growth, reproduction, behavior, immune system, and tumorigenesis, of various species including human. Some of these effects by POPs have been observed in animals from laboratory works and epidemiological studies in the wild, and/or in human populations. However, not all POPs cause toxicity through alteration of the endocrine system, and mechanisms for eliciting certain effects are not fully understood for the time being. Nevertheless, subchronic or chronic toxicities are also one of the major considerations in the POPs listing scheme or of concerns.
- Based on the criteria set by UNEP, to be defined as POPs, the following properties of a POP should be considered or met: log K_{ow} >5, bioconcentration factor >5000, half-life in water >2 months, half-life in sediment or soil >6 months, and half-life in air >2 days. However, it should be noted that these criteria are used for the screening or reference purpose and are not necessary prerequisites for a POP to be legally defined and listed. A chemical deemed persistent or bioaccumulative may not meet the values, as those prescribed in the POP criteria, but would still merit additional considerations in the reviewing process.

2. Types and sources of pollution

Initially, when the Stockholm Convention was enacted globally, there were the 12 worst POPs on the convention list, commonly known as the "dirty dozen," including eight organic pesticides and PCBs (polychlorinated biphenyls) for total elimination in production and use, DDT for malaria control only, and dioxins/furans for mitigation in inadvertent production. After eight amendments, a total of 30 substances are currently regulated under the Stockholm Convention as of 2018, and another two (dicofol and pentadecafluorooctanoic acid) have been recommended and are being reviewed for listing. In this chapter, we categorized POPs into three groups, namely pesticides, industrial chemicals, and unintentional production, mainly based on the convention. A complete list of regulated POPs is shown in Table 5.1 along with some of their physical and chemical properties related to characteristics of POPs. Some basic information on these POPs is briefly described in the following sections.

2.1 Pesticide

Currently, there are 16 organochlorine pesticides on the POP list, i.e., aldrin, chlordane, DDT, dieldrin, endrin, hexachlorobenzene, heptachlor, mirex, toxaphene, chlordecone, hexachlorocyclohexane (both α and β forms), pentachlorobenzene, lindane, endosulfan, and pentachlorophenol (PCP). Except for DDT, which is permitted to be used only in disease vector control by the convention and as an intermediate to produce dicofol and other compounds, the rest of these pesticides are totally banned for production or use in order to eliminate these contaminants in the environment and biota.

First synthesized in 1874, DDT is probably the most infamous and controversial POPs, which was widely used during World War II to protect soldiers and civilians from malaria, typhus, and other diseases spread by insects. DDT also had an agricultural application, especially for cotton. After the war, DDT continued to be applied against mosquitoes in several countries to control malaria. Because of the discovery of the insecticidal nature of DDT in 1939 and its contribution for saving millions of lives from vector diseases, the Swiss Chemist Paul Hermann Müller received the 1948 Nobel Prize in Physiology or Medicine. However, after the *"Silent Spring"* published in 1962, the scientific community and public started to worry about the persistence of DDT and other counterparts in the environment and their adverse ecological and health effects. The best-known toxic effect of DDT (or its breakdown product DDE, dichlorodiphenyldichloroethylene) is "eggshell thinning" found in birds, especially birds of prey species. The mechanism for such effect appeared to link with reduction of the transport of calcium carbonate from the blood into

Table 5.1: Persistent organic pollutants listed in the Stockholm Convention.

Type	Chemical name	Chemical formula	Molecule weight	Vapor pressure (mmHg)	Water solubility (μg/L)	Log K_{ow}	Note
Pesticide	Aldrin	$C_{12}H_8Cl_6$	364.92	2.31×10^{-5}	17–180	5.17–7.4	Elimination
	Chlordane	$C_{10}H_6Cl_8$	409.78	10^{-6}	56	6.00	Elimination
	DDT	$C_{14}H_9Cl_5$	354.49	1.6×10^{-7}	1.2–5.5	4.89–6.914	Restriction for production or use
	Dieldrin	$C_{12}H_8Cl_6O$	380.91	1.78×10^{-7}	140	3.692–6.2	Elimination
	Endrin	$C_{12}H_8Cl_6O$	380.92	7×10^{-7}	220–260	5.1	Elimination
	Hexachlorobenzene	C_6Cl_6	284.78	1.089×10^{-5}	40	3.03–6.42	Elimination, unintentional releases
	Heptachlor	$C_{10}H_5Cl_7$	373.32	3×10^{-4}	180	4.40–5.5	Elimination
	Mirex	$C_{10}Cl_{12}$	545.55	3×10^{-7}	6.8	6.9	Elimination
	Toxaphene	$C_{10}H_{10}Cl_8$	413.82	9.8×10^{-7}	550	3.23–5.50	Elimination
	Chlordecone	$C_{10}Cl_{10}O$	490.64	2.25×10^{-7}	2.70	5.41	Elimination
	α, β-Hexachlorocyclohexane	$C_6H_6Cl_6$	290.83	$0.25(\alpha), 0.053(\beta)$	$0.33(\alpha), 1.44(\beta)$	3.9	Elimination
	Pentachlorobenzene	C_6HCl_5	250.34	0.0065	3.46	5.17	Elimination
	Lindane	$C_6H_6Cl_6$	290.83	4.2×10^{-5}	8.2	3.20–3.89	Restriction, only to be produced and used as pharmaceutical for control of head lice and scabies as second line treatment
	Endosulfan	$C_9H_6Cl_6O_3S$	406.96	1.05×10^{-3}	0.33	4.7	Restriction, only to be used in crop-pest complexes
	Pentachlorophenol (PCP) and its salts and esters	C_6HCl_5O	266.35	1.1×10^{-4}	14 mg/L	5.01	Restriction, only to be used for utility poles and crossarms

Continued

Table 5.1: Persistent organic pollutants listed in the Stockholm Convention.—cont'd

Type	Chemical name	Chemical formula	Molecule weight	Vapor pressure (mmHg)	Water solubility (μg/L)	Log K_{ow}	Note
Industrial chemical	Polychlorinated biphenyl (PCB)	$C_{12}H_xCl_y$, 209 substances	291.99 (PCB77)	2.3×10^{-6} (PCB77)	0.175 (PCB77)	6.36 (PCB77)	Banned for production and use, only unintentional releases
	Hexabromocyclododecane (HBCDD)	$C_{12}H_{18}Br_6$	641.70	4.7×10^{-7}	3.4	5.6	Banned for production and use, only unintentional releases
	Polybrominated biphenyl (PBB)	$C_{12}H_xBr_y$, 209 substances	627.58	6.9×10^{-6}	3–11	6.39	OOnly hexa-BB is listed in stockholm Convention for elimination
	Hexachlorobutadiene (HCBD)	C_4Cl_6	260.76	0.15	3.2 mg/L	4.78	Banned for production and use, only unintentional releases
	Polybrominated diphenyl ether (PBDE)	$C_{12}H_xBr_yO$ 209 substances	485.79 (tetra) ~ 722.48 (hepta)	2.5×10^{-4} (tetra) ~ 3.3×10^{-10} (hepta)	0.9 (hexa) ~ 11 (tetra)	6.5–8.4 (penta)	Only tetra ~ deca-BDEs are listed in stockholm Convention for elimination
	Perfluorooctanesulfonic acid (PFOS) and it salts	$C_8F_{17}SO_3H$	500.13	2.0×10^{-3}	680 mg/L (potassium salt)	NM	Restriction in production and use
	Perfluorooctanesulfonyl fluoride (PFOSF)	$C_8F_{18}O_2S$	499.17	0.31	8.04	5.8	Restriction in production use

Unintentional production						
Short-chain chlorinated paraffins (SCCPs)	$C_xH_{(2x-y+2)}Cl_y$, x = 10 – 13; y = 3 – 12	320–500	2.8 to 0.028× 10^{-7} Pa	150–470.59% chlorine content at 20°C	4.5–7.4	Average chlorine content 40%–70% with the limiting molecular formulas set at $C_{10}H_{19}Cl_{13}$ and $C_{13}H_6Cl_{12}$ were listed in stockholm Convention for elimination
Polychlorinated dibenzo-p-dioxins and -furans	$C_{12}H_xCl_yO_2$ (75 dioxins), $C_{12}H_xCl_yO$ (135 furans)	321.97 (2.3,7,8-TCDD)	7.4×10^{-10} (2.3,7,8-TCDD)	7.9×10^{-3} – 3.2×10^{-1} (2.3,7,8-TCDD)	7.02 (2.3,7,8-TCDD)	Unintentional releases
Polycyclic aromatic hydrocarbons (PAHs)	Over 650 PAHs in nature	NA	5.4×10^{-2} (naphthalene), 2.44×10^{-6} (benzo[a]pyrene)	1,2500–03,4000 (naphthalene), 13.3 (benzo[a]pyrene)	3.37 (naphthalene), 6.06 (benzo[a] pyrene)	Only chlorinated naphthalenes are listed in stockholm Convention, 16 PAHs are listed as priority pollutants

NA, not applicable; NM, not measurable.

the eggshell gland by DDT (Lundholm, 1997). The consequence of the decline in some bird populations found in the United States and Europe led to its bans in many countries during the 1970s. Due to bioaccumulation and biomagnification, food-borne DDT remains the greatest source of exposure for the general human population. Detection of DDT in breast milk from mothers in different regions, especially countries without restriction to its use, has also raised serious concerns about infant health. Due to worldwide prohibition, DDT and other organochlorine pesticide residues have declined steadily in wildlife and human over the last few decades (Solomon and Weiss, 2002; ATSDR, 1989), but which still be at low levels for decades, and controversy remains over the positive and negative effects for using (or not using) DDT in some areas to control malaria.

Beside DDT, dieldrin and heptachlor are the only two other organochlorine pesticide POPs used for both vector disease control and agricultural applications. For the latter purpose, dieldrin was used primarily to control termites and textile pests, while heptachlor was mainly used to kill soil insects and termites. Heptachlor has also been used more widely to kill cotton insects, grasshoppers, and other crop pests. The rest of pesticide POPs have different purposes in the agricultural sectors, but most of them are used as insecticides. One of the disadvantages of the agricultural chemicals is that they have a strong tendency to become environmental contaminants because of their open field application and usually used in large quantities. For POPs, once released into the environment regardless of sources and locations employed, their persistence will allow them to be distributed by various natural forces to remote areas with limited degradation and will continuously pose threats to whatever organisms encountered. Thus, measures to control their use on a global scale are necessary.

Other than agricultural application, some POPs also had other industrial usages. For hexachlorobenzene, it has been used mainly as a crop fungicide as early as 1945, and as a by-product of industrial chemicals such as tetrachlorocarbon or trichloroethylene. Mirex was used to control fire ants and termites, but mainly as fire retardants in plastics, rubber, paints, papers, or home appliances. Pentachlorobenzene was used in the production of PCBs, in dyestuff carriers, and as a fungicide and a flame retardant. It is also an intermediate in producing quintozene, as well as impurities in chemical products such as solvents or pesticides; however, there are not known natural sources of pentachlorobenzene. PCP and its salt and ester derivates are currently allowed only to be used as a wood preservative for utility poles and crossarms. PCP had been used as herbicide, insecticide, fungicide, algaecide, disinfectant, and as an ingredient in antifouling paint. Some PCP applications were in agricultural seeds (for nonfood uses), leather, masonry, wood preservation, cooling tower water, rope, and paper mills. PCP can be contaminated with relatively high concentrations of dioxins during the production process.

2.2 Industrial chemicals

At present, 13 POPs are assigned as industrial chemicals used for different purposes and regulated in the Stockholm Convention, which include polychlorinated biphenyl (PCB), hexabromocyclododecane (HBCDD), hexabromobiphenyl (HexBB), hexachlorobutadiene (HCBD), tetrabromodiphenylether (TeBDE), pentabromodiphenyl ether (PeBDE), hexabromodiphenyl ether (HexBDE), heptabromodiphenyl ether (HepBDE), octabromodiphenyl ether (OBDE), decabromodiphenyl ether (DeBDE), perfluorooctanesulfonic acid (PFOS), perfluorooctanesulfonyl fluoride (PFOSF), and short-chain chlorinated paraffins (SCCPs).

Polychlorinated biphenyls (PCBs), as one of the "thirty dozen," are a group of 209 chemicals with a similar structure having two benzene rings (biphenyl) at different degree of chlorination. Initially commercialized under the trade name of "Aroclor," which are the mixtures with different PCBs potions, by the Monsanto Company from 1930 to 1977, PCBs were used either in so-called "closed" or "open" applications. Examples of the former include coolants and insulating fluids for transformers and capacitors (transformer oil), hydraulic fluids, and lubricating and cutting oils. For the open application, PCBs were used as transfer agents in carbonless copy paper and as plasticizers in paints, cements, electrical cables and electronic components, pesticide extenders, reactive flame retardants, sealants, adhesives, wood floor finishing, dedusting agents, waterproofing compounds, and casting agents. PCBs can also be produced unintentionally during the combination of any chlorine and organic materials. Due to discoveries of its toxicity as early as in the 1940s and concerns over its persistence and widespread contamination in the environment starting from the 1960s, PCBs were initially banned in the open applications from the 1970s in different countries. There have also been several incidents worldwide associated with PCBs poisoning or spill/leakage resulting in environmental contamination before any regulation was enacted. Globally, it was estimated to have at least 2500 tonnes of PCBs from various sources released into the environment annually during its use, and the burning of PCBs containing products were considered the major source of pollution (ATSDR, 2000).

Brominated organic compounds listed in the POP list are mainly used as either reactive or additive flame retardants for fire protection in a wide array of products, such as in building materials, electronics, furnishings, motor vehicles, airplanes, plastics, polyurethane foams, and textiles. When used as additive flame retardants, these substances are supposed to slow down ignition and fire growth, but because of this process, it increases available time for them to escape from a fire. In addition, they are not chemically bonded to the material, which leads to easy detachment into the surroundings during use, aging, and wear of the end consumer products.

Hexachlorobutadiene (HCBD) is mainly generated as a by-product in the manufacture of chlorinated hydrocarbons or as the chlorinated butane derivative in the production of both carbon tetrachloride and tetrachloroethene. With a relatively high vapor pressure and Henry's law constant, HCBD can be easily volatilized from wet surfaces and water, and partition in the atmosphere and remain there for a substantial amount of time.

Perfluorooctanesulfonic acid (PFOS) is a fluorinated anion and is commonly commercialized with its salt or incorporated into polymers with high molecular weights for various applications. PFOS-related substances have been defined somewhat differently in different contexts, so there are numbers defined by different authorities ranging from 48 to 271 (OECD, 2002). PFOS and its related compounds are both hydrophobic and lipophobic. Therefore, they are used as surface-active agents in different applications. The extreme persistence of these substances makes them suitable for high temperature applications and for applications in contact with strong acids or bases, which include firefighting foams, carpets, leather/apparel, textiles/upholstery, paper and packaging, coatings and coating additives, industrial and household cleaning products, and pesticides and insecticides. Manufacturing processes are believed to constitute a major source of PFOS to the local environment (3M, 2000), and some volatile PFOS-related substances may be released to the atmosphere. Theoretically, any molecule containing the PFOS moiety could be a precursor to PFOS; therefore, to have a better estimation for its amount of releases from various sources and environmental loadings is difficult.

Perfluorooctanesulfonyl fluoride (PFOSF) is a synthetic perfluorinated compound with a sulfonyl fluoride functional group. It is solely used to make PFOS and PFOS-based compounds. In 1949, 3M began producing PFOS-based compounds from the synthetic precursor PFOSF. PFOSF has a variety of industrial and consumer uses, but as mentioned above, PFOSF-derived substances ultimately degrade to form PFOS. Because of environmental concerns over PFOS, 3M, the largest global producer before 2000, ceased PFOSF use in 2002 and resulted in a sharp decrease in global production. The peak annual global production was around 4500 tonnes before 3M's phase-out, and total historical global production was estimated to be at least 120,000 tonnes before 2009 (Paul et al., 2009). Even on the POP list, both PFOS and PFOSF are not totally banned and has approved uses and exemptions, and with a program to encourage reduced production under the convention.

Short-chain chlorinated paraffins (SCCPs) are complex mixtures of polychlorinated n-alkanes with chain lengths ranging from C10 to C13 and chlorine contents between 30% and 70% by weight. SCCPs are mainly used in extreme pressure lubricants in the metal processing industry, in fillers or sealers, glues and coating materials used in the building industry, and in rubber and leather treatments. Some SCCPs are used as secondary plasticizers and flame retardants in plastics. The uses of chlorinated paraffin probably provide the major source of environmental contamination. Polymers containing chlorinated paraffin will act as sources of chlorinated paraffin for centuries after disposal, due to their resistance in degradation. Estimation of annual releases of SCCPs into the atmosphere and the water in European Union were 393.9 kg and 1784 tonnes, respectively (EU, 2000).

2.3 Unintentional production

The convention recognized PCBs, pentachlorobenzene, polychlorinated naphthalenes (PCNs), polychlorinated dibenzo-p-dioxins (PCDDs), and polychlorinated dibenzofurans (PCDFs) as the inadvertently produced compounds which could be formed naturally in different processes or activities. The first two were briefly described in Section 2.2 because of their also being commercial products, while the last three compounds (groups should be a more accurate term) have no commercial values whatsoever.

PCDDs and PCDFs (PCDD/F) are two groups of chemicals with structures akin to each other, and were collectively but inaccurately called "dioxins" when they have attracted global attention since the 1970s for their ubiquity and high acute toxicity of one of the congeners, 2,3,7,8-TCDD (tetrachloro-dibenzo-p-dioxins). The basic chemical structure of both PCDDs and PCDFs is two benzene rings joined by either one or two oxygens. The degrees of chlorination on the benzenes result in PCDDs and PCDFs having 75 and 135 congeners, respectively. Although the properties and toxicities vary among these 210 congeners, the focus is on the 12 most potent ones, which assigned with different dioxin toxic equivalent factors (TEFs) based on their relatively effective potency to 2,3,7,8-TCDD, the most toxic congener. The dioxin TEQ (toxic equivalent) reports the toxicity-weighted masses of mixtures containing PCDD/F and has been applied in risk assessment and regulatory control. This concept has also been used to evaluate the total toxicity of a mixture consisting of not only PCDD/F but also the so-called dioxin-like chemicals (DLCs), such as some PCBs.

PCDDs and PCDFs were found to be mainly formed as the by-product of incineration. Theoretically, any different kind of burning process of products containing organic compounds and chlorines, which are commonly found in nature, will produce these chemicals in suitable conditions. They are also generated in reactions not involving burning, such as chlorine bleaching fibers for paper or textiles, and in the manufacture of chlorinated phenols (Kulkarni et al., 2008). UNEP has classified the sources of PCDD/F into nine major categories, which not only include various types of incineration/burning but also include processes involving metals or mineral products (UNEP, 2005). A recent study showed that the total amount of global annual releases of PCDD/F into the environment was 100.04 kg-TEQ/year with regional variations from 1.8 (Oceania) to 47.1 (Asia) (Wang et al., 2016). Although there was a two-fold difference from the estimation by another study (Fiedler, 2016), which was 70.8 kg-TEQ/year, the emission of PCDD/F from various sources are believed to be greatly reduced from that before the beginning of the 21st century resulted from measurements and efforts by both national and international authorities. Nevertheless, due to their POP characteristics, nowadays, various levels of PCDD/F can still be detected in all humans with higher levels commonly found in persons living in more industrialized countries or with diets with higher lipid contents.

Polycyclic aromatic hydrocarbons (PAHs) are a group of over 100 hydrocarbons composed of multiple aromatic rings, within which naphthalene is the simplest with only two rings. If considered PAHs derivatives, the number of PAHs compounds can exceed 600. PAHs with five- or six-membered rings are the most common and are usually uncharged, nonpolar molecules with low water solubility and can be synthesized for commercial purposes. However, most of the PAHs are formed intentionally during the incomplete burning of coal, oil and gas, garbage, or other organic substances such as tobacco or charbroiled meat. They are also produced by the thermal decomposition of organic matter. Some PAHs are potent carcinogens as well as endocrine disruptors thus attracting attentions from health-related institutes or organizations. In 1976, US EPA (Environmental Protection Agency) listed 16 PAHs as hazardous priority pollutants for their health and environment relevancy and served as a standardized set of compounds to be analyzed and regulated accordingly. Currently, some PAHs other than the "sweet sixteen" have also been designated as priority PAHs and regulated by different agencies due to their carcinogenicity or genotoxicity, and/or ability to be measured based on available analytical technology (Table 5.2).

Environmental prevalence and bioaccumulation of PAHs are also of major concern. Fortunately, animals of the higher trophic level usually possess complex metabolic systems that allow most PAHs to be eliminated rapidly and not accumulate in their bodies. This results in relatively insignificant biomagnification for most of the PAHs (ATSDR, 1995). For this reason, although some can migrate to remote areas through airborne transportation, like POPs do, most PAHs are not considered as the POP substances, except PCNs (polychlorinated naphthalenes).

PCNs belong to chlorinated PAHs with 75 possible congeners. Since 1910, PCNs were commercialized as mixtures of several congeners and have ranged from thin liquids to hard waxes, and high melting point solids depending on the degrees of chlorination. The main uses of PCNs include in cable insulation, wood preservation, engine oil additives, electroplating masking compounds, capacitors, and refractive index testing oils and as dye intermediates; however, their production decreased in the late 1970s due to concerns of their dioxin-like toxicity. Some congeners of PCNs were assigned TEFs, as described above, to deal with their combined adverse effects due to exposure to PCN mixtures. The most potent PCNs include all penta-, hexa-, and hepta-chloronaphthalenes, and their TEFs varied from 0.001 to 0.004 which were similar to those for some dioxin-like PCBs

Table 5.2: Polycyclic aromatic hydrocarbons (PAHs) regulated by various agencies/conventions and their classification for Carcinogenicity by IARC.

PAHs	Agency/Convention	IARC classification for carcinogenicity
Acenaphthene	US EPA, ATSDR	3
Acenaphthylene	US EPA, ATSDR	NA
Anthracene	US EPA, ATSDR	3
Benzo[*a*]anthracene	US EPA, ATSDR, EFSA	2B
Benzo[*b*]fluoranthene	US EPA, ATSDR, EFSA, LRTAP	2B
Benzo[*j*]fluoranthene	US ATSDR, EFSA	2B
Benzo[*k*]fluoranthene	US EPA, ATSDR, EFSA, LRTAP	2B
Benzo[*c*]fluorene	EFSA	3
Benzo[*g,h,i*]perylene	EPA, ATSDR, EFSA	3
Benzo[*a*]pyrene	EPA, ATSDR, EFSA, LRTAP	1
Benzo[*e*]pyrene	ATSDR	3
Chrysene	US EPA, ATSDR, EFSA	2B
Coronene	ATSDR	NA
Cyclopenta[*c,d*]pyrene	EFSA	2A
Dibenzo[*a,h*]anthracene	EPA, ATSDR, EFSA	2A
Dibenzo[*a,e*]pyrene	EFSA	3
Dibenzo[*a,h*]pyrene	EFSA	2B
Dibenzo[*a,i*]pyrene	EFSA	2B
Dibenzo[*a,l*]pyrene	EFSA	2A
Fluoranthene	US EPA, ATSDR	3
Fluorene	US EPA, ATSDR	3
Indeno[*1,2,3-c,d*]pyrene	US EPA, ATSDR, EFSA, LRTAP	2B
5-Methylchrysene	EFSA	2B
Naphthalene	US EPA	2B
Phenanthrene	US EPA, ATSDR	3
Pyrene	US EPA, ATSDR	3

ATSDR, Agency for Toxic Substances and Disease Registry; *EFSA*, European Food Safety Authority; *IARC*, International Agency for Research on Cancer; *LRTAP*, Convention on Long-Range Transboundary Air Pollution, UNECE. IARC Classification for Carcinogenicity: 1. carcinogenic to human; 2A. probably carcinogenic to human; 2B. possibly carcinogenic to human; 3. not classifiable as to its carcinogenicity to humans.

(Van De Plassche and Schwegler, 2002). There is no commercial use of PCNs anymore, and they are unintentionally generated during some high-temperature industrial processes in the presence of chlorines. Currently, the major sources of release of PCNs, sometimes concomitantly with PCBs, into the environment are likely to be from waste incineration and disposal of items containing chlorinated naphthalenes to landfills (WHO, 2001).

One of the obstacles to deal with POPs in respect to environmental management is that they rarely occur as a single compound and are usually from multiple sources. In fact, in most of the environmental samples, different POPs with various levels may be present simultaneously as a POP mixture. Besides, individual field studies are insufficient to provide compelling and comprehensive evidence to define a substance or a group of isomers to be listed as one of the POPs. Furthermore, humans and wildlife are exposed to a mixture of POPs which act contextually with a wide range effect, of which the interactions between them are yet another area of research to be explored. Nevertheless, based on the precautionary principle and progressively accumulated information regarding environmental contamination and impacts of listed or potential POPs, better management strategies will be deployed to reduce the risk posed by this type of chemicals.

3. Occurrence in the aquatic environment and biota

For the initial "dirty dozen" POPs (legacy POPs), due to attention from the public and intervention and responding measures collaborated internationally through various authorities from back to 1970s, tremendous information has been gathered regarding their environmental pollution and ecological consequences, although discrepancies exist in occurrences of legacy POPs in different areas due to several factors, such as chemical characteristics, distribution, degradation patterns, environmental conditions, amounts and histories of application, etc. In general, the decline of their level in different kinds of environmental compartments was observed worldwide, specially in the Arctic system and biota, which serve as the indicators for POP contamination on the global scale (Riget et al., 2010, 2016, 2019; Braune et al., 2005; Brown et al., 2018; Evans et al., 2005; Yogui and Sericano, 2009).

In contrast, for the newly listed POPs (emerging POPs), most of the investigation started from the 1990s mainly due to the later restrictions and advance of analytical methods compared to those of legacy POPs. Although data from environmental monitoring studies have been collected relatively rapidly for emerging POPs in recently years, conclusions still cannot be drawn regarding their globally temporal trends of contamination (Houde et al., 2011; Xu et al., 2013; Ahrens, 2011; Kiesling et al., 2019; Leonel et al., 2014; Oros et al., 2007; Llorca et al., 2017; Riget et al., 2016). This may be because of some POPs being complex mixtures of isomers instead of a single compound with difficulty to define during the process of selection as a POP candidate and still not enough temporal and spatial data for meaningful statistical analysis. Nevertheless, the occurrences of POPs in various environments or biota are beyond the scope of this chapter. Several reviews on the occurrences of either legacy or emerging POPs global and regional contamination in the aquatic environments are well documented and can be served as excellent references for further studies in trend, spatial, or temporal investigation (Riget et al., 2019; Llorca et al., 2017; Ahrens, 2011; Xu et al., 2013; Braune et al., 2005; Houde et al., 2011; Wenning and Martello, 2014; Li et al., 2016; Meng et al., 2019; Han and Currell, 2017; Van Mourik et al., 2016; Tanabe and Minh, 2010; Law et al., 2014).

4. Toxic effects on aquatic animals

One of the major concerns for POPs is their wide variety of adverse chronic effects on species caused by long-term exposure and accumulation. Such toxic effects are species-dependent and not limited to human and mammals, upon which most of the research and information were focusing (El-Shahawi et al., 2010; Ashraf, 2017; Harmon, 2015; Desforges et al., 2016; Alharbi et al., 2018; Akortia et al., 2016). Major toxic effects of POPs observed in human or animal studies are outlined in Table 5.3. Since POPs covered a variety of different substances and could produce different adverse outcomes with high-species dependency but with only limited information available for aquatic animals, in the following discussion, we focus on endocrine-disrupting, reproductive and developmental, and carcinogenic effects and briefly generalize from the standpoint of molecular mechanisms of action with most of the information not limited to aquatic species.

4.1 Endocrine disruption

Some POPs may interfere with the body's endocrine system through mimicking or antagonizing the action of hormones (such as estrogen, androgen, and thyroid hormone). Basically, any system in the body controlled by hormones can be derailed by xenobiotics even at very low levels of exposure, thus being the target for insult from these substances. Alteration of normal endocrine functions along the hypothalamic—pituitary—gonadal (HPG) axis may consequently produce abnormal developmental, reproductive, neurological, and immune functions in both humans and wildlife, as well as in fish (Paul Grillasca, 2012). For arthropods, the ecdysteroid system is used by crustaceans and other arthropods as the major endocrine-signaling molecules, which will be affected by antiecdysteroidal POPs for delaying the molting process (Soin et al., 2009).

Different chemicals exert their endocrinal effects via different mechanisms of action, mainly by interacting with the effects of hormones and/or with their synthesis or signaling, and/or interfering with the corresponding steroid receptors in vertebrates, namely estrogen receptor (ER), androgen receptor, progesterone receptor, glucocorticoid receptor, constitutive androstane receptor, rodent pregnane X receptor, mineralocorticoid receptor, and thyroid hormone receptor. In arthropods, the ecdysone receptor will bind to and is activated by ecdysteroids, thus is also the target for disruption leading to their abnormal developmental as well as reproductive functions.

Other less well-explored or understood mechanisms of action are direct or indirect effects on genes or their regulatory sequences on DNA responsible for transcription/translation of certain hormones or their receptors (Wuttke et al., 2010; Yang et al., 2015). Nevertheless, it is now believed that the endocrine-disruptive effects on animals are determined by dose-dependent fashion and involved in cross-communication between different groups of

Table 5.3: Toxicities of persistent organic pollutants listed in the Stockholm Convention.

Chemical name	Endocrine disruption (human/wildlife)[a]	Reproductive/ developmental toxicity	Carcinogenicity[b]	Genotoxicity	Immunotoxicity	Neurotoxicity/ behavior	Major molecular mechanisms of toxicity identified in animals
Aldrin	2/2	√	3	√	√	√	Competitive binding to AR, stimulation of ER production
Chlordane	1/2	√	2B	√	√	√	Competitive binding to AR, antiestrogenic effect, inhibition of estradiol binding, epigenetic for carcinogenicity
DDT	1/1	√	2A	√	√	√	Competitive binding to AR, activation of androgen-sensitive cell proliferation, stimulation of ER production, ER agonist and PR antagonist
Dieldrin	2/2	√	3	√	√	√	Competitive binding to AR, stimulation of ER production
Endrin	2/2	√	3	X	X	√	Competitive binding to AR
Hexachlorobenzene	1/3	√	2B	√	√	√	DLC, weak ligand of AhR, induce alterations in IGF signaling pathway, inhibition of uroporphyrinogen decarboxylase

Heptachlor	2/3	√	2B	X	√	√	Noncompetitively blocks neurotransmitter action at GABA, carcinogenic effects by activating key kinases in signaling pathways and inhibiting apoptosis
Mirex	1/2	√	2B	X	√	√	Inhibit sodium/potassium-transporting ATPases, impairing energy-dependent cellular processes
Toxaphene	1/2	√	2B	√	√	√	Cholinesterase or acetylcholinesterase (AChE) inhibitor, antagonistic effects on ER
Chlordecone	1/2	√	2B	√	√	√	Competitive binding to the estrogen and androgen receptors
α, β-Hexachlorocyclohexane	3B/3B	√	2B	√	√	√	Antagonist for AR and ER, antagonist for PR, inhibit sodium/potassium-transporting ATPase and magnesium ATPase
Pentachlorobenzene	1/3B	√	3	X	X	X	Inhibitor of oxidative phosphorylation, disturb retinoid and TR, phenobarbital-type inducer
Lindane	1/2	√	2B	X	√	√	Antagonists of ARs, antagonist of inhibitory GABA and glycine synapses, inhibition of ligand biding to AR and ER

Continued

Table 5.3: Toxicities of persistent organic pollutants listed in the Stockholm Convention.—cont'd

Chemical name	Endocrine disruption (human/wildlife)[a]	Reproductive/ developmental toxicity	Carcinogenicity[b]	Genotoxicity	Immunotoxicity	Neurotoxicity/ behavior	Major molecular mechanisms of toxicity identified in animals
Endosulfan	2/2	√	3	X	√	√	ER-mediated, antagonists of AR and PR, antagonist of GABA
Pentachlorophenol	1/3B	√	2B	√	√	√	DLC, cholinesterase or acetylcholinesterase (AChE) inhibitor, competitive binding to AhR, competitive binding to ER and TR
Polychlorinated biphenyls (PCBs)	1/NA	√	1	√	√	√	DLC, competitive binding to AhR, PXR agonist, ER agonist or antagonist, affect TR metabolism, bind to thyroid transport proteins, TRs agonist and antagonist
Hexabromocyclododecane	√	√	ND	ND	ND	√	Antiandrogenic, antiprogesteronic and T3-potentiation, low binding of thyroxine to TR, PXR agonist, AR and PR antagonists
Hexabromobiphenyl	1/NA	√	2B	X	√	√	AhR-mediated, epigenetic effects

Chemical						Mode of action
Hexachlorobutadiene	X	√	3	√	X	√	Interaction of the reactive metabolite with the inner mitochondrial membrane
Decabromodiphenyl ether	NA/2	√	3	√	√	√	ER, AR, TR antagonists, competitive binding to AhR
Perfluorooctanesulfonic acid and perfluorooctanesulfonyl fluoride	√	√	2B	X	√	√	Peroxisome proliferator–activated receptor mediated, inhibit TR transportation and metabolism
Short-chain chlorinated paraffins	1/NA	√	2B	X	X	X	Interfere TR transportation and metabolism, peroxisome proliferator, activation of CAR signaling pathway
Polychlorinated dibenzo-p-dioxins and -furans	1/NA	√	1	X	√	√	Competitive binding to AhR, causing TR insufficient by affecting TR signaling
Chlorinated naphthalenes	√	√	X	X	X	√	DCL, AhR mediated

Note should be taken that for chemicals as a group, especially for PCBs, dioxins, polychlorinated naphthalenes, short-chain chlorinated paraffin, and perfluorooctanesulfonic acid, information summarized in the table not necessarily indicates every chemical in that group exhibiting such effect(s). √, positive; AhR, aryl hydrocarbon receptor; AR, androgen receptor; CAR, constitutive androstane receptor; DLC, dioxin-like chemical; ER, estrogen receptor; GABA, gamma-amino butyric acid receptor; IGFs, insulin-like growth factors; NA, not available; NC, not conclusive; ND, no data; PR, progesterone receptor; PXR, pregnane X receptor; TR, thyroid hormone receptor; TTR, transthyretin; X, negative.

aCategorization of endocrine-disruptive capability based on EUs-Strategy for Endocrine Disruptors. Category 1: evidence of endocrine-disrupting activity in at least one species using intact animals; category 2: at least some in vitro evidence of biological activity to endocrine disruption; category 3: no evidence of endocrine-disrupting activity or no data available, category 3 with 3 subcategories, A: data available on wildlife relevant and/or mammal relevant endocrine effects; B: some data are available, but the evidence is insufficient for identification; C: data available indicating no scientific basis for inclusion in the list.

bClassification of carcinogenicity based on the IARC system. See notes in Table 5.2.

nuclear receptors (Goksoyr, 2006; Mrema et al., 2013; Maqbool et al., 2016). Two reviews have summarized and covered endocrine-disruptive effects on both aquatic vertebrates and invertebrates in details (Zou, 2019; Ottinger et al., 2018). Table 5.3 lists POPs demonstrating endocrine-disrupting capability in literature.

4.2 Reproductive and developmental toxicity

POPs with reproductive and/or developmental toxicities, whether through alteration of the endocrine system or not, may have a profound impact not only to an individual but also to its inhabiting community. These community-level effects may result from disturbance in reproduction success of a species population, such as changes in sex ratio or decline in numbers, which will ultimately change the intrinsic dynamics in an ecosystem. Two cases during the 1970 and 1990s, respectively, were well known and documented for such cascade effects of POPs and also raised public awareness and concerns over the whole "endocrine-disrupting compounds (EDCs)" issue since. The first one was organochlorine pesticide, PCBs, and dioxin pollution in the Great Lakes basin led to decline in native fish (Baumann and Michael Whittle, 1988; Devault et al., 1986; Stow, 1995), as well as in bird populations (Fry, 1995; Colborn, 1991). The second case involved a dicofol (containing DDT and its metabolites) spill in Lake Apopka, Florida, US, in 1979. In this incident, the fecundity of resident alligators (Alligator mississippiensis) was compromised (Guillette et al., 1994), along with males with elevated estrogen levels, females with abnormal hormone levels, and genetic mutations of both sexes, which led to a local population crash (Semenza et al., 1997; Guillette et al., 2000; Woodward et al., 2011; Milnes and Guillette, 2008).

Interferences of HPG axis activity at any point and/or on the normal function of hormones, not necessarily reproductive, during the growth of a progeny are the major modes of action for POPs producing developmental disorders for higher aquatic organisms (fish, amphibians, reptiles, mammals). Such effects include morphological deformity, delayed developmental process, malfunction, teratogenesis, embryotoxicity, and fetotoxicity, which will, in most of the time, result in decreases in survivorship and changes in the population of a species. Two cases of contamination by POPs mentioned above have also provided valuable information for our understanding of some POP developmental effects in the wild. For instance, DDT-exposed male juvenile alligators in Lake Apopka were observed with a reduction in penis size (Guillette et al., 1996), and in Great Lakes water, snapping turtles with deformities of the carapace, head, toe, and tail were also reported (De Solla et al., 2008).

Details of modes of action of POPs producing reproductive/developmental effects on both vertebrates and invertebrates through endocrine disruption have been well reviewed (Zou, 2019; Ottinger et al., 2018). However, some POPs eliciting reproductive and developmental toxicities to animals may not be via alteration of normal endocrine functions, but through oxidative stress and/or apoptosis during different stages of development from formation of embryo to adult (Mrema et al., 2013). In addition, involvement of the aromatic hydrocarbon receptor (AhR) in the reproduction health of fish, birds, and mammals has been confirmed (King-Heiden et al., 2012; Barron et al., 1995; Eisler, 1985; White and Birnbaum, 2009). Although not a hormonal receptor, AhR, a cytosolic ligand—activated transcription factor and a key regulator of the cellular response to exogenous and endogenous chemicals (Denison and Nagy, 2003), also plays a role in reproduction. It is known to be involved in ovarian function, establishment of an optimum environment for fertilization, nourishing the embryo and maintaining pregnancy, as well as in regulating reproductive lifespan and fertility (Hernández-Ochoa et al., 2009). Moreover, in developing vertebrates, AhR also involves cellular proliferation and differentiation of many developmental pathways, including hematopoiesis, immune system, and hepatocytes, which is vital for normal growth of an offspring into adulthood (Larigot et al., 2018; Bock, 2018). It is now commonly recognized that the AhR-mediated processes are responsible for the majority, if not all, of toxicities produced by dioxin-like POPs or DLCs across all the species responsive upon exposure (Bock, 2018).

In recent years, epigenetic effects, impairing chromosomal stability and gene expression while not directly affecting changes in DNA sequence, have been proved as a possible mechanism of some of the POPs' toxicities. These include reproductive and developmental effects in vitro and in lab animals (Cupp et al., 2003; Anway and Skinner, 2006; Anway et al., 2006), but with very few reports on aquatic species (Vandegehuchte and Janssen, 2011). Epigenetics is an emerging field and yet not been fully exploited for ecotoxicological study (Head et al., 2012). POPs with reproductive/developmental toxicity are indicated in Table 5.3.

4.3 Genotoxicity/carcinogenicity

All the POPs listed have been already well reviewed for their carcinogenicity and some have been recognized as human carcinogens with different degrees of potency, such as 2,3,7,8-TCDD, some PAHs (Table 5.2), PCBs, DDT, PBDEs, etc., and are noted in Table 5.3. The majority of carcinogenesis in animals are the consequences of DNA aberrations (genotoxicity), including mutation of specific genes, changes in the nucleotide sequence of genomic DNA, and chromosomal alteration. Some POPs are known to produce genetic damage, such as PCBs, DDT, chlordecone, hexachlorocyclohexane, PCP, and hexachlorobutadiene. However, the epigenetic effects mentioned above are considered as one of the causes for carcinogenesis, which some POPs are proved to produce such effects in humans, alligators, and fish (Pizzorno and Katzinger, 2013; Best et al., 2018; Rusiecki et al., 2008; Moore et al., 2010). It is now recognized that although genotoxic events are fundamentally the essential and critical step in carcinogenesis and/or reproductive disorders, the genotoxic and epigenetic events are usually associated with each other, and the transgenerational epigenetic role of carcinogens could also be important during carcinogenesis.

Endocrine disruption is also one of the mechanisms that leads to carcinogenesis presumably through free radical generation or biomolecule damage, which may lead to dysregulation of hormones. Whether it is receptor-dependent or receptor-independent, the hormonal carcinogenesis often involves cross-talk between hormones and sometimes coupling with complex signaling cascades (Kim and Lee, 2011). Theoretically, a substance with the capability to alter reproductive and growth hormone functions have the potential to be a carcinogenic initiator or as a role of promotor, and some may also exert such effects transgenerationally.

The role of AhR in carcinogenesis has been explored and documented, as reviewed by different researchers (Feng et al., 2013; Safe et al., 2013; Xun and Wei, 2019). It seems that AhR possesses both tumorigenesis and tumor suppression properties depending on ligands, target sites, magnitudes of activation or deactivation, signaling pathways, and other physiological factors. The proposed mechanisms of action in carcinogenesis involve multiple aspects, including inhibition of the functional expression of key antioncogenes, promotion of stem cells transforming and angiogenesis, influence in cell cycle, apoptosis, cell–cell communication, metabolism and remodel of cell matrix, and interruption of cross-talking with the signaling pathways of ER and inflammation (Feng et al., 2013).

Recently, it was proposed that POPs may be able to activate telomerase or other telomere-elongating mechanisms, which may result in cancer (Joyce and Hou, 2015). Activation of telomerase is a critical, early event in carcinogenesis for the majority of cancer types, but the exact nature of this role remains to be explored (Barrett et al., 2015).

The aim of traditional ecotoxicological studies is to determine the health status of sentinel organisms and their aquatic environments, but limitations exist since conventional methodology or approaches can only observe specific effects or outcomes at different biological levels in an organism without a systemic appreciation of the mechanisms and physiological processes involved and clarification of all the interactions within. In recent years, the application of "omics" techniques, which consent the simultaneous and comprehensive evaluation of a broad number of biomolecules, allows a better understanding of the whole picture. Among the "omics," metabolomics is now a well-established scientific field in the systems biology and has demonstrated to be a high-throughput approach with notable potential in the field of aquatic toxicology (Bundy et al., 2008; Francesco et al., 2010; Lee et al., 2018; Long et al., 2015). In the following discussions, we will highlight some employments of metabolomics in the investigation of POPs' effects on aquatic animals.

5. Application of metabolomics in aquatic animals for the profiling of POPs

Metabolomics is the scientific study of chemical processes involving metabolites, the small molecule intermediates and products of metabolism. Specifically, metabolomics is the "systematic study of the unique chemical fingerprints that specific cellular processes leave behind" (Tyagi et al., 2010) and, simply speaking, is the study of small-molecule metabolite profiles. The metabolome represents the complete set of metabolites in a biological cell, tissue, organ, or organism. Generally, the mRNA gene expression data and proteomic analyses reveal the set of gene products being produced in a cell, and only represent one aspect of the cellular function. In contrast, metabolic profiling can give an instantaneous snapshot of the physiology of the cell, which provides a direct "functional readout of the physiological state" of an organism. Since metabolomics usually focuses on a global profile of small molecular weight metabolites, the researcher can reveal the metabolic responses induced by pollutant through the comparative profiling of metabolomes in organisms.

Aquatic animals are vulnerable for chemical pollution due to most or all of their lifetime being in the freshwater or seawater environments, where most chemicals from different land sources will end up and concentrate. In this section, we mainly discuss the application of metabolomics in four categories of aquatic animals, namely bivalve, amphibian, teleostei, and mammalia, for their responses to POPs. This selection was done because studies associated with these animals were more commonly found.

5.1 Application of metabolomics profiling for bivalves exposed to POPs

Bivalve is a class of aquatic molluscs that have compressed bodies enclosed by a shell. Animals in this class include clams, oysters, cockles, mussels, scallops, and numerous other families that live in saltwater, as well as in freshwater. The use of bivalve is especially beneficial in ecotoxicological studies because they are easy to handle, can accumulate toxins, and be transplanted to other sites with less complexity. In addition, due to their filter-feeding and sessile nature, bivalves are ideal biomonitoring species for POP contamination in the aquatic system. There has been a tremendous amount of studies that utilized bivalves as target organisms for researching POPs. Some of them applied metabolomics approaches using different species such as *Mytilus galloprovincialis* (Bonnefille et al., 2018; Ji et al., 2013, 2014, 2016; Fasulo et al., 2012), *Perna viridis* (Song et al., 2016a,b, 2017; Qiu et al., 2016), *Pinctada martensii* (Chen et al., 2016, 2018a, 2018b), *Dreissena polymorpha* (Watanabe et al., 2015), *Ruditapes philippinarum* (Zhang et al., 2011), *and Mytilus edulis* (Tuffnail et al., 2009).

Mussels are widely distributed from oceanic waters to estuarine systems and can be found as a dominant intertidal species on polluted rocky shores. Apart from their abundance and wide distribution within local waterbodies, their sedentary and filter-feeding characteristics ensure them to be an ideal pollution indicator species.

Gill, digestive gland, gonad, and adductor muscle are the main tissues in mussel where POPs deposit. Benzo(a)pyrene (BaP) and DDT that belong to two distinct groups of POPs have shown to induce tissue-specific metabolic responses in mussels. Chen et al. (2018b) performed a laboratory research of *P. martensii* exposed to BaP via NMR-based metabolomics and found that at 10 μg/L, BaP disturbed energy metabolism and increased osmotic stress in the digestive glands and the gills of the mussels (Chen et al., 2018b). In comparison, Song et al. (2016a, 2017) obtained similar results for the gill and digestive glands of *P. viridis* treated with BaP. They also concluded that the male gonad was more sensitive to DDT than BaP (Song et al., 2016a, 2017). On the other hand, some POPs can induce the gender-specific metabolic responses in bivalves. For example, Ji et al. (2016) investigated the gender-specific responses in *M. galloprovincialis* exposed to tetrabromobisphenol A; interestingly, they found that the changes of metabolites in female gills were much more sensitive than those in male's.

Regarding field works, Cappello and his group applied metabolomics on mussels in a field study by using ^{1}H NMR spectroscopy. In this case, *M. galloprovincialis* response to PAHs exposure at an anthropogenic-impacted site was studied (Cappello et al., 2013, 2017). They found that PAH-exposed mussels exhibited significant changes in amino acids, energy metabolites, osmolytes, and neurotransmitters (Table 5.4). Although not related to POPs, it is worthily noted that Campillo and his colleagues conducted a remarkable study in Mediterranean coastal lagoon areas using caged *Ruditapes decussatus* exposed to agriculture and urban runoff possibly containing endocrine-disrupting compounds to observe their metabolomics responses with the results outlined in Table 5.4 (Campillo et al., 2015). These studies mentioned above illustrate the advantage of metabolomics in the field, especially using bivalves species, which can quickly reflect the changes of organism's condition induced by pollutants (e.g., POPs) (Table 5.4).

5.2 Application of metabolomics profiling for teleost exposed to POPs

The teleost is by far the largest infraclass in the class Actinopterygii, the ray-finned fishes, and makes up 96% of all existent species of fish. Teleost ranges from giant oarfish to the minute male anglerfish, dominates the seas from pole to pole, and inhabits the ocean depths, estuaries, rivers, lakes, and even swamps. Human activities have strongly and adversely affected stocks of many species of fishes by mostly overfishing and pollution for

Table 5.4: Summary of metabolites observed in aquatic animal in response to persistent organic pollutant stress.

Stressor	Organism	Metabolic responses	References
Bivalve			
Diclofenac	*Mytilus galloprovincialis* (digestive glands)	↑Norepinephrine sulfate, ↑normetanephrine, ↑5-hydroxy-L-tryptophan, ↑serotonin, ↑5-hydroxyindoleacetic acid, ↑n-methyltryptamine, ↑17α-hydroxypregnenolone, ↑18-hydroxycorticosterone, ↓5,6-dihydroxyindole, ↓vanillactic acid, ↓3, 4-dihydroxybenzeneacetic acid, ↓metanephrine, ↓3, 4-dihydroxymandelic acid, ↓gentisate aldehyde, ↓gentisic acid, ↓3-hydroxyanthranilic acid, ↓2-oxoadipic acid, ↓3-methyldioxindole	Bonnefille et al. (2018)
Benzo(a)pyrene	*Perna viridis* (male gills)	↓BCAAs, ↓dimethylamine, ↓dimethylglycine	Song et al. (2016)
2.4′-DDT	*P. viridis* (male gills)	↑Succinate, ↓aspartate, ↓arginine	Song et al. (2016)
Benzo(a)pyrene + 2.4′-DDT	*P. viridis* (male gills)	Have no obvious metabolite changes	Song et al. (2016)
Benzo(a)pyrene	*P. viridis* (female gonads)	Have no obvious metabolite changes	Qiu et al. (2016)
2.4′-DDT	*P. viridis* (female gonads)	↑Valine, ↑leucine, ↑succinate	Qiu et al. (2016)
Benzo(a)pyrene + 2.4′-DDT	*P. viridis* (female gonads)	↑Valine, ↑leucine, ↑succinate	Qiu et al. (2016)
Benzo(a)pyrene	*P. viridis* (male gonads)	There was no significant change of metabolites	Song et al. (2016c)
2.4′-DDT	*P. viridis* (male gonads)	↑Valine, ↑leucine, ↑isoleucine, ↑threonine, ↑alanine, ↑glutamate, ↑glycine, ↑homarine, ↓arginine, ↓dimethylamine, ↓unknown metabolite (3.53 ppm)	Song et al. (2016c)
Benzo(a)pyrene + 2.4′-DDT	*P. viridis* (male gonads)	↑BCAAs, ↑alanine, ↑glutamate, ↑glycine, ↑succinate, ↑4-aminobutyrate, ↑homarine, ↑dimethylglycine, ↓arginine, ↓an unknown metabolite (3.53 ppm), ↓ATP	Song et al. (2016c)

Continued

Table 5.4: Summary of metabolites observed in aquatic animal in response to persistent organic pollutant stress.—cont'd

Stressor	Organism	Metabolic responses	References
Benzo(a)pyrene	*P. viridis* (male hepatopancreas)	↑Glycine, ↑BCAAs, ↑alanine, ↑glutamate, ↑succinate, ↓acetoacetate, ↓glucose, ↓glycogen	Song et al. (2017)
2.4′-DDT	*P. viridis* (male hepatopancreas)	↑Malonate, ↑succinate, ↓3-aminoisobutyrate	Song et al. (2017)
Benzo(a)pyrene + 2.4′-DDT	*P. viridis* (male hepatopancreas)	↑Hypotaurine, ↑glycine, ↑glutamate, ↑alanine, ↑π-histidine, ↓3-aminoisobutyrate, ↓unknown 5 (1.30 ppm), ↓unknown 4(4.20 ppm), ↓unknown 3 (4.02 ppm)	Song et al. (2017)
Benzo(a)pyrene	*Pinctada martensii* (gill)	10 ug/L exposure: ↑arginine, ↑acetoacetate, ↑malonate, ↑phosphocholine, ↓BCAAs, ↓threonine, ↓glutamine, ↓glucose	Chen et al. (2018a)
Benzo(a)pyrene	*P. martensii* (digestive glands)	1 ug/L exposure: ↑glutamate, ↓BCAAS, ↓taurine, ↓glycine, ↓tyrosine, ↓histidine, ↓phenylalanine, ↓glucose, ↓inosine10 ug/L exposure: ↑Homarine, ↑ATP, ↓BCAAS, ↓taurine, ↓glycine, ↓tyrosine, ↓histidine, ↓phenylalanine, ↓glucose, ↓inosine, ↓arginine, ↓threonine, ↓glutamate	Chen et al. (2018b)
Benzo(a)pyrene	*P. martensii* (gill)	1 ug/L exposure: ↑alanine, ↑glucose, ↑glutamate, ↑ATP, ↓dimethylglycine, ↓betaine, ↓glycine, ↓phosphocholine, ↓acetoacetate, ↓malonate, ↓inosine10 ug/L exposure: ↑dimethylglycine, ↑ATP, ↑glucose, ↓betaine, ↓phosphocholine, ↓acetoacetate, ↓malonate, ↓inosine, ↓glutamate	Chen et al. (2018b)

Continued

Table 5.4: Summary of metabolites observed in aquatic animal in response to persistent organic pollutant stress.—cont'd

Stressor	Organism	Metabolic responses	References
Benzo(a)pyrene	*P. martensii* (digestive gland)	↑Phosphocholine, ↑glycine, ↑NADP⁺, ↓branched chain amino acids (BCAAs), ↓threonine, ↓alanine, ↓glutamate, ↓hypotaurine, ↓dimethyl-glycine, ↓malonate, ↓acetoacetate	Chen et al. (2016)
Benzo(a)pyrene	*Ruditapes philippinarum* (gill)	0.2uM after exposure for 24h ↑valine, ↑leucine, ↑isoleucine, ↑alanine, ↑succinate, ↑malonate, ↑phosphocholine, ↑homarine, ↓dimethylamine, ↓betaine, ↓taurine, ↓glucose, ↓glycogen	Zhang et al. (2011)
2,2′,4,4′-tetrabromodiphenylether (BDE 47)	*M. galloprovincialis* (gill)	Female versus male: ↑glutamate, ↑hypotaurine, ↓arginine	Ji et al. (2013)
PAH-contaminated water exposure (field study)	*M. galloprovincialis* (digestive gland)	↑Valine, ↑lysine, ↑phenylalanine, ↑acetoacetate, ↑adenine, ↑thymidine	Fasulo et al. (2012)
Chlorpyrifos	*M. galloprovincialis* (mediterranean mussel)	↑Adenosine, ↑ATP/ADP, ↑choline, ↑glutamine, ↑guanine, ↑homarine, ↓arginine, ↓glutamate, ↓hippurate, ↓isoleucine, ↓lactate, ↓leucine, ↓phenylacetylglycine, ↓phenylalanine, ↓proline, ↓tyrosine, ↓valine	Jones et al. (2008)
PAHs	Caged mussels *M. galloprovincialis* (gill)	↑Isoleucine, ↑leucine, ↑valine, ↑alanine, ↑arginine, ↑glutamate, ↑glutamine, ↑aspartate, ↑glycine, ↑acetate, ↑acetoacetate, ↑succinate, ↑malonate, ↑glucose, ↑glycogen, ↑ATP, ↑ADP, ↑hypotaurine, ↑taurine, ↑betaine, ↑homarine, ↑unknown #1, ↑unknown #2, ↓tyrosine, ↓acetylcholine	Cappello et al. (2013)

Continued

Table 5.4: Summary of metabolites observed in aquatic animal in response to persistent organic pollutant stress.—cont'd

Stressor	Organism	Metabolic responses	References
Teleost			
Benzo[a]pyrene	*Oncorhynchus mykiss* (muscle)	↑Proline, ↑L-serine phosphoethanolamine, ↑hypotaurine, ↑taurine, ↑betaine, ↓branched chain amino acids, ↓dimethylamine, ↓betaine, ↓taurine	Roszkowska et al. (2018)
PCB95	*Danio rerio* (embryo)	↑Serine, ↑threonine, ↑cysteine, ↑tryptophan, ↑betaine, ↑choline, ↑aspartate acid, ↑phenylalanine, ↑glutamine, ↑tyrosine, ↑γ-aminobutyrate, ↑N-acetylornithine, ↑IMP. These results not only indicate that (−)- and (+)-PCB95 have opposite effects on alanine and threonine but also predict that (−)-PCB95 plays a more important role when the two enantiomers are administered simultaneously	Xu et al. (2016)
Perfluorooctanesulfonate	*D. rerio* (embryo)	↑Glucuronolactone, ↑neopterin, ↑tyrosine, ↑phenylalanine, ↑nicotinamide, ↑maltose, ↑coumarate, ↑inosine, ↑hypotaurine, ↑uracil, ↑phenylalanine, ↑tyrosine, ↑maltose, ↓inosine monophosphate, ↓asparagine, ↓oxoglutarate	Ortiz-Villanueva et al. (2018)
Benz[a]anthracene	*D. rerio* (developing)	↑GSH, ↑serine, ↑threonine, ↑cystathionine, ↑S-adenosylmethionine, ↑S-adenosylhomocysteine, ↑methionine, ↑uanosine, ↑hypoxanthine, ↑inosine, ↑xanthine, ↑phenylalanine (phe), ↑tyrosine (Tyr), ↑tryptophan	Elie et al. (2015)

Continued

Table 5.4: Summary of metabolites observed in aquatic animal in response to persistent organic pollutant stress.—cont'd

Stressor	Organism	Metabolic responses	References
PCB (field study)	*Salvelinus alpinus* (plasma)	↓Alanine, ↓aspartate, ↓glutamate, ↓glycine, ↓serine, ↓threonine, ↓arginine	Gauthier et al. (2018)
Crude oil (PAHs)	*Oncorhynchus tshawytscha*	↑Taurine, ↑glutamate, ↓lactate, ↓ATP, ↓histidine, ↓glycine	Van Scoy et al. (2010)
Amphibian			
A mixture of diclofenac, metformin, and valproic acid	*Limnodynastes peronii* tadpoles	Pronounced differences were observed in leucine, acetate, glutamine, citrate, glycogen, tyrosine, arginine, purine nucleotides, and an unidentified metabolite at 6.53 ppm	Melvin et al. (2018)
A mixture of the drugs metformin, atorvastatin, and bezafibrate	*L. peronii* tadpoles (striped marsh frog)	↑Lactic acid, ↑branched-chain amino acids	Melvin et al. (2017)
Mammalia			
PCBs, OCPs, PBDEs, PFAS-contaminated water exposure (field study)	Male polar bear (liver)	ARA, glycerophospholipid, and several amino acid metabolic pathways were identified as different between subpopulations	Morris et al. (2019)
Organohalogenated compound—contaminated water exposure (field study)	Female polar bears (*Ursus maritimus*)	Glucose and lactate were the main metabolites driving seasonal metabolome segregation with higher concentrations of glucose and lactate in feeding	Tartu et al. (2017)
PCBs	Gray seal pups *Halichoerus grypus* (blubber)	↑Lactate, ↑glucose	Robinson et al. (2018)

the past decades, and global warming at present. Chemical pollution, especially in rivers, estuaries, and coastal waters, has harmed teleost when various types of chemicals such as pesticides, herbicides, or other chemical substances enter the water. Some POPs with endocrine-disrupting property can interfere with fish reproduction by altering their normal endocrine function. For instance, river pollution caused by some endocrine disrupters has induced the intersex condition, an individual's gonads containing both cells, in roach and some other freshwater fish. POP pollution has also caused local extinction of some fish populations in many northern European lakes during the second half of the 20th century (Wootton and Smith, 2014).

Numerous POP metabolomics studies were reported using different fish species such as *Oncorhynchus mykiss* (Roszkowska et al., 2018), *Danio rerio* (Elie et al., 2015; Ortiz-Villanueva et al., 2017, 2018; Yoon et al., 2017; Xu et al., 2016), *Carassius auratus* (Liu et al., 2015), *Acipenser persicus* (Hajirezaee et al., 2017), *Sparus aurata* (Ziarrusta et al., 2018), freshwater carp (Kokushi et al., 2012), *Oncorhynchus tshawytscha* (Viant et al., 2006), *O. mykiss* (Samuelsson et al., 2011), *Atherinops affinis* (Van Scoy et al., 2012), *Cyprinus carpio* (Kokushi et al., 2012), and even *Salvelinus alpinus* (Gauthier et al., 2018) (Table 5.4). Teleosts are also good bioindicators to detect the presence of certain contaminates in water. Thus, they are suitable to be used in ecotoxicological studies, especially when POPs are of concern (Samuelsson et al., 2006). Although several fish species (both marine and freshwater) have been used in different omics studies, *D. rerio* has been the most popular one maybe because of its different life stages including embryo, developing larvae, and adult (Table 5.4). For example, Ortiz-Villanueva et al. (2018) investigated endocrine disruptor's effects on zebrafish (*D. rerio*) embryos by untargeted LC-HRMS metabolomics analysis. They found that TBT and BPA treatments induced changes in the concentrations of about 50 metabolites in *D. rerio* embryos, and PFOS in 25 metabolites. Further analysis of the corresponding metabolic changes suggested a similar underlying mode of actions of BPA, TBT, and PFOS for affecting the metabolism of glycerophospholipids, amino acids, purines, and 2-oxocarboxylic acids (Table 5.4) (Ortiz-Villanueva et al., 2018). Ortiz-Villanueva's team also noted that different metabolic profiles were observed in zebrafish embryos treated with Bisphenol A with the same dosages but at different dosing times (Ortiz-Villanueva et al., 2017, 2018). This observation also raises a question regarding how the timing of exposure affects the metabolomics profile of an individual, especially during its developing stage, and simultaneously reemphasizes the importance and urgency of standardization of the metabolomics method for expanding its application in ecotoxicological studies. Among metabolomics studies for POPs, the study performed by Xu and collaborators is unique for its role in understanding the enantioselective effects of PCB95 in zebrafish embryos (Xu et al., 2016). Their accomplishment also highlights the importance of enantiomer-specific structures of PCB95 in understanding its toxicity and similar structure–toxicity relationship should be considered for other POPs' ecotoxicities.

Fish is not only an invaluable test model species in environmental toxicology for determination of lethal and sublethal effects of POPs, and monitoring biochemical and histopathological changes, it is also highly sensitive and accurate when using metabolomics in field studies. For example, Gauthier et al. (2018) conducted a field study of metabonomics experiment using LC-MS on the Arctic char exposed to PCBs at the remote Norwegian island of Bjørnøya. They found that lifelong exposure to PCBs in Ellasjøen char disrupts their plasma metabolome and may impair the adaptive metabolic response to stressors, leading to a reduced fitness (Gauthier et al., 2018).

5.3 Application of metabolomics profiling for amphibian exposed to POPs

Amphibians are ectothermic, tetrapod vertebrates belonging to the class Amphibian, with most species living within terrestrial, fossorial, arboreal, or freshwater aquatic ecosystems and spending their life cycles in intermediate zones of wet and dry environments. Amphibians typically start as larvae living in water, but some species have developed behavioral adaptations to bypass this. A young generally undergo metamorphosis from larva with gills to an adult air-breathing form with lungs. With their complex reproductive needs and permeable skins, amphibians are often ecological food indicators for deterioration of an aquatic ecosystem. In recent decades, there has been a dramatic decline in amphibian populations for many species around the globe. Despite amphibians play an important role in the aquatic ecosystems, metabolomics research with subjects from this phylum comprises only a small part of environmental metabolomics researches, leaving alone POPs.

5.4 Application of metabolomics profiling for aquatic mammalian exposed to POPs

Aquatic mammals are comprised of a diverse group of species that dwell partly or entirely in different marine and freshwater environments around the world. The level of dependence on aquatic life is vastly different among species. For example, the Amazonian manatee and river dolphins are aquatic and fully dependent on aquatic ecosystems, whereas the Baikal seal feeds underwater but rests, molts, and breeds on land. The capybara and hippopotamus can venture in and out of water in search of food. Like other aquatic animals, aquatic mammals are also at risk of POP pollution. However, at present, examinations of aquatic mammal species are very scarce. For wild mammals, polar bears *Ursus maritimus* and gray seals *Halichoerus grypus* have been studied utilizing metabolomics approaches (Morris et al., 2019; Tartu et al., 2017; Robinson et al., 2018) (Table 5.4). Interestingly, Morris et al. (2019) found that levels of perfluorinated alkyl substances, polybrominated diphenyl ethers, p,p′-DDE, and some highly chlorinated orthopolychlorinated biphenyl congeners were greater in the bears from southern Hudson Bay and were consistently inversely correlated with discriminating acylcarnitines and phosphatidylcholines between the subpopulations. This is surprising that dietary variation was also an important factor in the differentiation of the subpopulations (Morris et al., 2019). Furthermore, Tartu et al. (2017) assessed multiple-stressor effects for polar bears and found that several organohalogen contaminants (OHCs) affected lipid biosynthesis and catabolism in females. In addition to these findings, they also revealed that these effects were more pronounced when combined with reduced sea ice extent and thickness. The information also suggested that climate could drive the sea ice decline, and OHCs may have synergistic negative effects on polar bears. Robinson et al. (2018) used blubber explants from wild gray seal (*H. grypus*) pups to examine impacts of intrinsic tissue's POP burden and acute experimental POP exposure on adipose metabolic characteristics. Their data showed that POP burdens are high enough in seal pups to alter adipose function early in life, when fat deposition and mobilization are vital. Such POP-induced alterations to adipose metabolic properties may significantly alter energy balance regulation in marine top predators, with the potential for long-term impacts on fitness and survival.

The observations of these changes in metabolome occurred in field aquatic animals being stable are very promising, given that metabolomics studies may ultimately require the utilization of wild animals. Besides, these results should be applicable to monitor the aquatic animal's well-being or to assess environmental health rather than those from experiments mimicking environmental conditions in a laboratory setting.

6. Future challenges and possibilities

Metabolomics method can describe a more detailed and actual health condition of organisms. It has been also proven to be a sensitive, reliable, accurate, and high throughput technology. Analysis of metabolome also provides a unique opportunity to study factors on which the endogenous metabolic state of organisms will be influenced, such as genetic variation, disease, applied treatment, or diet. Although the area of human toxicology might also benefit from metabolomics research, applications in ecotoxicological studies have been limited (not to mention in aquatic animals), mainly due to wide variety of species and confounding factors involved, higher technical threshold and limitations, massive analysis works, and capacities in bioinformatics and complexity of analyses. Despite these obstacles, other gaps in environmental metabolomics/ecometabolomics prevent its application in the field of ecological/ecotoxicological sciences or specifically in the study of effects of endocrine disruptors or POPs on aquatic organism.

Among these gaps, lack of communication and interaction between researchers from different respective disciplines, such as chemistry, biology, ecology, mathematics, and computer science, is often observed. This is very common for an innovative multidisciplinary technology; especially for the ecological system, which itself is essentially multisubject oriented. To this aspect, it will be a challenge for environmental metabolomics to be fully applied in ecotoxicological research.

Current metabolomics application mainly depends on NMR or MS analysis, which still needs to be improved in order to better identify and quantify metabolites of concerns (biomarkers) within a single sample. Standardized procedures, compatible software, and a database recognized by the scientific community are equally important for validation and interpretation of results.

The major concern of metabolomics for application on wide-ranging fields is the efficiency for recognition of metabolites, because in most cases, most metabolites detected by NMR and MS have an unknown chemical nature. According to a study, reference databases for metabolomics currently provide information for hundreds of thousands of compounds, barely 5% of these known small molecules have experimental data from pure standards (Frainay et al., 2018). Unfortunately, it is still unknown how well-existing mass spectral libraries cover the biochemical landscape of prokaryotic and eukaryotic organisms. Moreover, in toxicological studies, the distinction between metabolic changes due to the toxicant or the consequences of damage, and variability between individuals, and for one individual under different conditions is sometimes technically difficult under current condition. Besides, the presence of xenobiotics and/or their metabolites is making the issue more complicated. Furthermore, to rule out the effects of environmental factors including biological (e.g., viral infection), chemical (e.g., pH or other toxicants), and physical (e.g., light or temperature) on the homeostasis, in which any change will definitely result in the shift in metabolite profile, is a real challenge for metabolomics used in field-based studies.

For environmental metabolomics research on the toxic effects of chemicals of concern mentioned in this chapter, i.e., POPs, transformation products or metabolites (e.g., hydroxylated or methylsulfonyl metabolites) are of concern in a review by Blaha and Holoubek (Blaha and Holoubek, 2013). These biotransformed compounds are more polar than the parent compound, but may be more resistant to further detoxification with more harmful effects. Whether they can be characterized by metabolomics or served as biomarkers to better understand the adverse consequence for animals exposed to POPs remains to be explored. For some emerging POPs, there is a growing number of other chemical classes that have been shown to have POP properties, but information regarding their ecotoxicity is comparatively limited (Muir and de Wit, 2010). Comparing with other environmental pollutants, to study the outcomes of wild species exposed to POPs, especially for those living in the Arctic and possessing many unique biological and physiological properties, will be much more ecologically significant and realistic. This is due to, firstly it is difficult to examine their condition in a controlled environment, such as in a laboratory; and secondly, they are usually more vulnerable by the insults than the others due to higher POP exposure. Despite the environmental and internal factors influencing the analysis as mentioned above, the fieldwork constitutes more limitations and difficulties and requires more effort when metabolomics approach is applied, thus remaining an open challenge for researchers.

The major challenge in environmental metabolomics is the lack of basic studies for most of the aquatic organisms to establish baseline information and to expand the contents of database regarding metabolite information, possibly because of their wide-ranging diversity. Although many environmental metabolomics studies using various fish species, such as rainbow trout, Chinook salmon, flatfish, fathead minnows, zebrafish, topsmelt, freshwater carp, Atlantic salmon, whale sharks, and Japanese medaka, present significant results (reviewed by Lankadurai et al., 2013), most fish examined were prototypical laboratory species and are hardly representative for all the aquatic organisms. Besides, those studies were rarely involved with POPs. Remarkably, despite the importance of arthropods to aquatic ecosystems, metabolomics research working with species from this phylum consists only a small portion of environmental metabolomics studies. For instance, *Daphnia* species are commonly used in ecological monitoring or toxicity tests, but there have been very few studies using this species with the metabolomics approach (Lankadurai et al., 2013). Nevertheless, with a greater onset of targeted metabolomics, which required comprehensive appreciation of a massive array of metabolic enzymes, their kinetics, end products, and the known biochemical pathways to which they contribute, the field of environmental metabolomics has grown significantly over the last decade, and the role of metabolomics for studying physiological responses in aquatic organisms will undoubtedly continue to expand.

It is a personal belief that in the future it will become less important to identify individual genes, proteins, and metabolites as biomarkers in species or even in humans responding to disturbances by environmental factors/pollutants; instead, to identify altered or modified cell pathways resulting from stressors will be more important from a toxicological point of view. However, before that goal can be achieved, we need to have a better picture of how different molecular biomarkers interact with each other and how they relate to the adverse effects observed in an organism. Metabolomics methods will be an excellent aid to answer those questions. It is also possible to serve as an early warning indicator for potential ecosystem changes. There is no doubt that metabolomics will have a bright future. It is also essential to incorporate it with other -omics technologies, such as genomics, transcriptomics, proteomics, etc. Only in this way, it can ultimately allow us to comprehensively understand the impacts of various environmental stressors on organisms and to some extent and the environmental health status under various pressures.

References

3M, 20003M, 2000. Sulfonated Perfluorochemicals in the Environment, Sources, Dispersion, Fate and Effects. 3M Company, St Paul, MN. AR226−0545.

Ahrens, L., 2011. Polyfluoroalkyl compounds in the aquatic environment: a review of their occurrence and fate. J. Environ. Monit. 13, 20−31.

Akortia, E., Okonkwo, J.O., Lupankwa, M., Osae, S.D., Daso, A.P., Olukunle, O.I., Chaudhary, A., 2016. A review of sources, levels, and toxicity of polybrominated diphenyl ethers (PBDEs) and their transformation and transport in various environmental compartments. Environ. Rev. 24, 253−273.

Alharbi, O.M.L., Basheer, A.A., Khattab, R.A., Ali, I., 2018. Health and environmental effects of persistent organic pollutants. J. Mol. Liq. 263, 442−453.

Anway, M.D., Skinner, M.K., 2006. Epigenetic transgenerational actions of endocrine disruptors. Endocrinology 147, s43−s49.

Anway, M.D., Leathers, C., Skinner, M.K., 2006. Endocrine disruptor vinclozolin induced epigenetic transgenerational adult-onset disease. Endocrinology 147, 5515−5523.

Ashraf, M.A., 2017. Persistent organic pollutants (POPs): a global issue, a global challenge. Environ. Sci. Pollut. R. 24, 4223−4227.

ATSDR, 1989. Toxicological Profile for DDT, DDE, and DDD. Agency for Toxic Substances and Disease Registry, U.S. Public Health Service, Atlanta, GA.

ATSDR, 1995. Toxicological profile for Polycyclic Aromatic Hydrocarbons (PAHs). Department of Health and Human Services, Public Health Service, Agency for Toxi Substances and Diease Registry, Atlanta, GA.

ATSDR, 2000. Toxicological Profile for Polychlorinated Biphenyls (PCBs). U.S. Dept. of Health and Human Services, Public Health Service, Agency for Toxic Substances and Disease Registry, Atlanta, GA.

Barrett, J.H., Iles, M.M., Dunning, A.M., Pooley, K.A., 2015. Telomere length and common disease: study design and analytical challenges. Hum. Genet. 134, 679−689.

Barron, M.G., Galbraith, H., Beltman, D., 1995. Comparative reproductive and developmental toxicology of PCBs in birds. Comp. Biochem. Physiol. C. 112, 1−14.

Baumann, P.C., Michael Whittle, D., 1988. The status of selected organics in the Laurentian Great Lakes: an overview of DDT, PCBs, dioxins, furans, and aromatic hydrocarbons. Aquat. Toxicol. 11, 241−257.

Best, C., Ikert, H., Kostyniuk, D.J., Craig, P.M., Navarro-Martin, L., Marandel, L., Mennigen, J.A., 2018. Epigenetics in teleost fish: from molecular mechanisms to physiological phenotypes. Comp. Biochem. Physiol. B 224, 210−244.

Blaha, L., Holoubek, I., 2013. Emerging issues in ecotoxicology: persistent organic pollutants (POPs). In: Ferard, J.F., BLAISE, C. (Eds.), Encyclopedia of Aquatic Ecotoxicology. Springer Netherlands., Dordrecht, pp. 429−436.

Bock, K.W., 2018. From TCDD-mediated toxicity to searches of physiologic AHR functions. Biochem. Pharmacol. 155, 419−424.

Bonnefille, B., Gomez, E., Alali, M., Rosain, D., Fenet, H., Courant, F., 2018. Metabolomics assessment of the effects of diclofenac exposure on *Mytilus galloprovincialis*: potential effects on osmoregulation and reproduction. Sci. Total Environ. 613, 611−618.

Braune, B.M., Outridge, P.M., Fisk, A.T., Muir, D.C.G., Helm, P.A., Hobbs, K., Hoekstra, P.F., Kuzyk, Z.A., Kwan, M., Letcher, R.J., Lockhart, W.L., Norstrom, R.J., Stern, G.A., Stirling, I., 2005. Persistent organic pollutants and mercury in marine biota of the Canadian Arctic: an overview of spatial and temporal trends. Sci. Total Environ. 351−352, 4−56.

Brown, T.M., Macdonald, R.W., Muir, D.C.G., Letcher, R.J., 2018. The distribution and trends of persistent organic pollutants and mercury in marine mammals from Canada's Eastern Arctic. Sci. Total Environ. 618, 500−517.

Bundy, J.G., Davey, M.P., Viant, M.R., 2008. Environmental metabolomics: a critical review and future perspectives. Metabolomics 5, 3.

Campillo, J.A., Sevilla, A., Albentosa, M., Bernal, C., Lozano, A.B., Cánovas, M., León, V.M., 2015. Metabolomic responses in caged clams, Ruditapes decussatus, exposed to agricultural and urban inputs in a Mediterranean coastal lagoon (Mar Menor, SE Spain). Sci. Total Environ. 524, 136–147.

Cappello, T., Mauceri, A., Corsaro, C., Maisano, M., Parrino, V., Paro, G.L., Messina, G., Fasulo, S., 2013. Impact of environmental pollution on caged mussels *Mytilus galloprovincialis* using NMR-based metabolomics. Mar. Pollut. Bull. 77, 132–139.

Cappello, T., Maisano, M., Mauceri, A., Fasulo, S., 2017. H-1 NMR-based metabolomics investigation on the effects of petrochemical contamination in posterior adductor muscles of caged mussel *Mytilus galloprovincialis*. Ecotoxicol. Environ. Saf. 142, 417–422.

Carson, R., Darling, L., 1962. Silent Spring. Houghton Mifflin, Boston.

Chen, H., Song, Q., Diao, X., Zhou, H., 2016. Proteomic and metabolomic analysis on the toxicological effects of Benzo[a]pyrene in pearl oyster *Pinctada martensii*. Aquat. Txicol. 175, 81–89.

Chen, H., Diao, X., Wang, H., Zhou, H., 2018a. An integrated metabolomic and proteomic study of toxic effects of Benzo[a]pyrene on gills of the pearl oyster *Pinctada martensii*. Ecotoxicol. Environ. Saf. 156, 330–336.

Chen, H., Diao, X., Zhou, H., 2018b. Tissue-specific metabolic responses of the pearl oyster Pinctada martensii exposed to benzo[a]pyrene. Mar. Pollut. Bull. 131, 17–21.

Colborn, T., 1991. Epidemiology of Great lakes bald eagles. J. Toxicol. Environ. Health 33, 395–453.

Cupp, A.S., Uzumcu, M., Suzuki, H., Dirks, K., Phillips, B., Skinner, M.K., 2003. Effect of transient embryonic in vivo exposure to the endocrine disruptor methoxychlor on embryonic and postnatal testis development. J. Androl. 24, 736–745.

De Solla, S.R., Fernie, K.J., Ashpole, S., 2008. Snapping turtles (*Chelydra serpentina*) as bioindicators in Canadian Areas of Concern in the Great Lakes Basin. II. Changes in hatching success and hatchling deformities in relation to persistent organic pollutants. Environ. Pollut. 153, 529–536.

Denison, M.S., Nagy, S.R., 2003. Activation of the aryl hydrocarbon receptor by structurally diverse exogenous and endogenous chemicals. Annu. Rev. Pharmacol. Toxicol. 43, 309–334.

Desforges, J.P.W., Sonne, C., Levin, M., Siebert, U., Dietz, R., 2016. Immunotoxic effects of environmental pollutants in marine mammals. Environ. Int. 86, 126–139.

Devault, D.S., Willford, W.A., Hesselberg, R.J., Nortrupt, D.A., Rundberg, E.G.S., Alwan, A.K., Bautista, C., 1986. Contaminant trends in Lake trout (*Salvelinus namaycush*) from the Upper Great Lakes. Arch. Environ. Contam. Toxicol. 15, 349–356.

Eisler, R., 1985. Dioxin hazards to fish, wildlife, and invertebrates: a synoptic review. Fish and Wildlife Service, U.S. Dept. of Interior. 85, 8.

El, Shahawi, M.S., Hamza, A., Bashammakh, A.S., Al-Saggaf, W.T., 2010. An overview on the accumulation, distribution, transformations, toxicity and analytical methods for the monitoring of persistent organic pollutants. Talanta 80, 1587–1597.

Elie, M.R., Choi, J., Nkrumah-Elie, Y.M., Gonnerman, G.D., Stevens, J.F., Tanguay, R.L., 2015. Metabolomic analysis to define and compare the effects of PAHs and oxygenated PAHs in developing zebrafish. Environ. Res. 140, 502–510.

European Chemicals Bureau, 2000. European Union Risk Assessment Report, vol. 4. Institute for Health and Consumer Protection, pp. C10–C13.

Evans, M.S., Muir, D., Lockhart, W.L., Stern, G., Ryan, M., Roach, P., 2005. Persistent organic pollutants and metals in the freshwater biota of the Canadian Subarctic and Arctic: an overview. Sci. Total Environ. 351–352, 94–147.

Fasulo, S., Iacono, F., Cappello, T., Corsaro, C., Maisano, M., D'agata, A., Giannetto, A., De Domenico, E., Parrino, V., LoParo, G., Mauceri, A., 2012. Metabolomic investigation of *Mytilus galloprovincialis* (Lamarck 1819) caged in aquatic environments. Ecotoxicol. Environ. Saf. 84, 139–146.

Feng, S., Cao, Z., Wang, X., 2013. Role of aryl hydrocarbon receptor in cancer. Biochim. Biophys. Acta 1836, 197–210.

Fiedler, H., 2016. Release inventories of polychlorinated dibenzo-p-dioxins and polychlorinated dibenzofurans. In: Alafe, M. (Ed.), Dioxin and Related Compounds: Special Volume in Honor of Otto Hutzinger. Springer International Publishing., Cham, pp. 1–27.

Frainay, C., Schymanski, E., Neumann, S., Merlet, B., Salek, R., Jourdan, F., Yanes, Ó., 2018. Mind the Gap: Mapping Mass Spectral Databases in Genome-Scale Metabolic Networks Reveals Poorly Covered Areas.

Francesco, I., Cappello, T., Corsaro, C., Branca, C., Maisano, M., Giuseppina, G., De Domenico, E., Mauceri, A., Salvatore, F., 2010. Environmental metabolomics and multibiomarker approaches on biomonitoring of aquatic habitats. Comp. Biochem. Physiol. A 157 (1), 550.

Fry, D.M., 1995. Reproductive effects in birds exposed to pesticides and industrial chemicals. Environ. Health Perspect. 103 (Suppl. 7), 165–171.

Gauthier, P.T., Evenset, A., Christensen, G.N., Jorgensen, E.H., Vijayan, M.M., 2018. Lifelong exposure to PCBs in the remote Norwegian arctic disrupts the plasma stress metabolome in arctic charr. Environ. Sci. Technol. 52, 868–876.

Goksoyr, A., 2006. Endocrine disruptors in the marine environment: mechanisms of toxicity and their influence on reproductive processes in fish. J. Toxicol. Environ. Health A. 69, 175–184.

Guillette Jr., L.J., Gross, T.S., Masson, G.R., Matter, J.M., Percival, H.F., Woodward, A.R., 1994. Developmental abnormalities of the gonad and abnormal sex hormone concentrations in juvenile alligators from contaminated and control lakes in Florida. Environ. Health Perspect. 102, 680–688.

Guillette Jr., L.J., Pickford, D.B., Crain, D.A., Rooney, A.A., Percival, H.F., 1996. Reduction in penis size and plasma testosterone concentrations in juvenile alligators living in a contaminated environment. Gen. Comp. Endocrinol. 101, 32–42.

Guillette, J.L.J., Crain, D.A., Gunderson, M.P., Kools, S.A.E., Milnes, M.R., Orlando, E.F., Rooney, A.A., Woodward, A.R., 2000. Alligators and endocrine disrupting contaminants: a current perspective 1. Am. Zool. 40, 438–452.

Hajirezaee, S., Mirvaghefi, A., Farahmand, H., Agh, N., 2017. NMR-based metabolomic study on the toxicological effects of pesticide, diazinon on adaptation to sea water by endangered Persian sturgeon, Acipenser persicus fingerlings. Chemosphere 185, 213–226.

Han, D., Currell, M.J., 2017. Persistent organic pollutants in China's surface water systems. Sci. Total Environ. 580, 602–625.

Harmon, S.M., 2015. Chapter 18 – the toxicity of persistent organic pollutants to aquatic organisms. In: Zeng, E.Y. (Ed.), Comprehensive Analytical Chemistry. Elsevier., Amsterdam, pp. 587–613.

Head, J.A., Dolinoy, D.C., Basu, N., 2012. Epigenetics for ecotoxicologists. Environ. Toxicol. Chem. 31, 221–227.

Hernández, Ochoa, I., Karman, B.N., Flaws, J.A., 2009. The role of the aryl hydrocarbon receptor in the female reproductive system. Biochem. Pharmacol. 77, 547–559.

Houde, M., De Silva, A.O., Muir, D.C., Letcher, R.J., 2011. Monitoring of perfluorinated compounds in aquatic biota: an updated review. Environ. Sci. Technol. 45, 7962–7973.

Ji, C., Wu, H., Wei, L., Zhao, J., Yu, J., 2013. Proteomic and metabolomic analysis reveal gender-specific responses of mussel *Mytilus galloprovincialis* to 2,2',4,4'-tetrabromodiphenyl ether (BDE 47). Aquat. Toxicol. 140, 449–457.

Ji, C., Wei, L., Zhao, J., Wu, H., 2014. Metabolomic analysis revealed that female mussel *Mytilus galloprovincialis* was sensitive to bisphenol A exposures. Environ. Toxicol. Pharmacol. 37, 844–849.

Jones, O.A., Dondero, F., Viarengo, A., Griffin, J.L., 2008. Metabolic profiling of *Mytilus galloprovincialis* and its potential applications for pollution assessment. Mar. Ecol. Prog. Ser. 369, 169–179.

Joyce, B.T., Hou, L., 2015. Organic pollutants and telomere length: a new facet of carcinogenesis. EBioMedicine 2, 1854–1855.

Kiesling, R.L., Elliott, S.M., Kammel, L.E., Choy, S.J., Hummel, S.L., 2019. Predicting the occurrence of chemicals of emerging concern in surface water and sediment across the U.S. portion of the Great Lakes Basin. Sci. Total Environ. 651, 838–850.

Kim, H.S., Lee, B.M., 2011. Chapter 70 — mutagenicity and carcinogenicity: human reproductive cancer and risk factors. In: Gupta, R.C. (Ed.), Reproductive and Developmental Toxicology. Academic Press., San Diego, pp. 913—922.

King, Heiden, T.C., Mehta, V., Xiong, K.M., Lanham, K.A., Antkiewicz, D.S., Ganser, A., Heideman, W., Peterson, R.E., 2012. Reproductive and developmental toxicity of dioxin in fish. Mol. Cell. Endocrinol. 354, 121—138.

Kokushi, E., Uno, S., Harada, T., Koyama, J., 2012. 1H NMR-based metabolomics approach to assess toxicity of bunker a heavy oil to freshwater carp, *Cyprinus carpio*. Environ. Toxicol. 27, 404—414.

Kulkarni, P.S., Crespo, J.G., Afonso, C.A.M., 2008. Dioxins sources and current remediation technologies — a review. Environ. Int. 34, 139—153.

Lankadurai, B., G.Nagato, E., Simpson, M., 2013. Environmental metabolomics: an emerging approach to study organism responses to environmental stressors. Environ. Rev. 21 (3), 180—205.

Larigot, L., Juricek, L., Dairou, J., Coumoul, X., 2018. AhR signaling pathways and regulatory functions. Biochimie Open 7, 1—9.

Law, R.J., Covaci, A., Harrad, S., Herzke, D., Abdallah, M.A.E., Fernie, K., Toms, L.M.L., Takigami, H., 2014. Levels and trends of PBDEs and HBCDs in the global environment: status at the end of 2012. Environ. Int. 65, 147—158.

Lee, S.W., Chatterjee, N., Im, J.E., Yoon, D., Kim, S., Choi, J., 2018. Integrated approach of eco-epigenetics and eco-metabolomics on the stress response of bisphenol-A exposure in the aquatic midge Chironomus riparius. Ecotoxicol. Environ. Saf. 163, 111—116.

Leonel, J., Sericano, J.L., Secchi, E.R., Bertozzi, C., Fillmann, G., Montone, R.C., 2014. PBDE levels in franciscana dolphin (Pontoporia blainvillei): temporal trend and geographical comparison. Sci. Total Environ. 493, 405—410.

Li, J., Chen, C., Li, F., 2016. Status of POPs accumulation in the Yellow River Delta: from distribution to risk assessment. Mar. Pollut. Bull. 107, 370—378.

Liu, Y., Chen, T., Li, M.H., Xu, H.D., Jia, A.Q., Zhang, J.F., Wang, J.S., 2015. 1H NMR based metabolomics approach to study the toxic effects of dichlorvos on goldfish (*Carassius auratus*). Chemosphere 138, 537—545.

Llorca, M., Farre, M., Eljarrat, E., Diaz-Cruz, S., Rodriguez-Mozaz, S., Wunderlin, D., Barcelo, D., 2017. Review of emerging contaminants in aquatic biota from Latin America: 2002—2016. Environ. Toxicol. Chem. 36, 1716—1727.

Long, S.M., Tull, D.L., Jeppe, K.J., De Souza, D.P., Dayalan, S., Pettigrove, V.J., Mcconville, M.J., Hoffmann, A.A., 2015. A multi-platform metabolomics approach demonstrates changes in energy metabolism and the transsulfuration pathway in Chironomus tepperi following exposure to zinc. Aquat. Toxicol. 162, 54—65.

Lundholm, C.E., 1997. DDE-induced eggshell thinning in birds: effects of p,p′-DDE on the calcium and prostaglandin metabolism of the eggshell gland. Comp. Biochem. Physiol. C. 118, 113—128.

Maqbool, F., Mostafalou, S., Bahadar, H., Abdollahi, M., 2016. Review of endocrine disorders associated with environmental toxicants and possible involved mechanisms. Life Sci. 145, 265—273.

Melvin, S.D., Habener, L.J., Leusch, F.D., Carroll, A.R., 2017. 1H NMR-based metabolomics reveals sub-lethal toxicity of a mixture of diabetic and lipid-regulating pharmaceuticals on amphibian larvae. Aquat. Toxicol. 184, 123—132.

Melvin, S.D., Jones, O.A., Carroll, A.R., Leusch, F.D., 2018. 1H NMR-based metabolomics reveals interactive effects between the carrier solvent methanol and a pharmaceutical mixture in an amphibian developmental bioassay with *Limnodynastes peronii*. Chemosphere 199, 372—381.

Meng, Y., Liu, X., Lu, S., Zhang, T., Jin, B., Wang, Q., Tang, Z., Liu, Y., Guo, X., Zhou, J., Xi, B., 2019. A review on occurrence and risk of polycyclic aromatic hydrocarbons (PAHs) in lakes of China. Sci. Total Environ. 651, 2497—2506.

Milnes, M.R., Guillette, J.L.J., 2008. Alligator tales: new lessons about environmental contaminants from a sentinel species. Bioscience 58, 1027—1036.

Moore, B.C., Kohno, S., Cook, R.W., Alvers, A.L., Hamlin, H.J., Woodruff, T.K., Guillette, L.J., 2010. Altered sex hormone concentrations and gonadal mRNA expression levels of activin signaling factors in hatchling alligators from a contaminated Florida lake. J. Exp. Zool. Part. A. 313, 218–230.

Morris, A., Letcher, R., Dyck, M., Chandramouli, B., Cosgrove, J., 2019. Concentrations of legacy and new contaminants are related to metabolite profiles in Hudson Bay polar bears. Environ. Res. 168, 364–374.

Mrema, E.J., Rubino, F.M., Brambilla, G., Moretto, A., Tsatsakis, A.M., Colosio, C., 2013. Persistent organochlorinated pesticides and mechanisms of their toxicity. Toxicology 307, 74–88.

Muir, D.C.G., De, Wit, C.A., 2010. Trends of legacy and new persistent organic pollutants in the circumpolar arctic: overview, conclusions, and recommendations. Sci. Total Environ. 408, 3044–3051.

Oecd, 2002. OECD Report Hazard Assessment of Perfluorooctane Sulfonate (PFOS) and Its Salts.

Oros, D.R., Ross, J.R.M., Spies, R.B., Mumley, T., 2007. Polycyclic aromatic hydrocarbon (PAH) contamination in San Francisco Bay: a 10-year retrospective of monitoring in an urbanized estuary. Environ. Res. 105, 101–118.

Ortiz-Villanueva, E., Navarro-Martin, L., Jaumot, J., Benavente, F., Sanz-Nebot, V., Pina, B., Tauler, R., 2017. Metabolic disruption of zebrafish (*Danio rerio*) embryos by bisphenol A. An integrated metabolomic and transcriptomic approach. Environ. Pollut. 231, 22–36.

Ortiz-Villanueva, E., Jaumot, J., Martinez, R., Navarro-Martin, L., Pina, B., Tauler, R., 2018. Assessment of endocrine disruptors effects on zebrafish (*Danio rerio*) embryos by untargeted LC-HRMS metabolomic analysis. Sci. Total Environ. 635, 156–166.

Ottinger, M.A., Dean, K., Russart, K.L.G., 2018. Vertebrate Endocrine Disruption. Reference Module in Life Sciences. Elsevier.

Paul, A.G., Jones, K.C., Sweetman, A.J., 2009. A first global production, emission, and environmental inventory for perfluorooctane sulfonate. Environ. Sci. Technol. 43, 386–392.

Paul Grillasca, J., 2012. Impact of endocrine disrupting chemicals [EDCs] on hypothalamic-pituitary-gonad-liver [HPGL] Axis in fish. World J. Fish Mar. Sci. 4 (1), 14–30.

Pizzorno, J., Katzinger, J., 2013. Clinical implications of persistent organic pollutants – epigenetic mechanisms. Journal of Restorative Medicine 2, 4–13.

Qiu, L., Song, Q., Jiang, X., Zhao, H., Chen, H., Zhou, H., Han, Q., Diao, X., 2016. Comparative gonad protein and metabolite responses to a binary mixture of 2,4′-DDT and benzo (a) pyrene in the female green mussel *Perna viridis*. Metabolomics 12, 140.

Riget, F., Bignert, A., Braune, B., Stow, J., Wilson, S., 2010. Temporal trends of legacy POPs in Arctic biota, an update. Sci. Total Environ. 408, 2874–2884.

Riget, F., Vorkamp, K., Bossi, R., Sonne, C., Letcher, R.J., Dietz, R., 2016. Twenty years of monitoring of persistent organic pollutants in Greenland biota. A review. Environ. Pollut. 217, 114–123.

Riget, F., Bignert, A., Braune, B., Dam, M., Dietz, R., Evans, M., Green, N., Gunnlaugsdóttir, H., Hoydal, K.S., Kucklick, J., Letcher, R., Muir, D., Schuur, S., Sonne, C., Stern, G., Tomy, G., Vorkamp, K., Wilson, S., 2019. Temporal trends of persistent organic pollutants in Arctic marine and freshwater biota. Sci. Total Environ. 649, 99–110.

Robinson, K.J., Hall, A.J., Debier, C., Eppe, G., Thomé, J.P., Bennett, K.A., 2018. Persistent organic pollutant burden, experimental POP exposure, and tissue properties affect metabolic profiles of blubber from gray seal pups. Environ. Sci. Technol. 52, 13523–13534.

Roszkowska, A., Yu, M., Bessonneau, V., Bragg, L., Servos, M., Pawliszyn, J., 2018. Metabolome profiling of fish muscle tissue exposed to benzo [a] pyrene using in vivo solid-phase microextraction. Environ. Sci. Technol. Lett. 5, 431–435.

Rusiecki, J.A., Baccarelli, A., Bollati, V., Tarantini, L., Moore, L.E., Bonefeld-Jorgensen, E.C., 2008. Global DNA hypomethylation is associated with high serum-persistent organic pollutants in Greenlandic Inuit. Environ. Health Perspect. 116, 1547–1552.

Safe, S., Lee, S.O., Jin, U.H., 2013. Role of the aryl hydrocarbon receptor in carcinogenesis and potential as a drug target. Toxicol. Sci. 135, 1–16.

Samuelsson, L.M., Förlin, L., Karlsson, G., Adolfsson-Erici, M., Larsson, D.J., 2006. Using NMR metabolomics to identify responses of an environmental estrogen in blood plasma of fish. Aquat. Toxicol. 78, 341–349.

Samuelsson, L.M., Björlenius, B., Förlin, L., Larsson, D.J., 2011. Reproducible 1H NMR-based metabolomic responses in fish exposed to different sewage effluents in two separate studies. Environ. Sci. Technol. 45, 1703–1710.

Semenza, J.C., Tolbert, P.E., Rubin, C.H., Guillette Jr., L.J., Jackson, R.J., 1997. Reproductive toxins and alligator abnormalities at Lake Apopka, Florida. Environ. Health Perspect. 105, 1030–1032.

Soi, T., Verslycke, T., Janssen, C., Smagghe, G., 2009. Ecdysteroids and their importance in endocrine disruption research. In: Smagghe, G. (Ed.), Ecdysone: Structures and Functions. Springer Netherlands., Dordrecht, pp. 539–549.

Solomon, G.M., Weiss, P.M., 2002. Chemical contaminants in breast milk: time trends and regional variability. Environ. Health Perspect. 110, A339–A347.

Song, Q., Chen, H., Li, Y., Zhou, H., Han, Q., Diao, X., 2016a. Toxicological effects of benzo (a) pyrene, DDT and their mixture on the green mussel *Perna viridis* revealed by proteomic and metabolomic approaches. Chemosphere 144, 214–224.

Song, Q., Zheng, P., Qiu, L., Jiang, X., Zhao, H., Zhou, H., Han, Q., Diao, X., 2016b. Toxic effects of male *Perna viridis* gonad exposed to BaP, DDT and their mixture: a metabolomic and proteomic study of the underlying mechanism. Toxicol. Lett. 240, 185–195.

Song, Q., Zhou, H., Han, Q., Diao, X., 2017. Toxic responses of *Perna viridis* hepatopancreas exposed to DDT, benzo (a) pyrene and their mixture uncovered by iTRAQ-based proteomics and NMR-based metabolomics. Aquat. Toxicol. 192, 48–57.

Stow, C.A., 1995. Factors associated with PCB concentrations in Lake Michigan fish. Environ. Sci. Technol. 29, 522–527.

Tanabe, S., Minh, T.B., 2010. Dioxins and organohalogen contaminants in the Asia-Pacific region. Ecotoxicology 19, 463–478.

Tartu, S., Lille-Langøy, R., Størseth, T.R., Bourgeon, S., Brunsvik, A., Aars, J., Goksøyr, A., Jenssen, B.M., Polder, A., Thiemann, G.W., Torget, V., Routti, H., 2017. Multiple-stressor effects in an apex predator: combined influence of pollutants and sea ice decline on lipid metabolism in polar bears. Sci. Rep-U. K. 7, 16487.

Tuffnail, W., Mills, G.A., Cary, P., Greenwood, R., 2009. An environmental 1H NMR metabolomic study of the exposure of the marine mussel *Mytilus edulis* to atrazine, lindane, hypoxia and starvation. Metabolomics 5, 33–43.

Tyagi, S., Raghvendra, S.U., Kalra, T., Munjal, K., 2010. Applications of metabolomics-a systematic study of the unique chemical fingerprints: an overview. Int. J. Pharmaceut. Sci. Rev. Res. 3, 83–86.

UNEP, 2005. Standardized Toolkit for Identification and Quantification of Dioxins and Furan Releases. UNEP Chemicals, Geneva, Switzerland.

Van De Plassche, E., Schwegler, A., 2002. Polychlorinated Naphthalenes, Preliminary Risk Profile.

Van Mourik, L.M., Gaus, C., Leonards, P.E.G., De Boer, J., 2016. Chlorinated paraffins in the environment: a review on their production, fate, levels and trends between 2010 and 2015. Chemosphere 155, 415–428.

Van Scoy, A.R., Lin, C.Y., Anderson, B.S., Philips, B.M., Martin, M.J., Mccall, J., Todd, C.R., Crane, D., Sowby, M.L., Viant, M.R., 2010. Metabolic responses produced by crude versus dispersed oil in Chinook salmon pre-smolts via NMR-based metabolomics. Ecotoxicol. Environ. Saf. 73, 710–717.

Van Scoy, A.R., Anderson, B.S., Philips, B.M., Voorhees, J., Mccann, M., De Haro, H., Martin, M.J., Mccall, J., Todd, C.R., Crane, D., 2012. NMR-based characterization of the acute metabolic effects of weathered crude and dispersed oil in spawning topsmelt and their embryos. Ecotoxicol. Environ. Saf. 78, 99–109.

Vandegehuchte, M.B., Janssen, C.R., 2011. Epigenetics and its implications for ecotoxicology. Ecotoxicology 20, 607–624.

Viant, M.R., Pincetich, C.A., Tjeerdema, R.S., 2006. Metabolic effects of dinoseb, diazinon and esfenvalerate in eyed eggs and alevins of Chinook salmon (*Oncorhynchus tshawytscha*) determined by 1H NMR metabolomics. Aquat. Toxicol. 77, 359−371.

Wang, B., Fiedler, H., Huang, J., Deng, S., Wang, Y., Yu, G., 2016. A primary estimate of global PCDD/F release based on the quantity and quality of national economic and social activities. Chemosphere 151, 303−309.

Watanabe, M., Meyer, K.A., Jackson, T.M., Schock, T.B., Johnson, W.E., Bearden, D.W., 2015. Application of NMR-based metabolomics for environmental assessment in the Great Lakes using zebra mussel (*Dreissena polymorpha*). Metabolomics 11, 1302−1315.

Wenning, R.J., Martello, L., 2014. Chapter 8 − POPs in marine and freshwater environments. In: O,Sullivan, G., Sandau, C. (Eds.), Environmental Forensics for Persistent Organic Pollutants. Elsevier., Amsterdam, pp. 357−390.

White, S.S., Birnbaum, L.S., 2009. An overview of the effects of dioxins and dioxin-like compounds on vertebrates, as documented in human and ecological epidemiology. J. Environ. Sci. Health C Environ. Carcinog. Ecotoxicol. Rev. 27, 197−211.

WHO, 2001. Polychlorinated naphthalenes. International Programme on Chemical Safety CICAD 34.

Woodward, A.R., Percival, H.F., Rauschenberger, R.H., Gross, T.S., Rice, K.G., Conrow, R., 2011. Abnormal alligators and organochlorine pesticides in Lake Apopka, Florida. In: Elliott, J.E., Bishop, C.A., Morrissey, C.A. (Eds.), Wildlife Ecotoxicology: Forensic Approaches. Springer New York., New York, NY, pp. 153−187.

Wootton, R.J., Smith, C., 2014. Reproductive Biology of Teleost Fishes. John Wiley & Sons.

Wuttke, W., Jarry, H., Seidlova,Wuttke, D., 2010. Definition, classification and mechanism of action of endocrine disrupting chemicals. Hormones (Basel) 9 (1), 9−15.

Xu, F.L., Jorgensen, S.E., Shimizu, Y., Silow, E., 2013. Persistent organic pollutants in fresh water ecosystems. The. Scientific. World. J. 2013, 303815.

Xu, N., Mu, P., Yin, Z., Jia, Q., Yang, S., Qian, Y., Qiu, J., 2016. Analysis of the enantioselective effects of PCB95 in zebrafish (*Danio rerio*) embryos through targeted metabolomics by UPLC-MS/MS. PloS One 11, e0160584.

Xun, C., Wei, D., 2019. Aryl hydrocarbon receptor: its regulation and roles in transformation and tumorigenesis. Curr. Drug Targets 20, 1−10.

Yang, O., Kim, H.L., Weon, J.I., Seo, Y.R., 2015. Endocrine-disrupting chemicals: review of toxicological mechanisms using molecular pathway analysis. J. Cancer. Prev. 20, 12−24.

Yogui, G.T., Sericano, J.L., 2009. Polybrominated diphenyl ether flame retardants in the U.S. marine environment: a review. Environ. Int. 35, 655−666.

Yoon, C., Yoon, D., Cho, J., Kim, S., Lee, H., Choi, H., Kim, S., 2017. 1H-NMR-based metabolomic studies of bisphenol A in zebrafish (*Danio rerio*). J. Environ. Sci. Heal. B. 52, 282−289.

Zhang, L., Liu, X., You, L., Zhou, D., Wang, Q., Li, F., Cong, M., Li, L., Zhao, J., Liu, D., Yu, J., Wu, H., 2011. Benzo(a)pyrene-induced metabolic responses in Manila clam Ruditapes philippinarum by proton nuclear magnetic resonance (H-1 NMR) based metabolomics. Environ. Toxicol. Pharmacol. 32, 218−225.

Ziarrusta, H., Mijangos, L., Picart-Armada, S., Irazola, M., Perera-Lluna, A., Usobiaga, A., Prieto, A., Etxebarria, N., Olivares, M., Zuloaga, O., 2018. Non-targeted metabolomics reveals alterations in liver and plasma of gilt-head bream exposed to oxybenzone. Chemosphere 211, 624−631.

Zou, E., 2019. Aquatic invertebrate endocrine disruption. In: Encyclopedia of Animal Behavior, pp. 112−123.

CHAPTER 6

Metabolomics in plant protection product research and development: discovering the mode(s)-of-action and mechanisms of toxicity

Konstantinos A. Aliferis[1,2]

[1]*Pesticide Science Laboratory, Agriculture University of Athens, Athens, Greece;* [2]*Department of Plant Science, McGill University, Sainte-Anne-de-Bellevue, QC, Canada*

Chapter Outline

Environmental Metabolomics. https://doi.org/10.1016/B978-0-12-818196-6.00006-6

1. Introduction

The significant increase of crop yields that was observed during the "Green Revolution" was largely based on the discovery and development of chemicals as plant protection products (PPPs) (Tilman, 1998; Evenson and Gollin, 2003; Pingali, 2012). Currently, however, the agrochemical and agri-food sectors are facing great challenges. The increasing concerns of consumers, organizations, and governments on the effects of PPPs on human health, the environment, and nontarget organisms (Delorenzo et al., 2001; Grandjean and Landrigan, 2014), high levels of PPPs residues in the food and the environment (Wilkowska and Biziuk, 2011; Regueiro et al., 2015), and the existence of fungal diseases for which no effective PPPs currently exist (Klosterman et al., 2009) are among the major ones. The situation becomes even more complex taking into account other important factors such as the sporadic and unpredictable outbreaks of plant diseases (Fisher et al., 2012), the development of resistant to PPP populations of weeds, phytopathogens, and insects (Delp and Dekker, 1985; Sparks et al., 2012; Délye et al., 2013; Fisher et al., 2018), and the banning of soil fumigants (Mouttet et al., 2014).

In response to some of the above challenges, international organizations have introduced new registration requirements (EFSA, 2017), which has resulted in the significant reduction of the available PPPs. The latter, in combination with the ineffectiveness to develop PPPs exhibiting novel mode(s)-of-action (MoA) (Delp and Dekker, 1985; Dayan et al., 2009; Cantrell et al., 2012; Duke, 2012), has intensified the issues that the plant protection sector is facing.

Within this context, the discovery and assessment of new or alternative sources of bioactivity, and their exploitation in crop protection could provide valuable solutions. However, in the PPP's research and development (R&D) pipeline (Fig. 6.1), the discovery of the MoA of a vast number of natural or chemically synthesized bioactive compounds and their toxicological risk assessment represent major bottlenecks (Aliferis and Chrysayi-Tokousbalides, 2011; Aliferis and Jabaji, 2011), being time-consuming and costly tasks.

Currently, based on the advances in the fields of natural products' and synthetic chemistry, scientists have access to a vast number of bioactive compounds with diverse physicochemical properties, structures, and bioactivities (Dayan et al., 2009; Cantrell et al., 2012; Gerwick and Sparks, 2014). Such compounds need to be screened based on their toxicity to human and nontarget organisms, their efficacy in plant protection, and persistence in the environment. Furthermore, they should provide solutions to issues such as the development of pest and pathogen resistance to PPPs by the introduction, among others, to new MoA. Taking into account that 103 different MoA of PPPs have been reported, 54 for fungicides (Fig. 6.2), 20 for herbicides, and 29 for insecticides (Fig. 6.3) (Aliferis and Jabaji, 2011; Sparks and Nauen, 2015; FRAC, 2019; HRAC, 2019; IRAC,

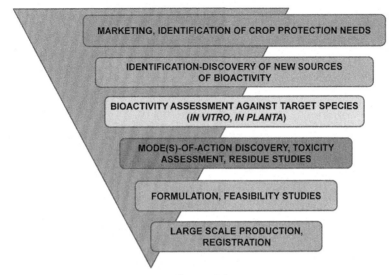

Figure 6.1

A representative pipeline for the research and development of active ingredients (a.i.) as plant protection products. The discovery of the mode(s)-of-action (MoA) and the assessment of the toxicity of an a.i. represent bottlenecks. The development of robust metabolomics models is highly foreseen to accelerate the process reducing at the same time the corresponding costs and improving the selection process.

2019), it is evident that the employment of high-throughput validated tools are required in order to facilitate the above endeavor in a timely and cost-effective manner. The early discovery of the MoA of superior bioactive compounds and the assessment of their toxicity could greatly accelerate the R&D of new/alternative plant protection agents.

PPPs exert their bioactivity by interfering with a primary biochemical system (primary MoA) of the target organism, as well as in many cases with the secondary ones (secondary MoA) (Aliferis and Jabaji, 2011), leading to the disturbance of its metabolic state and possibly, finally, to its elimination. For the majority of the active ingredients (a.i.) of PPPs, the secondary MoAs are largely unknown. Both the primary and the secondary MoA determine the bioactivity and toxicological profiles of a given a.i.

The exposure of a biological system to bioactive compounds causes a general disturbance of their metabolism, which could be reversible or not. The latter depends on factors such as the time of exposure, the metabolism of the a.i., their MoA, dosage, and environmental conditions. Therefore, the monitoring of the global metabolism of an organism and its regulation in response to treatments with bioactive agents could reveal evidence on their MoA and toxicity leading to the discovery of corresponding biomarkers of effect. Nonetheless, the discovery of validated biomarkers (individual metabolites, sets of metabolites, or whole metabolic profiles) of PPPs' toxicity is challenging. Upon their

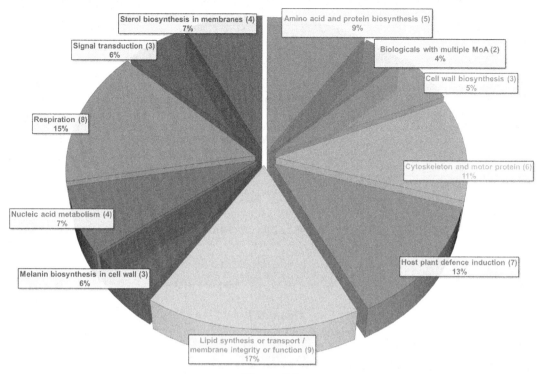

Figure 6.2

Grouping of the mode(s)-of-action (MoA) of commercially developed active ingredients of fungicides as plant protection products based on the targeted physiological functions. The corresponding number of MoA is displayed in the parenthesis following the physiological function. *Data were retrieved from the Fungicide Resistance Action Committee (FRAC, http://www.frac.info/ home). (Accessed November 2019).*

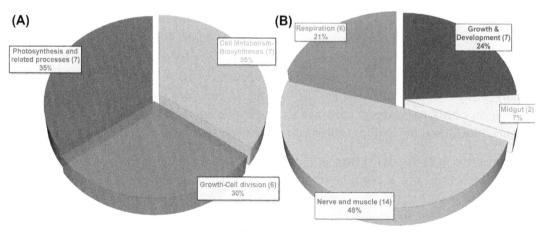

Figure 6.3

Grouping of the mode(s)-of-action (MoA) of commercially developed as plant protection products active ingredients of herbicides (A) and insecticides (B) based on the targeted physiological functions of plants and insects, respectively. The corresponding number of MoA is displayed in the parenthesis following the physiological function. *Data were retrieved from the Herbicide Resistance Action Committee (HRAC, http://www.hracglobal.com/) and the Insecticide Resistance Action Committee (https://www.irac-online.org/). (Accessed April 2019).*

discovery, such biomarkers could serve as a reliable high-throughput tool in the early screening of bioactive molecules based on their MoA and toxicological profiles, with profound benefits for their R&D.

In contrast to other "omics" such as genomics, transcriptomics, or proteomics, being closest to the phenome, metabolomics best describes phenotypes, correlating them with the undergoing changes at the metabolome level. In theory, the monitoring of the global metabolic network of a biological system and its perturbation in response to stimuli, such as exposure to bioactive compounds, can provide valuable insights into their MoA and mechanisms of toxicity (Fig. 6.4). Within this context, and based on its capacities, metabolomics could greatly assist the R&D of PPPs by reducing the required times and costs for the discovery of the MoA of bioactive compounds and by eliminating at the same time, at early stages of research, those exhibiting unfavorable MoA or toxicological profiles.

Metabolomics is a robust bioanalytical tool, whose concept was introduced in the begging of the previous decade (Fiehn et al., 2000; Trethewey, 2001), that enables the high-throughput monitoring of the metabolism of a biological system and its regulation in response to stimuli. To date, it has been successfully applied in almost every scientific discipline: the study, diagnosis, and prognosis of human diseases (Spratlin et al., 2009; Madsen et al., 2010; Griffin et al., 2015; Wishart, 2016; Mccartney et al., 2018), clinical chemistry and toxicology (Robertson et al., 2010; Roux et al., 2011), and nutritional research (Jones et al., 2012; Gibbons et al., 2015). Focusing on the agricultural and environmental sciences, it has been applied in the discovery of the MoA of bioactive substances (Araníbar et al., 2001; Guo et al., 2009; Liu et al., 2010; Aliferis and Jabaji, 2011), plant sciences (Saito and Matsuda, 2010; Urano et al., 2010; Aliferis et al., 2014; Bowne et al., 2018), the study of the environmental impact of bioactive agents on nontarget organisms (Miller, 2007; Bundy et al., 2009; Ahuja et al., 2010; Viant and Sommer, 2013), microbiology (Aldridge and Rhee, 2014; Barkal et al., 2016), food science (Wishart, 2008a; Cevallos-Cevallos et al., 2009; Antignac et al., 2011; Herrero et al., 2012; Castro-Puyana and Herrero, 2013; Xu, 2017), and the risk assessment of genetically modified organisms (Manetti et al., 2006; García-Villalba et al., 2008; Leon et al., 2009). Nonetheless, although it exhibits a great potential in PPPs' R&D (Aliferis and Chrysayi-Tokousbalides, 2011; Aliferis and Jabaji, 2011), that remains yet partially exploited.

Liquid chromatography (LC) and gas chromatography (GC) hyphened with various mass spectrometry (MS) detectors and nuclear magnetic resonance (NMR) spectrometers are the most commonly employed in metabolomics studies (Dunn and Ellis, 2005; Wishart, 2008b; Viant and Sommer, 2013; Markley et al., 2017). The topics of instrumentation, analytical methodologies, and data processing and analyses for metabolomics have been

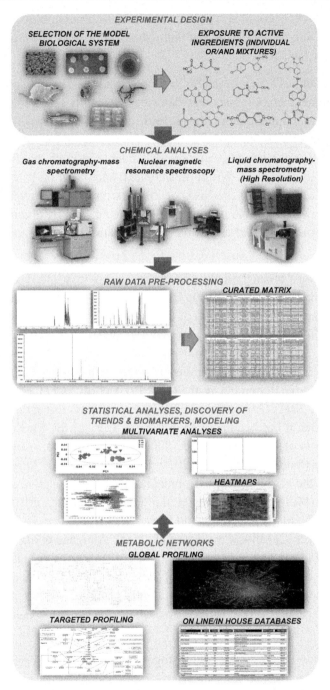

Figure 6.4

Sequential steps of a typical metabolomics experiment for the discovery of the mode(s)-of-action and the assessment of the toxicity of bioactive compounds. Metabolomics is a multidisciplinary bioanalytical tool. A solid experimental design and robust protocols for sample preparation and chemical analyses, data processing, and bioinformatics are prerequisites for the validity of analyses and the discovery of biomarkers of effect. In a final step, the use of databases, online repositories, and tools will assist in the biological interpretation of the obtained large-scale data sets.

recently thoroughly reviewed (Kim et al., 2010; Hendriks et al., 2011; Dunn et al., 2012; Haug et al., 2012; Tautenhahn et al., 2012). Thus, here, the current state-of-the-art of metabolomics in the discovery and study of the MoA and mechanisms of toxicity of bioactive compounds is being presented. Additionally, the conceptual application of metabolomics in the development of the new generation PPPs is being proposed.

2. Discovery and study of the MoA of PPPs and assessment of their toxicity using model biological systems employing metabolomics

The choice of the biological system to be used in a metabolomics study for the development of predictive models is crucial for the robust and successful discovery of the MoA of PPPs. Initially, the bioactivity of the a.i. under investigation should be tested to the targeted biological system or the closest available one based on its effect on the metabolic function. As a rule of thumb, biological systems that exhibit features such as minimum requirements for handling and space produce uniform populations in relatively short time and are economical in their maintenance, and preferably those that at the same time have sequenced genomes, are favorable for use in such studies. The latter facilitates the study of an organism's metabolism within a systems biology context. Several plant species, microorganisms, aquatic organisms, earthworms, rats and mice, and cell cultures have been used as model biological systems performing metabolomics for the discovery and study of the MoA of PPPs and, additionally, in the assessment of their toxicity, as discussed below. Furthermore, the application of metabolomics in such systems is rapidly expanding as it is indicated by the exponential increase of the corresponding published research (Fig. 6.5).

2.1 Metabolomics pipeline

Metabolomics is multidisciplinary concept that requires the integration of aspects from different fields such as experimental design and execution (Dunn et al., 2012; Hounoum et al., 2016), sample preparation (Kim and Verpoorte, 2010), chemical analyses, data processing and bioinformatics (Van Den Berg et al., 2006; Katajamaa and Orešič, 2007; Hendriks et al., 2011), database utilization/development (Kopka et al., 2004; Kind et al., 2009; Tautenhahn et al., 2012), online repositories and tools (Haug et al., 2012; Xia and Wishart, 2016), and experimental quality control (Fiehn et al., 2007). A representative metabolomics pipeline for the discovery of the MoA and mechanisms of toxicity of bioactive compounds is presented in Fig. 6.4. The thorough discussion on the topic is not within the scope of this chapter; however, selected main components and considerations regarding the application of metabolomics in the study of the MoA and toxicity mechanisms of PPPs are being presented.

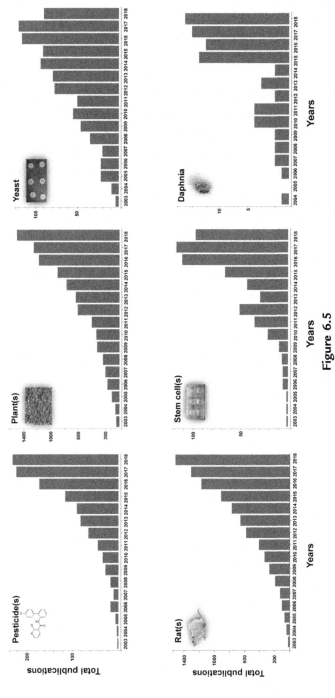

Figure 6.5

Diagrams displaying the total number of publications for the years 2003–18. Searches performed using the tools of the ISI Web of Knowledge[SM] acquiring for the terms "metabolomics" and the displayed biological systems.

The development of metabolomics models exhibiting high predictive capabilities and the discovery of validated biomarkers of effect or sets of biomarkers require the execution of large-scale experiments with a large number of biological replications. Additionally, pooling of samples is desirable in order to reduce the inner-experimental variation and the inclusion of quality control samples, composed of aliquots of the different biological replications, as well as quality control samples composed of aliquots from all samples of an experiment are highly recommended. When feasible, multiplatform analyses should be performed in order to achieve the widest possible coverage of the metabolite composition of the samples being analyzed. Combining gas chromatography/mass spectrometry (GC/MS) with liquid chromatography-mass spectrometry (LC/MS) and/or NMR platforms could provide wide metabolome coverage as well as confidence in metabolite identification (Fig. 6.4).

The identical experimental and analytical conditions for all treatments and samples are fundamental for the validity of a metabolomics analyses and the biological interpretation of results. The obtained raw data could be processed using a variety of online or locally installed bioanalytical software that enable their preprocessing (alignment, normalization) as well as their analyses (trend and biomarker discovery). Guidelines that have been proposed for the standardization of metabolomics data processing and reporting should be followed (Fiehn et al., 2007). The discovery of the MoA of a bioactive compound could be performed by modeling, using a.i. of known MoA for machine learning or by the discovery of biomarkers of effect that are closely linked to its target site (Aliferis and Jabaji, 2011). The latter requires the comprehensive monitoring of the global metabolism of a biological system. Overall, the use of advanced bioinformatics tools is required for the biological interpretation of metabolomics results (Shannon et al., 2003; Caspi et al., 2015).

2.2 Model biological systems used in the discovery and study of the MoA of PPPs and the assessment of their toxicity applying metabolomics

2.2.1 Plants

Plants represent the immediate targets (weeds) or the intermediates (crops) of PPPs, and therefore they should be used as models in studies on the MoA of bioactive compounds and their metabolism applying metabolomics. For the majority of PPPs, knowledge on their secondary MoA if nonexistent is largely fragmented. Therefore, of particular interest is the study of the effects of groups of PPPs such as fungicides and insecticides on plants' metabolism, in order to avoid undesirable effects on their growth or develop superior PPPs that could additionally trigger plants' defense mechanisms or promote their growth (Lydon and Duke, 1989; García et al., 2003). Fungicides have been reported not only to promote the biosynthesis of phytoalexins or other plant defense—related responses (Lydon and Duke, 1989; Ronchi et al., 1997; García et al., 2003) and growth (Amaro et al., 2018) but also to exert toxicity on the photosynthetic apparatus (Petit et al., 2012) or negatively affect plant growth (Mohamed and Akladious, 2017). On the other hand, although the metabolism of insecticidal a.i. in the plants has been studied (Rattan, 2010), their effects on their global metabolism is yet largely unknown, and to the best of our knowledge, no comprehensive metabolomics studies exist on the topic. Therefore, efforts targeting the discovery and study of the secondary MoA of PPPs applying metabolomics are highly expected to provide valuable insights that could be further exploited toward the optimization or the development of improved ones. Discussion on the metabolism of PPPs in plants is beyond the scope of this chapter.

Plant metabolomics is the firstly developed and one of the most advanced fields of metabolomics, with a great potential and a vast array of applications (Schauer and Fernie, 2006; Kim et al., 2011; Sumner et al., 2015; Hall, 2018; Maroli et al., 2018). Nonetheless, plants exhibit enormously complex metabolomes, composed of a vast number of metabolites with diverse physicochemical properties and unique structures (Last et al., 2007; Viant et al., 2017), which makes the coverage of their global metabolomes challenging. Therefore, the employment of more than one analyzers combined with bioinformatics analyses and use of metabolite databases is being required.

The majority of the commercially developed herbicides and plant growth regulators target functions of the plants that are vital for their survival and development, exhibiting 20 different MoA (Fig. 6.3A) (Aliferis and Jabaji, 2011; HRAC, 2019). The concept of using model plant species for the development of robust predictive metabolomics models in the discovery of the MoA of phytotoxic compounds was introduced in the beginning of the previous decade using maize (*Zea mays* L.) as the model organism (Araníbar et al., 2001). In this study, ^{1}H NMR spectroscopy metabolomics combined with artificial neural network analysis was applied for the rapid discrimination and classification of herbicidal compounds with various MoA based on the resulting metabolic plant profiles. Applying a similar approach, ^{1}H NMR metabolomics were applied in maize following treatments of the plants with herbicides exhibiting 19 MoA (Ott et al., 2003). Application of artificial neural network analyses revealed a strong discrimination among the recorded profiles, confirming the applicability of the applied protocol for the robust discovery of the MoA of phytotoxic compounds. In addition to the study of the MoA of synthetic herbicides, ^{1}H NMR metabolomics has been applied in the study of the MoA of natural phytotoxins using the weed species wild oat (*Avena sterilis* L.) as model (Aliferis and Chrysayi-Tokousbalides, 2006). Results of multivariate analyses revealed that the phytotoxins exhibited distinct MoA, differing from those of the herbicides being studied, making them interesting molecules for further experimentation and development.

Arabidopsis (*Arabidopsis thaliana* L.) is probably the most widely used plant species in various "omics" studies. Its sequenced genome (Kaul et al., 2000) and the existence of advanced online resources (Garcia-Hernandez et al., 2002; Mueller et al., 2003) are advantages for high-throughput metabolomics and system biology approaches. The potential of the plant as a model for the study of the MoA of herbicides was explored applying Fourier-transform ion cyclotron resonance—mass spectrometry (FT-ICR/MS) metabolomics (Oikawa et al., 2006). Treatments with the 10 selected a.i., exhibiting four different MoA, resulted in distinct metabolic profiles confirming the applicability of the approach in the high-throughput study of the MoA of phytotoxic compounds.

The aquatic microphyte *Lemna* (*Lemna* sp.) is another model plant species with high potential in the development of robust metabolomics models for the study of the MoA and toxicity mechanisms of bioactive compounds (Aliferis et al., 2009; Grossmann et al., 2012; Kostopoulou et al., 2020). It produces, fast and relatively inexpensively, uniform populations, requiring minimum laboratory space, which are major advantages for metabolomics modeling. It is a testing organism used in standardized bioassays for the toxicological risk assessment of xenobiotics in the environment (Brain et al., 2004; Michel et al., 2004; OECD, 2006; Aliferis et al., 2009; Mkandawire et al., 2014).

Lemna has early attracted the research interest as a model organism for the robust study of the MoA of bioactive compounds (Brain et al., 2004; Michel et al., 2004); however, such studies were focused on the assessment of the toxicities based on phenotypic traits. In the first report on the use of *Lemna minor* L. as the model for the study of the MoA and toxicity of phytotoxic compounds applying metabolomics, ^{1}H NMR metabolomics was applied in plant extract following exposure to selected phytotoxic compounds exhibiting various MoA (Aliferis et al., 2009). Results of multivariate analyses revealed discrete metabolic profiles following the various treatments, highlighting the potential of *Lemna* in the study of the MoA and toxicity of xenobiotics applying metabolomics. Following the same concept, in a large-scale study, *Lemna paucicostata* plants were exposed to a.i. of herbicides exhibiting different MoA (Grossmann et al., 2012). Combining information from the phenome and metabolome levels, the developed model could be applied for the discovery of bioactive compounds acting on plastoquinone, and auxin or very long—chain fatty acid biosyntheses. The above studies confirm the potential and applicability of *Lemna* metabolomics in the study and discovery of the MoA and toxicities of phytotoxic compounds in a timely and cost-effective fashion.

2.2.2 Microorganisms

Fungal soil-borne and foliar plant pathogens cause severe yield losses annually. The banning of several a.i. of fungicides, together with the development of fungal resistance to a wide range of applied fungicides, dictate the need for the acceleration of the process for the discovery and assessment of new sources of bioactivity.

The direct fungitoxic or fungistatic activity and MoA of bioactive compounds intended to be developed as fungicides can be investigated using the target species or other microorganisms as the models. Fungal metabolomics is a well-established field of metabolomics with a vast array of applications and great potential (Mashego et al., 2007; Barkal et al., 2016; Kalampokis et al., 2018). The existence of online metabolite repositories and bioinformatics tools such as those of the BioCyc (Caspi et al., 2015), combined with the complete sequencing of major species, could greatly support the development of robust metabolomics models for the study, prediction, and discovery of the MoA and toxicity mechanisms of bioactive compounds.

Yeast (*Saccharomyces cerevisiae*) represents an ideal biological system for metabolomics studies. It exhibits similar advantages for metabolomics modeling as the cell cultures (see below) that can be employed for the development of high-throughput metabolomics protocols aiming to the discovery of the MoA of bioactive compounds and the study of their toxicity. The developed yeast metabolite databases and corresponding bioinformatics tools (Cherry et al., 1998; Jewison et al., 2011; Ramirez-Gaona et al., 2016), as well as the well-established protocols for the analysis of its exo- and endo-metabolome (Canelas et al., 2009; Karamanou and Aliferis, 2020), facilitate the high-throughput metabolomics analyses and the in silico modeling of the MoA discovery and toxicity assessment. Additionally, the advances in the genetic engineering of yeast with the construction of numerous mutants available to the research community, and the technical advances in the workflows of yeast analyses protocols for metabolomics (Ewald et al., 2009), further strengthen the notion of yeast metabolomics employment for the purposes of PPPs' R&D.

The effects of yeast exposure to PPPs have been investigated in numerous studies at the genome level (toxicogenomics) (Dos Santos and Sá-Correia, 2015), related to the involvement of yeast transporters in multidrug resistance (Cabrito et al., 2011), and mechanisms of resistance to various a.i. of fungicides (Dias et al., 2010). Nonetheless, although numerous yeast metabolomics studies have been conducted to date on a wide range of research topics (Mülleder et al., 2016; Mendes et al., 2017), only in a handful of studies the effect PPPs on yeast has been investigated. The effect of bioactive compounds on yeast, such as sterol biosynthesis and respiratory chain inhibitors, have been studied applying direct-infusion MS metabolic fingerprinting and footprinting for their classification based on their MoA (Allen et al., 2004). The applied machine learning strategy was confirmed as an approach of high potential in the robust discovery of the MoA of bioactive compounds using yeast as the model organism.

The application of metabolomics on fungal plant pathogens for the study of PPPs and their MoA is largely unexploited. For the discovery of the metabolic basis of the resistance of carbendazim-resistant *Fusarium graminearum* isolates, ^{1}H NMR metabolomics modeling was applied, leading to the discovery of toxicity biomarkers and the grouping of isolates in distinct groups, separated from isolates exhibiting known mutations (Sevastos et al., 2018). Among others, a strong correlation was discovered between the carbendazim-resistance levels of the strains and their content in the amino acids of the aromatic and pyruvate families. In another study, using the model species *Aspergillus nidulans*, metabolomics has been employed to dissect the role of nucleobase transporters in its resistance to the fungicide boscalid. The effect of mutations in the metabolism of the fungus provided evidence of their involvement in the observed resistance and insights that could be further exploited for the development or improvement of plant protection agents (Kalampokis et al., 2018). Following a similar concept, performing LC/MS metabolomics, the pathogenic bacterium *Staphylococcus aureus* was used as a model organisms for the discovery of the MoA of the antibiotic berberine (Yi et al., 2007) or natural antibiotics (Yu et al., 2007).

The above studies highlight the potential of fungal metabolomics in the R&D of PPPs for the study of the MoA of bioactive compounds and the mining of the mechanisms of fungal resistance, which, however, is partially exploited.

2.2.3 Aquatic organisms

Since the aquatic environments are major endpoints of the applied PPPs in the agricultural practice, the use of aquatic organisms as models in the study of their toxicity ensures the development of those exhibiting improved toxicological profiles. Additionally, it provides an integrated pipeline for the assessment of the ecotoxicological risk of the applied PPPs (Fig. 6.4). Among the aquatic organisms, the fish species zebrafish (*Danio rerio*), the crustacean Daphnia (*Daphnia* sp.), and the microphyte *Lemna* sp. (discussed in Section 2.2.1) are the most studied organisms performing metabolomics for the assessment of the toxicity of bioactive compounds.

By applying LC/MS metabolomics, the effect of selected PPPs on the concentration of neurotransmitters and precursors in zebrafish has been recorded (Tufi et al., 2016a). Results confirmed the applicability of the approach for the assessment of the mechanisms of toxicity of PPPs in aquatic environments. In another study, the effects of the insecticide chlorpyrifos on the metabolite composition of muscles of the individuals being exposed were indicative of a muscular deterioration (Gómez-Canela et al., 2017). Similarly, the effects of various PPPs such as those of the fungicide difenoconazole (Teng et al., 2018), the fungicides iprodione, pyrimethanil, pyraclostrobin, and the insecticide acetamiprid, individually or in binary mixtures (Wang et al., 2018), have been studied on zebrafish. The applied a.i. affected functions such as the energy, amino acid, and lipid metabolism, cell apoptosis, and triggered oxidative stress responses.

In addition to zebrafish, the effect of PPPs applying metabolomics has been studied on species such as Chinook salmon (*Oncorhynchus tshawytscha*), goldfish (*Carassius auratus*), and the freshwater snail (*Lymnaea stagnalis*). Application of NMR and LC/UV (liquid chromatography with ultraviolet detection) metabolomics on eyed eggs and alevins of Chinook salmon (*O. tshawytscha*) following their exposure to the insecticides dinoseb, diazinon, and esfenvalerate revealed a MoA-dependent fluctuation of their metabolism (Viant et al., 2006). Similarly, ^{1}H NMR–based metabolomics has been applied for the dissection of lambda-cyhalothrin's toxicity to goldfish (*C. auratus*) (Li et al., 2014). Several toxicity biomarkers (e.g., leucine, isoleucine, valine, lactate, and taurine) were discovered, providing evidence of the applicability and potential of metabolomics in the risk assessment of PPPs in aquatic environments. The latter was further supported by results of a metabolomics study using the freshwater snail *L. stagnalis* as the model organism, which was exposed to individual a.i. or their mixtures (combined toxicity) (Tufi et al., 2016b). Although the study of combined toxicity is challenging, the work highlighted the potential of metabolomics in the assessment of the toxicity of mixtures of a.i. in aquatic environments.

The minute freshwater crustacean *Daphnia* sp. (water flea, Daphniidae) is an OECD test species (OECD, 2004, 2012) widely used as a model in the assessment of the toxicity of bioactive compounds in aquatic environments. Although its use has a great potential in the study of the MoA of bioactive compounds, such application is yet largely unexploited. For chronic toxicity reproduction, *Daphnia* is being used as a phenotypic endpoint, whereas for acute toxicity testing, the mortality of neonates is being assessed. Other toxicity parameters include its algae feeding rates (Grintzalis et al., 2017) and respiration as surrogate measurements of physiology. Two species the *Daphnia magna* and *Daphnia pulex* are the most commonly used ones, with the genome of the latter being 200 megabases containing at a minimum 30,907 genes (Colbourne et al., 2011).

Although *Daphnia* has been extensively used in studies related to pesticide toxicity (Olmstead and Leblanc, 2003; Barata et al., 2004; Jansen et al., 2011; Silva et al., 2018), there is only a handful of related studies applying metabolomics (Taylor et al., 2009, 2010, 2018; Poynton et al., 2011; Nagato et al., 2016; Toyota et al., 2016). Robust metabolomics protocols were introduced less than a decade ago (Taylor et al., 2009). In this study, the high-throughput classification of bioactive compounds based on their MoA was proposed by applying direct-infusion (DI) FT-ICR metabolomics on *D. magna*. In another research from the same group, DI FT-ICR/MS was applied for the study of the toxicity of the bioactive compounds cadmium, fenvalerate, dinitrophenol, and propranolol, which exhibit different MoA (Taylor et al., 2010). Results provided insights into the underlying mechanisms of toxicity and the potential of metabolomics in the study of the MoA of bioactive compounds using *Daphnia* as the model organism. In a recent study, differences have been reported between the metabolic responses of neonates and adults of *D. magna* following exposure to sublethal doses of bioactive compounds, including the herbicide atrazine (Wagner et al., 2017). In another recent study, the potential of metabolomics in the regulatory toxicity testing applying *Daphnia* metabolomics was highlighted when exposed to cadmium, 2,4-dinitrophenol, and propranolol (Taylor et al., 2018). The potential of the in silico modeling by the development of quantitative structure—activity relationship models for a.i. of PPPs using data sets of *D. magna* has been recently highlighted (Silva et al., 2018). Such approach could be further improved by using data sets and biomarkers from comprehensive metabolomics experiments on *Daphnia*.

All the above confirm the potential of *Daphnia* metabolomics in the research related to the modeling for the robust discovery of the MoA of bioactive compounds and toxicity mechanisms; however, the concept has not yet been implemented as a routine tool in the R&D of PPPs.

2.2.4 Earthworms

Earthworms represent another biological system with potential in the study of the MoA and toxicity of PPPs. Although the latter has been extensively studied for various pollutants applying metabolomics, there are few studies related to the MoA of PPPs using earthworms as the model biological system.

Application of GC/MS-based metabolomics on the earthworm *Metaphire posthuma* identified several metabolites biomarkers of toxicity, following exposure to the insecticide carbofuran, including various amino acids and carbohydrates (Mudiam et al., 2013). Using the same species, the toxicity of the synthetic pyrethroid cypermethrin has been studied applying metabolomics, with results being indicative of the disturbance of the neural system metabolism (Ch et al., 2015). In another study, the metabolism of *Lumbricus rubellus* was monitored following its exposure to the insecticides imidacloprid, thiacloprid, and chlorpyrifos, and also to nickel. The results of this study indicated distinct patterns of the metabolism regulation for the different MoA of the compounds being tested (Baylay et al., 2012).

The species *Eisenia fetida* has been used in the study of the effects of the herbicide trifluralin and the insecticide endosulfan, following treatments with sublethal concentrations performing 1D and 2D NMR metabolomics (Yuk et al., 2011). The same species has been employed as the model in the study of the toxicity of the insecticides dichlorodiphenyltrichloroethane (DDT) and endosulfan applying ^{1}H NMR and GC/MS metabolomics (Mckelvie et al., 2009). Analyses led to the discovery of biomarkers of the corresponding toxicities (e.g., alanine, glycine), further highlighting the potential of the approach in investigating the toxicity mechanisms of bioactive compounds. In a recent study, ^{1}H NMR and GC/MS metabolomics were applied on the coelomic fluid of *E. fetida* following exposure to chloroacetanilide herbicides (Griffith et al., 2018). Results of the study highlighted the applicability of using the celomic fluid instead of the whole organism for the study of the toxicity of herbicides applying metabolomics. Application of

^{1}H NMR metabolomics on the same species has led to the discovery of several amino acids, carbohydrates, and carboxylic acids as biomarkers of the toxicities of metalaxyl and metalaxyl-M, and to the conclusion that the two a.i. cause distinct enantiospecific disturbance of the metabolism of *E. fetida* (Zhang and Zhou, 2019).

Overall, although the use of earthworms in the study of the MoA and toxicity of PPPs applying metabolomics is in its infancy, results of the above studies provide evidence of its potential in the study of the effects of PPPs on the metabolism of earthworms and possibly the development of robust predictive models of their MoA.

2.2.5 Rats and mice

Rats (*Rattus* sp.) and mice (*Mus musculus*) are key model organisms for the toxicological risk assessment of bioactive compounds. The application of metabolomics following treatments of the animals with bioactive compounds could provide insights into their mechanisms of toxicity and MoA based on the discovery of biomarkers of effect and effects on the metabolism regulation. Both organisms can be used in the R&D of PPPs for the animal-to-human extrapolation of unintended effects. The existence of online databases and bioinformatics tools such as those of MouseCyc (Evsikov et al., 2009) provide great assistance to metabolomics analyses and the interpretation of results. In a recent study, based on this concept, rats were exposed to the insecticidal neurotoxic a.i. permethrin, deltamethrin, imidacloprid, carbaryl, and fipronil, which exhibit various MoA (Moser et al., 2015). Metabolomics of blood serum and plasma led to the discovery of distinct sets of biomarkers for the various toxicities, providing evidence that the applied metabolomics protocol could be employed in the robust assessment of the mechanisms of toxicity of neurotoxic compounds to rats. Application of ^{1}H NMR metabolomics of urine combined with GC coupled to flame ionization detector (GC/FID) analyses of the plasma, liver, and brain of rats following their exposure to a mixture of PPPs caused a general disturbance of their metabolism (Bonvallot et al., 2018). The biomarkers that were discovered (e.g., amino acids, carboxylic acids, fatty acids) participate in main biosynthetic pathways such as the Krebs cycle, the metabolism of amino acids, lipids, and carbohydrates, and energy equilibrium, which is indicative of the induced oxidative stress of the animals. Applying also ^{1}H NMR metabolomics in blood serum, liver, and testis of animals treated with the fungicides myclobutanil and triadimefon (triazoles), toxicity biomarkers (e.g., choline, phosphocholine, betaine, lactate, pyruvate) mainly involved in the cycle of methionine, creatine, and creatinine were discovered (Ekman et al., 2006). In another recent study, metabolomics was employed in order to gain insights into one of the most debatable modern issues for regulatory agencies and the agrochemical sector, the toxicity of glyphosate-based herbicides (Mesnage et al., 2017). Applying a systems biology approach combining results of metabolomics and proteomics on rat liver, several biomarkers were discovered (e.g., N-methyl proline, N-acetyl-beta-alanine, nicotinamide riboside), with the interpretation of results suggesting the activation of free radical scavenging systems, accompanied by liver dysfunction. The toxicity of a mixture of PPPs on rats was studied applying ^{1}H NMR metabolomics in rat's blood plasma leading to the discovery of metabolites such as glucose, lactate, choline, glycerophosphocholine, and phosphocholine as biomarkers of toxicity (Demur et al., 2013). Based on the interpretation of the results, authors concluded that the toxicity caused by mixtures of PPPs cannot always be predicted taking into account the cumulative effects of their individual components. ^{1}H NMR metabolomics has been also employed in the analyses of rat urine following injection of the insecticides chlorpyrifos, carbaryl, and their mixtures (Wang et al., 2011a). Interestingly, although no histopathological changes were detected in organs such as the liver or kidney, various biomarkers of toxicity were discovered (e.g., creatine, glycine, glutamine, succinate, alanine, lactate, glucose), including metabolites which are involved in the energy metabolism in liver mitochondria.

The results of the above studies confirm the applicability and potential of metabolomics in the study and discovery of toxicity mechanisms and MoA of bioactive compounds using rats as the model organism.

2.2.6 Cell cultures

Cell cultures (e.g., stem cell cultures) represent an excellent in vitro biological system that exhibits many advantages over the use of whole organisms in metabolomics studies regarding the MoA and toxicity of bioactive compounds. They have minimal space requirements, are easy to handle and maintain with relatively low cost, and they produce uniform populations. The above features make them ideal systems for the development of robust predictive metabolomics models; however, their potential in such applications is largely unexploited to date.

Following such approach, metabolomics was applied in the study of the exo-metabolome of human embryonic stem (hES) cell cultures following exposure to selected chemicals of the United States Environmental Protection Agency's (US EPA) ToxCast project (Kleinstreuer et al., 2011). Several biosynthetic pathways were affected by treatments and the developed model predicted developmental toxicity highly similar to that predicted by animal data. This study demonstrates the applicability and potential of metabolomics in the modeling and understanding of the developmental toxicity of bioactive compounds and as such can greatly complement and assist the R&D of PPPs in early stages. In another large-scale study, the metabolism of human liver cell line HepG2 was monitored following treatments with 35 bioactive compounds, including many a.i. of PPPs. The resulting exo- and endo-metabolomes were analyzed by LC/MS/MS and GC/MS and results confirmed the applicability of the protocol to reliably identify liver toxicity mechanisms, minimizing the need for using experimental animals (Ramirez et al., 2018). The HepG2 line has been also used in the monitoring of their endo-metabolome following exposure to the organochlorine insecticides endosulfan, lindane, DDT, and aldrin, applying GC/MS-TOF metabolomics (Zuluaga et al., 2016). Applying multivariate analyses, the corresponding metabolite-biomarkers were discovered, which are involved in the energy metabolism of the cell, which is indicative of the toxicity caused by the applied active ingredients. In addition, the effects of the pyrethroid insecticide permethrin and those of the organophosphate malathion on the metabolism of an immortalized rat neuronal cell line were studied employing GC/MS metabolomics (Hayton et al., 2017). Results of the study further confirmed the applicability and potential of metabolomics in the study of PPPs' toxicity using mammalian cell cultures as model biological system.

The above confirm the potential of metabolomics modeling using cell cultures as the model biological system in the discovery and study of the MoA and toxicity of bioactive compounds, complementing results acquired by other bioassays and helping at the same time toward minimizing the required number of experimental animals.

3. R&D of the next-generation PPPs: focusing on the potential and contribution of metabolomics in the research on bioelicitors and nano-PPPs

The ineffectiveness to introduce PPPs exhibiting new MoA, combined with the various challenges that the agrochemical sector is facing, dictates the need for the discovery of new sources of bioactivity and their assessment and development as new, improved plant protection agents. In this context, bioelicitors, including endophytic microorganisms and algal extracts, and nanomaterials have a great potential. Nonetheless, their bioactivities are yet largely unexplored and their MoA mostly unknown. These are disciplines that metabolomics could greatly assist toward the development of the improved next-generation PPPs.

3.1 Bioelicitors: endophytes and algal extracts

Bioelicitors represent one of the latest advances in the R&D of PPPs toward the development of alternative PPPs exhibiting new MoA, improved bioactivity, and toxicological profiles by modulating the plants' defense mechanism (Walters et al., 2013; Sharma et al., 2014). A variety of biotic and abiotic agents could serve as bioelicitors (Walters et al., 2013; Sharma et al., 2014). The endophytic microorganisms and algal extracts have recently attracted the research interest, and therefore, they will be further discussed below. Since bioelicitors mainly exert their bioactivity indirectly through their effects on the plants' metabolism, plant metabolomics is an ideal tool for the assessment of their effectiveness and the discovery of their MoA, which currently are largely unknown.

Being among the most advanced field of metabolomics, the existence of online repositories and bioinformatics tools for plant metabolomics (Schläpfer et al., 2017) could greatly accelerate the R&D related to bioelicitors.

3.1.1 Endophytes

An endophyte is any microorganism that lives into the plant tissues and establishes a symbiotic relationship without causing symptoms (Brader et al., 2017). They represent a great source of bioactivity (Strobel and Daisy, 2003; Aly et al., 2010; Brader et al., 2014; Nisa et al., 2015) and exhibit an outstanding potential for applications in plant protection. This is based on their abilities to induce plant growth and/or to make them more resistant to biotic and abiotic stresses (Gagne-Bourgue et al., 2013; Du Jardin, 2015; Brader et al., 2017). Furthermore, their potential in combating diseases for which no PPPs currently exists, such as vascular wilt diseases, represents another exciting perspective (Eljounaidi et al., 2016). Nonetheless, the exact mechanisms by which they exert such action are partially known.

The research on the study of the interactions of endophytes with their host plants is still in its infancy. To date, research has focused on the biosynthesis of secondary metabolites by endophytes (Gagne-Bourgue et al., 2013; Jalgaonwala et al., 2017), which could be further exploited for applications in medicine, the agri-food sector, and the industry (Kumar and Kaushik, 2012; Nisa et al., 2015; Jalgaonwala et al., 2017).

Early research on the exploitation of endophytes in plant protection, with main focus on bacterial species, has revealed their potential in controlling plant diseases (Benhamou et al., 2000; Hsieh et al., 2005; Mercado-Blanco and Bakker, 2007; Wang et al., 2009). *Bacillus* sp. is one of the mostly studied endophytic species (Gagne-Bourgue et al., 2013; Gond et al., 2015) with several of its strains already commercially developed. *Bacillus* sp. isolates have been reported to be directly mycotoxic and/or act as plant growth—promoting endophytes (Wang et al., 2009; Gagne-Bourgue et al., 2013). Species such as the *Serratia plymuthica* (strain R1GC4) has been reported to stimulate the defense mechanism of cucumber (*Cucumis sativus*) seedlings resulting in an improved defense against *Pythium ultimum* infections (Benhamou et al., 2000). Furthermore, the species *Pantoea agglomerans* has been reported as an effective agent against the bacterial wilt of bean (Hsieh et al., 2005). On the other hand, research on fungal endophytes has focused on the use of non-pathogenic *Fusarium oxysporum* (Vu et al., 2006; Wang et al., 2011b) strains and *Trichoderma* sp. (Bailey et al., 2006; Samuels, 2006; Druzhinina et al., 2011). However, the exact mechanisms by which these endophytes exerted their bioactivity are largely unknown. Due to the complexity of plant metabolism and the biosynthesis of a vast number of diverse metabolites by the endophytes, the employment of advanced analyzers and bioanalytical protocols is a prerequisite to decode their interactions.

Metabolomics analyses have uncovered the changes in the metabolism of *Cirsium arvensis* in response to the endophyte *Chaetomium cochlioides*, with many novel oxylipins and a galactolipid–jasmonate conjugate being detected as biomarkers (Hartley et al., 2015). Other studies have focused on the involvement of lipopeptides, surfactins, and other growth-promoting metabolites in the endophyte–plant interactions. The presence of lipopeptides and other growth-promoting metabolites has been reported in switchgrass (*Panicum virgatum* L.) for various endophytes (Gagne-Bourgue et al., 2013), which is a plausible hypothesis for other plant–endophyte systems (Gond et al., 2015). In addition, the biosynthesis of gibberellins has been reported as another major mechanism by which endophytes regulate plants' metabolism (Shahzad et al., 2016). Endophytes alter significantly the metabolism of the plant (Gagné-Bourque et al., 2016), and results of such studies could further support the development of an endophyte as PPP.

Additionally, metabolomics approaches could greatly accelerate the discovery of natural products via a chemotaxonomy-based approach (Maciá-Vicente et al., 2018) or the in-depth analysis of endophytes aiming to the discovery of novel bioactive secondary metabolites (Tawfike et al., 2017). For example, the endophytic fungus *Eupenicillium parvum*, isolated from the stem tissues of *Azadirachta indica*, is biosynthesizing in vitro the insecticidal molecules azadirachtin and 3-tigloyl-azadirachtol. This example highlights the tremendous potential of endophytes of medicinal plants (Kaul et al., 2012; Alvin et al., 2014; Scott et al., 2018) in the discovery of new bioactive compounds.

Nonetheless, further research is required to adequately understand the complexity of plant–endophyte interactions and the underlying mechanisms, in order to further develop and optimize endophytic species as plant protection agents.

3.1.2 Algal extracts

In addition to endophytes, marine algae represent another source of bioactivity with high potential in plant protection, with polysaccharides such as glucans (laminarin), sulfated fucans, carrageenans, and ulvans to attract the research interest (Jaulneau et al., 2010; Trouvelot et al., 2014; De Borba et al., 2019). Fucans and carrageenans are sulfated polysaccharides produced by the brown and red algae, respectively (Mercier et al., 2001; Klarzynski et al., 2003). Sulfated oligosaccharides obtained from fucan and carrageenans have shown to induce the salicylate signaling pathway (Klarzynski et al., 2003; Mercier et al., 2001). Nonetheless, the MoA of algal extracts is partially known, which represents a major bottleneck toward their further development and optimization.

The sulfated heteropolysaccharide, ulvan, has been shown to reduce the disease severity of *Fusarium* wilt (*F. oxysporum* f. sp. *phaseoli*) in beans, when applied with foliar spraying (De Borba et al., 2019). Application of ulvan has shown to affect plant hormones including signaling compounds (Jaulneau et al., 2010), with results being indicative of its potential as a plant protection agent. Although more solid evidence is required, the metabolites ulvan, carrageenan, alginate, and laminarin seem to have also a potential in the plant protection against verticillium wilt of olive (*Verticillium dahliae*) (Ben Salah et al., 2018). Similarly, the sulfated laminarin (PS3) provides improved protection of grapevine (*Vitis vinifera* cv. *Marselan*) against downy mildew (*Plasmopara viticola*), part of such action is attributed to the stimulation of both callose biosynthesis and the jasmonic acid pathway (Trouvelot et al., 2008). In another study using tobacco and Arabidopsis cell cultures as the model systems, it was concluded that the chain length of laminarin and the presence of sulfur groups in glucans are highly correlated with their bioactivity (Ménard et al., 2004).

The above confirm the tremendous potential of algal extracts for applications in plant protection; however, more solid evidence on their MoA is required and metabolomics studies are largely foreseen to assist toward this direction.

3.2 Nanomaterials and nano-plant protection products

Taking into account that the discovery of a.i. with new MoA is limited (Duke, 2012), the development of new forms of PPPs (a.i. and/or formulations) could provide solutions to issues that the plant protection sector is facing, and additionally it could give an add-on value to old a.i. Nanotechnology is among the most significant developments toward this direction (Arruda et al., 2015; Kah, 2015), being considered one of the major advances in the science and technology sector of the century. It can provide solutions through its application in the formulation of nano-PPPs or the use of nanoparticles per se as a.i. (Arruda et al., 2015). The formulation of nanomaterials with a nanocarrier could provide significant advantages to the resulting formulated products such as controlled release of the a.i., improved selectivity, efficacy, and toxicity. Focusing on the efficacy, a recent literature survey indicates that the efficacy of nano-PPPs is approximately 20%−30% greater than that of the corresponding conventional product (Kah et al., 2018). The nanoformulation provides a significantly improved release rate, loading capacity, and efficacy, while the individual nanomaterials exhibit substantially improved bioactivity against bacteria and fungi (Chhipa, 2017).

Within this context, the development and use of nano-PPPs is highly foreseen that will play an important role in plant protection in the near future (Kah, 2015) toward the securing of food quality and supply to the exponentially increasing human population. In a recent study, the application of the $Cu(OH)_2$ nanopesticide did not have a negative impact on the yield and microbial activity under conventional agriculture regime; however, applying low-input agricultural practice, its application affected microbial-mediated soil processes (Simonin et al., 2018).

In addition to the potential risks that nano-PPPs pose to the environment and nontarget organisms, their effects on the metabolism of the plants are another important factor, information on which is largely unknown. Nanoformulations and nanomaterials exhibiting low phytotoxicity, high bioactivity against the target organisms, and potentially promoting the growth and the defense mechanisms of the plants will be ideal candidates to be developed as commercial nano-PPPs. Most of the studies on the effects of nanomaterials on plants have focused on the $Cu(OH)_2$ nano-PPPs. The application of GC/TOF/MS metabolomics on lettuce following foliar treatments with nano $Cu(OH)_2$ revealed a disturbance of the plants' metabolism. Among others, nicotianamine, which is a copper chelator, was discovered as a major biomarker (Zhao et al., 2016). In another research, using again nano $Cu(OH)_2$, its effects on the metabolism of Corn (*Z. mays*) and cucumber (*C. sativus*) were investigated applying metabolomics (Zhao et al., 2017). Results revealed distinct responses by the two species in response to the nano $Cu(OH)_2$ toxicity, with the energy metabolism, the shikimate-phenylpropanoid and arginine and proline biosynthetic pathways being the most affected ones.

Based on the above, it is evident that nano-PPPs have a great potential for applications in plant protection. However, knowledge on their MoA and mechanisms of toxicity, their unintended effects on nontarget organisms and fate in the environment, is still largely fragmented. The latter represents a significant obstacle toward their further development and application in the agricultural practice. Within this context, metabolomics is highly expected to provide valuable insights into the above topics in the near future further supporting their R&D. Furthermore, the nanoformulation of a.i. is anticipated to have a major impact on their environmental fate and toxicological profiles; thus, there is an urge for the adaptation of the existing exposure assessment protocols and regulatory framework (Kah et al., 2013; Kookana et al., 2014; Kah, 2015).

4. Summary and future trends

Based on its unique capabilities and recent advances, metabolomics could serve as a powerful routine bioanalytical tool for the discovery of the MoA of bioactive compounds and the assessment of their toxicity to nontarget organisms. Representing a bottleneck in the PPPs' R&D, the acceleration of the discovery of the MoA of bioactive compounds could greatly assist the process and reduce the corresponding costs. Additionally, the in-depth study of their toxicity to nontarget organisms may serve as a reliable screening tool for the early identification of candidate molecules that should be excluded from further research. The discovery of robust, reliable biomarkers or sets of biomarkers of effect through application of cross-platform analyses is highly anticipated to assist toward this direction. Nonetheless, the integration of metabolomics with other large-scale "omics"-derived data sets within a systems biology approach could provide a holistic overview and solid information on the MoA of a given bioactive compound. Furthermore, the application of metabolomics using appropriate model biological systems could provide valuable insights into the MoA and toxicity mechanisms of the next-generation PPPs, like nano-PPPs and bioelicitors, as well as assist regulatory agencies in policy making regarding the toxicity of mixtures of a.i. Nonetheless, although a significant progress has been made during the last decade in the employment of metabolomics in the PPPs' R&D, its potential in the era of the next-generation PPPs, although tremendous, is still largely unexploited.

References

Ahuja, I., De Vos, R.C., Bones, A.M., Hall, R.D., 2010. Plant molecular stress responses face climate change. Trends Plant Sci. 15, 664–674.

Aldridge, B.B., Rhee, K.Y., 2014. Microbial metabolomics: innovation, application, insight. Curr. Opin. Microbiol. 19, 90–96.

Aliferis, K.A., Chrysayi-Tokousbalides, M., 2006. Metabonomic strategy for the investigation of the mode of action of the phytotoxin (5S, 8R, 13S, 16R)-(-)-pyrenophorol using ^{1}H nuclear magnetic resonance fingerprinting. J. Agric. Food Chem. 54, 1687–1692.

Aliferis, K.A., Chrysayi-Tokousbalides, M., 2011. Metabolomics in pesticide research and development: review and future perspectives. Metabolomics 7, 35–53.

Aliferis, K.A., Faubert, D., Jabaji, S., 2014. A metabolic profiling strategy for the dissection of plant defense against fungal pathogens. PLoS One 9, e111930.

Aliferis, K.A., Jabaji, S., 2011. Metabolomics-a robust bioanalytical approach for the discovery of the modes-of-action of pesticides: a review. Pestic. Biochem. Physiol. 100, 105–117.

Aliferis, K.A., Materzok, S., Paziotou, G.N., Chrysayi-Tokousbalides, M., 2009. *Lemna minor* L. as a model organism for ecotoxicological studies performing ^{1}H NMR fingerprinting. Chemosphere 76, 967–973.

Allen, J., Davey, H.M., Broadhurst, D., Rowland, J.J., Oliver, S.G., et al., 2004. Discrimination of modes of action of antifungal substances by use of metabolic footprinting. Appl. Environ. Microbiol. 70, 6157–6165.

Alvin, A., Miller, K.I., Neilan, B.A., 2014. Exploring the potential of endophytes from medicinal plants as sources of antimycobacterial compounds. Microbiol. Res. 169, 483–495.

Aly, A.H., Debbab, A., Kjer, J., Proksch, P., 2010. Fungal endophytes from higher plants: a prolific source of phytochemicals and other bioactive natural products. Fungal Divers. 41, 1–16.

Amaro, A.C.E., Ramos, A.R.P., Macedo, A.C., Ono, E.O., Rodrigues, J.D., 2018. Effects of the fungicides azoxystrobin, pyraclostrobin and boscalid on the physiology of Japanese cucumber. Sci. Hortic. 228, 66–75.

Antignac, J.-P., Courant, F., Pinel, G., Bichon, E., Monteau, F., et al., 2011. Mass spectrometry-based metabolomics applied to the chemical safety of food. Trends Anal. Chem. 30, 292–301.

Araníbar, N., Singh, B.K., Stockton, G.W., Ott, K.-H., 2001. Automated mode-of-action detection by metabolic profiling. Biochem. Biophys. Res. Commun. 286, 150–155.

Arruda, S.C., Silva, A.L., Galazzi, R.M., Azevedo, R.A., Arruda, M.A., 2015. Nanoparticles applied to plant science: a review. Talanta 131, 693–705.

Bailey, B., Bae, H., Strem, M., Roberts, D., Thomas, S., et al., 2006. Fungal and plant gene expression during the colonization of cacao seedlings by endophytic isolates of four *Trichoderma* species. Planta 224, 1449–1464.

Barata, C., Solayan, A., Porte, C., 2004. Role of B-esterases in assessing toxicity of organophosphorus (chlorpyrifos, malathion) and carbamate (carbofuran) pesticides to *Daphnia magna*. Aquat. Toxicol. 66, 125–139.

Barkal, L.J., Theberge, A.B., Guo, C.-J., Spraker, J., Rappert, L., et al., 2016. Microbial metabolomics in open microscale platforms. Nat. Commun. 7, 10610.

Baylay, A., Spurgeon, D., Svendsen, C., Griffin, J., Swain, S.C., et al., 2012. A metabolomics based test of independent action and concentration addition using the earthworm *Lumbricus rubellus*. Ecotoxicology 21, 1436–1447.

Ben Salah, I., Aghrouss, S., Douira, A., Aissam, S., El Alaoui-Talibi, Z., et al., 2018. Seaweed polysaccharides as bio-elicitors of natural defenses in olive trees against verticillium wilt of olive. J. Plant Interact. 13, 248–255.

Benhamou, N., Gagné, S., Le Quéré, D., Dehbi, L., 2000. Bacterial-mediated induced resistance in cucumber: beneficial effect of the endophytic bacterium *Serratia plymuthica* on the protection against infection by *Pythium ultimum*. Phytopathology 90, 45–56.

Bonvallot, N., Canlet, C., Blas-Y-Estrada, F., Gautier, R., Tremblay-Franco, M., et al., 2018. Metabolome disruption of pregnant rats and their offspring resulting from repeated exposure to a pesticide mixture representative of environmental contamination in Brittany. PLoS One 13, e0198448.

Bowne, J., Bacic, A., Tester, M., Roessner, U., 2018. Abiotic stress and metabolomics. Annu. Rev. Plant Sci. Online 61–85.

Brader, G., Compant, S., Mitter, B., Trognitz, F., Sessitsch, A., 2014. Metabolic potential of endophytic bacteria. Curr. Opin. Biotechnol. 27, 30–37.

Brader, G., Compant, S., Vescio, K., Mitter, B., Trognitz, F., et al., 2017. Ecology and genomic insights into plant-pathogenic and plant-nonpathogenic endophytes. Annu. Rev. Phytopathol. 55, 61–83.

Brain, R.A., Johnson, D.J., Richards, S.M., Sanderson, H., Sibley, P.K., et al., 2004. Effects of 25 pharmaceutical compounds to *Lemna gibba* using a seven-day static-renewal test. Environ. Toxicol. Chem. 23, 371–382.

Bundy, J.G., Davey, M.P., Viant, M.R., 2009. Environmental metabolomics: a critical review and future perspectives. Metabolomics 5, 3.

Cabrito, T.R., Teixeira, M.C., Singh, A., Prasad, R., Sá-Correia, I., 2011. The yeast ABC transporter Pdr18 (ORF YNR070w) controls plasma membrane sterol composition, playing a role in multidrug resistance. Biochem. J. 440, 195–202.

Canelas, A.B., Ten Pierick, A., Ras, C., Seifar, R.M., Van Dam, J.C., et al., 2009. Quantitative evaluation of intracellular metabolite extraction techniques for yeast metabolomics. Anal. Chem. 81, 7379–7389.

Cantrell, C.L., Dayan, F.E., Duke, S.O., 2012. Natural products as sources for new pesticides. J. Nat. Prod. 75, 1231–1242.

Caspi, R., Billington, R., Ferrer, L., Foerster, H., Fulcher, C.A., et al., 2015. The MetaCyc database of metabolic pathways and enzymes and the BioCyc collection of pathway/genome databases. Nucleic Acids Res. 44, D471–D480.

Castro-Puyana, M., Herrero, M., 2013. Metabolomics approaches based on mass spectrometry for food safety, quality and traceability. Trends Anal. Chem. 52, 74–87.

Cevallos-Cevallos, J.M., Reyes-De-Corcuera, J.I., Etxeberria, E., Danyluk, M.D., Rodrick, G.E., 2009. Metabolomic analysis in food science: a review. Trends Food Sci. Technol. 20, 557–566.

Ch, R., Singh, A.K., Pandey, P., Saxena, P.N., Mudiam, M.K.R., 2015. Identifying the metabolic perturbations in earthworm induced by cypermethrin using gas chromatography-mass spectrometry based metabolomics. Sci. Rep. 5, 15674.

Cherry, J.M., Adler, C., Ball, C., Chervitz, S.A., Dwight, S.S., et al., 1998. SGD: *Saccharomyces* genome database. Nucleic Acids Res. 26, 73–79.

Chhipa, H., 2017. Nanofertilizers and nanopesticides for agriculture. Environ. Chem. Lett. 15, 15–22.

Colbourne, J.K., Pfrender, M.E., Gilbert, D., Thomas, W.K., Tucker, A., et al., 2011. The ecoresponsive genome of *Daphnia pulex*. Science 331, 555–561.

Dayan, F.E., Cantrell, C.L., Duke, S.O., 2009. Natural products in crop protection. Bioorg. Med. Chem. 17, 4022–4034.

De Borba, M.C., De Freitas, M.B., Stadnik, M.J., 2019. Ulvan enhances seedling emergence and reduces Fusarium wilt severity in common bean (*Phaseolus vulgaris* L.). Crop Protect. 118, 66–71.

Delorenzo, M.E., Scott, G.I., Ross, P.E., 2001. Toxicity of pesticides to aquatic microorganisms: a review. Environ. Toxicol. Chem. 20, 84–98.

Delp, C., Dekker, J., 1985. Fungicide resistance: definitions and use of terms. EPPO Bull. 15, 333–335.

Délye, C., Jasieniuk, M., Le Corre, V., 2013. Deciphering the evolution of herbicide resistance in weeds. Trends Genet. 29, 649–658.

Demur, C., Métais, B., Canlet, C., Tremblay-Franco, M., Gautier, R., et al., 2013. Dietary exposure to a low dose of pesticides alone or as a mixture: the biological metabolic fingerprint and impact on hematopoiesis. Toxicology 308, 74–87.

Dias, P.J., Teixeira, M.C., Telo, J.P., Sá-Correia, I., 2010. Insights into the mechanisms of toxicity and tolerance to the agricultural fungicide mancozeb in yeast, as suggested by a chemogenomic approach. OMICS 14, 211–227.

Dos Santos, S.C., Sá-Correia, I., 2015. Yeast toxicogenomics: lessons from a eukaryotic cell model and cell factory. Curr. Opin. Biotechnol. 33, 183–191.

Druzhinina, I.S., Seidl-Seiboth, V., Herrera-Estrella, A., Horwitz, B.A., Kenerley, C.M., et al., 2011. Trichoderma: the genomics of opportunistic success. Nat. Rev. Microbiol. 9, 749.

Du Jardin, P., 2015. Plant biostimulants: definition, concept, main categories and regulation. Sci. Hortic. 196, 3–14.

Duke, S.O., 2012. Why have no new herbicide modes of action appeared in recent years? Pest Manag. Sci. 68, 505–512.

Dunn, W.B., Ellis, D.I., 2005. Metabolomics: current analytical platforms and methodologies. Trends Anal. Chem. 24, 285–294.

Dunn, W.B., Wilson, I.D., Nicholls, A.W., Broadhurst, D., 2012. The importance of experimental design and QC samples in large-scale and MS-driven untargeted metabolomic studies of humans. Bioanalysis 4, 2249–2264.

EFSA, European Food Safety Agency, 2017 (Accessed December 2019). https://www.efsa.europa.eu/en/applications/pesticides.

Ekman, D.R., Keun, H.C., Eads, C.D., Furnish, C.M., Murrell, R.N., et al., 2006. Metabolomic evaluation of rat liver and testis to characterize the toxicity of triazole fungicides. Metabolomics 2, 63–73.

Eljounaidi, K., Lee, S.K., Bae, H., 2016. Bacterial endophytes as potential biocontrol agents of vascular wilt diseases—review and future prospects. Biol. Contr. 103, 62–68.

Evenson, R.E., Gollin, D., 2003. Assessing the impact of the green revolution, 1960 to 2000. Science 300, 758–762.

Evsikov, A.V., Dolan, M.E., Genrich, M.P., Patek, E., Bult, C.J., 2009. MouseCyc: a curated biochemical pathways database for the laboratory mouse. Genome Biol. 10, R84.

Ewald, J.C., Heux, S., Zamboni, N., 2009. High-throughput quantitative metabolomics: workflow for cultivation, quenching, and analysis of yeast in a multiwell format. Anal. Chem. 81, 3623–3629.

Fiehn, O., Kopka, J., Dörmann, P., Altmann, T., Trethewey, R.N., et al., 2000. Metabolite profiling for plant functional genomics. Nat. Biotechnol. 18, 1157.

Fiehn, O., Robertson, D., Griffin, J., Van Der Werf, M., Nikolau, B., et al., 2007. The metabolomics standards initiative (MSI). Metabolomics 3, 175–178.

Fisher, M.C., Hawkins, N.J., Sanglard, D., Gurr, S.J., 2018. Worldwide emergence of resistance to antifungal drugs challenges human health and food security. Science 360, 739–742.

Fisher, M.C., Henk, D.A., Briggs, C.J., Brownstein, J.S., Madoff, L.C., et al., 2012. Emerging fungal threats to animal, plant and ecosystem health. Nature 484, 186.

FRAC, Fungicide Resistance Action Committee, 2019 (Accessed December 2019). http://www.frac.info/home.

Gagné-Bourque, F., Bertrand, A., Claessens, A., Aliferis, K.A., Jabaji, S., 2016. Alleviation of drought stress and metabolic changes in timothy (*Phleum pratense* L.) colonized with *Bacillus subtilis* B26. Front. Plant Sci. 7, 584.

Gagne-Bourgue, F., Aliferis, K., Seguin, P., Rani, M., Samson, R., et al., 2013. Isolation and characterization of indigenous endophytic bacteria associated with leaves of switchgrass (*Panicum virgatum* L.) cultivars. J. Appl. Microbiol. 114, 836−853.

Garcia-Hernandez, M., Berardini, T., Chen, G., Crist, D., Doyle, A., et al., 2002. TAIR: a resource for integrated *Arabidopsis* data. Funct. Integr. Genom. 2, 239−253.

García-Villalba, R., León, C., Dinelli, G., Segura-Carretero, A., Fernández-Gutiérrez, A., et al., 2008. Comparative metabolomic study of transgenic versus conventional soybean using capillary electrophoresis−time-of-flight mass spectrometry. J. Chromatogr. A 1195, 164−173.

García, P.C., Rivero, R.M., Ruiz, J.M., Romero, L., 2003. The role of fungicides in the physiology of higher plants: implications for defense responses. Bot. Rev. 69, 162.

Gerwick, B.C., Sparks, T.C., 2014. Natural products for pest control: an analysis of their role, value and future. Pest Manag. Sci. 70, 1169−1185.

Gibbons, H., O'gorman, A., Brennan, L., 2015. Metabolomics as a tool in nutritional research. Curr. Opin. Lipidol. 26, 30–34.

Gómez-Canela, C., Prats, E., Piña, B., Tauler, R., 2017. Assessment of chlorpyrifos toxic effects in zebrafish (*Danio rerio*) metabolism. Environ. Pollut. 220, 1231−1243.

Gond, S.K., Bergen, M.S., Torres, M.S., White Jr., J.F., 2015. Endophytic *Bacillus* spp. produce antifungal lipopeptides and induce host defence gene expression in maize. Microbiol. Res. 172, 79−87.

Grandjean, P., Landrigan, P.J., 2014. Neurobehavioural effects of developmental toxicity. Lancet Neurol. 13, 330−338.

Griffin, J.L., Wang, X., Stanley, E., 2015. Does our gut microbiome predict cardiovascular risk? A review of the evidence from metabolomics. Circulation 8, 187−191.

Griffith, C.M., Morgan, M.A., Dinges, M.M., Mathon, C., Larive, C.K., 2018. Metabolic profiling of chloroacetanilide herbicides in earthworm coelomic fluid using ^{1}H NMR and GC-MS. J. Proteome Res. 17, 2611−2622.

Grintzalis, K., Dai, W., Panagiotidis, K., Belavgeni, A., Viant, M.R., 2017. Miniaturising acute toxicity and feeding rate measurements in *Daphnia magna*. Ecotoxicol. Environ. Saf. 139, 352−357.

Grossmann, K., Christiansen, N., Looser, R., Tresch, S., Hutzler, J., et al., 2012. Physionomics and metabolomics-two key approaches in herbicidal mode of action discovery. Pest Manag. Sci. 68, 494−504.

Guo, Q., Sidhu, J.K., Ebbels, T.M.D., Rana, F., Spurgeon, D.J., et al., 2009. Validation of metabolomics for toxic mechanism of action screening with the earthworm *Lumbricus rubellus*. Metabolomics 5, 72−83.

Hall, R.D., 2018. Plant metabolomics in a nutshell: potential and future challenges. Annu. Rev. Plant Sci. Online 1−24.

Hartley, S.E., Eschen, R., Horwood, J.M., Gange, A.C., Hill, E.M., 2015. Infection by a foliar endophyte elicits novel arabidopside-based plant defence reactions in its host, *Cirsium arvense*. New Phytol. 205, 816−827.

Haug, K., Salek, R.M., Conesa, P., Hastings, J., De Matos, P., et al., 2012. MetaboLights-an open-access general-purpose repository for metabolomics studies and associated meta-data. Nucleic Acids Res. 41, D781−D786.

Hayton, S., Maker, G.L., Mullaney, I., Trengove, R.D., 2017. Untargeted metabolomics of neuronal cell culture: a model system for the toxicity testing of insecticide chemical exposure. J. Appl. Toxicol. 37, 1481−1492.

Hendriks, M.M., Van Eeuwijk, F.A., Jellema, R.H., Westerhuis, J.A., Reijmers, T.H., et al., 2011. Data-processing strategies for metabolomics studies. Trends Anal. Chem. 30, 1685−1698.

Herrero, M., Simó, C., García-Cañas, V., Ibáñez, E., Cifuentes, A., 2012. Foodomics: MS-based strategies in modern food science and nutrition. Mass Spectrom. Rev. 31, 49−69.

Hounoum, B.M., Blasco, H., Emond, P., Mavel, S., 2016. Liquid chromatography−high-resolution mass spectrometry-based cell metabolomics: experimental design, recommendations, and applications. Trends Anal. Chem. 75, 118−128.

HRAC, Herbicide Resistance Action Committee, 2019 (Accessed December 2019). http://www.hracglobal.com/.

Hsieh, T., Huang, H., Erickson, R., 2005. Biological control of bacterial wilt of bean using a bacterial endophyte, Pantoea agglomerans. J. Phytopathol. 153, 608−614.

IRAC, Insecticide Resistance Action Committee, 2019 (Accessed 2019). http://www.irac-online.org.

Jalgaonwala, R.E., Mohite, B.V., Mahajan, R.T., 2017. A review: natural products from plant associated endophytic fungi. J. Microbiol. Biotechnol. Res. 1, 21−32.

Jansen, M., Coors, A., Stoks, R., De Meester, L., 2011. Evolutionary ecotoxicology of pesticide resistance: a case study in *Daphnia*. Ecotoxicology 20, 543−551.

Jaulneau, V., Lafitte, C., Jacquet, C., Fournier, S., Salamagne, S., et al., 2010. Ulvan, a sulfated polysaccharide from green algae, activates plant immunity through the jasmonic acid signaling pathway. BioMed Res. Int. 2010, 525291.

Jewison, T., Knox, C., Neveu, V., Djoumbou, Y., Guo, A.C., et al., 2011. YMDB: the yeast metabolome database. Nucleic Acids Res. 40, D815−D820.

Jones, D.P., Park, Y., Ziegler, T.R., 2012. Nutritional metabolomics: progress in addressing complexity in diet and health. Annu. Rev. Nutr. 32, 183−202.

Kah, M., 2015. Nanopesticides and nanofertilizers: emerging contaminants or opportunities for risk mitigation? Front. Chem. 3, 64.

Kah, M., Beulke, S., Tiede, K., Hofmann, T., 2013. Nanopesticides: state of knowledge, environmental fate, and exposure modeling. Crit. Rev. Environ. Sci. Technol. 43, 1823−1867.

Kah, M., Kookana, R.S., Gogos, A., Bucheli, T.D., 2018. A critical evaluation of nanopesticides and nanofertilizers against their conventional analogues. Nat. Nanotechnol. 13, 677.

Kalampokis, I.F., Kapetanakis, G.C., Aliferis, K.A., Diallinas, G., 2018. Multiple nucleobase transporters contribute to boscalid sensitivity in *Aspergillus nidulans*. Fungal Genet. Biol. 115, 52−63.

Karamanou, D.A., Aliferis, K.A., 2020. The yeast (*Saccharomyces cerevisiae*) YCF1 vacuole transporter: evidence on its implication into the yeast resistance to flusilazole as revealed by GC/EI/MS metabolomics. Pestic. Biochem. Physiol. https://doi.org/10.1016/j.pestbp.2019.09.013 (in press).

Katajamaa, M., Orešič, M., 2007. Data processing for mass spectrometry-based metabolomics. J. Chromatogr. A 1158, 318−328.

Kaul, S., Gupta, S., Ahmed, M., Dhar, M.K., 2012. Endophytic fungi from medicinal plants: a treasure hunt for bioactive metabolites. Phytochemistry Rev. 11, 487−505.

Kaul, S., Koo, H.L., Jenkins, J., Rizzo, M., Rooney, T., et al., 2000. Analysis of the genome sequence of the flowering plant *Arabidopsis thaliana*. Nature 408, 796.

Kim, H.K., Choi, Y.H., Verpoorte, R., 2010. NMR-based metabolomic analysis of plants. Nat. Protoc. 5, 536.

Kim, H.K., Choi, Y.H., Verpoorte, R., 2011. NMR-based plant metabolomics: where do we stand, where do we go? Trends Biotechnol. 29, 267−275.

Kim, H.K., Verpoorte, R., 2010. Sample preparation for plant metabolomics. Phytochem. Anal. 21, 4−13.

Kind, T., Wohlgemuth, G., Lee, D.Y., Lu, Y., Palazoglu, M., et al., 2009. FiehnLib: mass spectral and retention index libraries for metabolomics based on quadrupole and time-of-flight gas chromatography/mass spectrometry. Anal. Chem. 81, 10038−10048.

Klarzynski, O., Descamps, V., Plesse, B., Yvin, J.-C., Kloareg, B., et al., 2003. Sulfated fucan oligosaccharides elicit defense responses in tobacco and local and systemic resistance against tobacco mosaic virus. Mol. Plant Microbe Interact. 16, 115−122.

Kleinstreuer, N., Smith, A., West, P., Conard, K., Fontaine, B., et al., 2011. Identifying developmental toxicity pathways for a subset of ToxCast chemicals using human embryonic stem cells and metabolomics. Toxicol. Appl. Pharmacol. 257, 111−121.

Klosterman, S., Atallah, Z., Vallad, G., Subbarao, K., 2009. Diversity, pathogenicity, and management of Verticillium species. Annu. Rev. Phytopathol. 47, 39−62.

Kookana, R.S., Boxall, A.B., Reeves, P.T., Ashauer, R., Beulke, S., et al., 2014. Nanopesticides: guiding principles for regulatory evaluation of environmental risks. J. Agric. Food Chem. 62, 4227−4240.

Kopka, J., Schauer, N., Krueger, S., Birkemeyer, C., Usadel, B., et al., 2004. GMD@ CSB. DB: the Golm metabolome database. Bioinformatics 21, 1635−1638.

Kostopoulou, S., Ntatsi, G., Arapis, G., Aliferis, K.A., 2020. Assessment of the effects of metribuzin, glyphosate, and their mixtures on the metabolism of the model plant *Lemna minor* L. applying metabolomics. Chemosphere 239, 124582.

Kumar, S., Kaushik, N., 2012. Metabolites of endophytic fungi as novel source of biofungicide: a review. Phytochemistry Rev. 11, 507–522.

Last, R.L., Jones, A.D., Shachar-Hill, Y., 2007. Towards the plant metabolome and beyond. Nat. Rev. Mol. Cell Biol. 8, 167.

Leon, C., Rodriguez-Meizoso, I., Lucio, M., Garcia-Cañas, V., Ibañez, E., et al., 2009. Metabolomics of transgenic maize combining Fourier transform-ion cyclotron resonance-mass spectrometry, capillary electrophoresis-mass spectrometry and pressurized liquid extraction. J. Chromatogr. A 1216, 7314–7323.

Li, M., Wang, J., Lu, Z., Wei, D., Yang, M., et al., 2014. NMR-based metabolomics approach to study the toxicity of lambda-cyhalothrin to goldfish (*Carassius auratus*). Aquat. Toxicol. 146, 82–92.

Liu, Y., Wen, J., Wang, Y., Li, Y., Xu, W., 2010. Postulating modes of action of compounds with antimicrobial activities through metabolomics analysis. Chromatographia 71, 253–258.

Lydon, J., Duke, S.O., 1989. Pesticide effects on secondary metabolism of higher plants. Pestic. Sci. 25, 361–373.

Maciá-Vicente, J.G., Shi, Y.N., Cheikh-Ali, Z., Grün, P., Glynou, K., et al., 2018. Metabolomics-based chemotaxonomy of root endophytic fungi for natural products discovery. Environ. Microbiol. 20, 1253–1270.

Madsen, R., Lundstedt, T., Trygg, J., 2010. Chemometrics in metabolomics-a review in human disease diagnosis. Anal. Chim. Acta 659, 23–33.

Manetti, C., Bianchetti, C., Casciani, L., Castro, C., Di Cocco, M.E., et al., 2006. A metabonomic study of transgenic maize (*Zea mays*) seeds revealed variations in osmolytes and branched amino acids. J. Exp. Bot. 57, 2613–2625.

Markley, J.L., Brüschweiler, R., Edison, A.S., Eghbalnia, H.R., Powers, R., et al., 2017. The future of NMR-based metabolomics. Curr. Opin. Biotechnol. 43, 34–40.

Maroli, A.S., Gaines, T.A., Foley, M.E., Duke, S.O., Doğramacı, M., et al., 2018. Omics in weed science: a perspective from genomics, transcriptomics, and metabolomics approaches. Weed Sci. 66, 681–695.

Mashego, M.R., Rumbold, K., De Mey, M., Vandamme, E., Soetaert, W., et al., 2007. Microbial metabolomics: past, present and future methodologies. Biotechnol. Lett. 29, 1–16.

Mccartney, A., Vignoli, A., Biganzoli, L., Love, R., Tenori, L., et al., 2018. Metabolomics in breast cancer: a decade in review. Cancer Treat Rev. 67, 88–96.

Mckelvie, J.R., Yuk, J., Xu, Y., Simpson, A.J., Simpson, M.J., 2009. ^{1}H NMR and GC/MS metabolomics of earthworm responses to sub-lethal DDT and endosulfan exposure. Metabolomics 5, 84.

Ménard, R., Alban, S., De Ruffray, P., Jamois, F., Franz, G., et al., 2004. β-1, 3 glucan sulfate, but not β-1,3 glucan, induces the salicylic acid signaling pathway in tobacco and Arabidopsis. Plant Cell 16, 3020–3032.

Mendes, I., Sanchez, I., Franco-Duarte, R., Camarasa, C., Schuller, D., et al., 2017. Integrating transcriptomics and metabolomics for the analysis of the aroma profiles of *Saccharomyces cerevisiae* strains from diverse origins. BMC Genom. 18, 455.

Mercado-Blanco, J., Bakker, P.A., 2007. Interactions between plants and beneficial *Pseudomonas* spp.: exploiting bacterial traits for crop protection. Antonie van Leeuwenhoek 92, 367–389.

Mercier, L., Lafitte, C., Borderies, G., Briand, X., Esquerré-Tugayé, M.T., et al., 2001. The algal polysaccharide carrageenans can act as an elicitor of plant defence. New Phytol. 149, 43–51.

Mesnage, R., Renney, G., Séralini, G.-E., Ward, M., Antoniou, M.N., 2017. Multiomics reveal non-alcoholic fatty liver disease in rats following chronic exposure to an ultra-low dose of Roundup herbicide. Sci. Rep. 7, 39328.

Michel, A., Johnson, R.D., Duke, S.O., Scheffler, B.E., 2004. Dose-response relationships between herbicides with different modes of action and growth of *Lemna paucicostata*: an improved ecotoxicological method. Environ. Toxicol. Chem. 23, 1074–1079.

Miller, M.G., 2007. Environmental metabolomics: a SWOT analysis (strengths, weaknesses, opportunities, and threats). J. Proteome Res. 6, 540–545.

Mkandawire, M., Teixeira Da Silva, J.A., Dudel, E.G., 2014. The Lemna bioassay: contemporary issues as the most standardized plant bioassay for aquatic ecotoxicology. Crit. Rev. Environ. Sci. Technol. 44, 154–197.

Mohamed, H.I., Akladious, S.A., 2017. Changes in antioxidants potential, secondary metabolites and plant hormones induced by different fungicides treatment in cotton plants. Pestic. Biochem. Physiol. 142, 117–122.

Moser, V.C., Stewart, N., Freeborn, D.L., Crooks, J., Macmillan, D.K., et al., 2015. Assessment of serum biomarkers in rats after exposure to pesticides of different chemical classes. Toxicol. Appl. Pharmacol. 282, 161–174.

Mouttet, R., Escobar-Gutiérrez, A., Esquibet, M., Gentzbittel, L., Mugniéry, D., et al., 2014. Banning of methyl bromide for seed treatment: could *Ditylenchus dipsaci* again become a major threat to alfalfa production in Europe? Pest Manag. Sci. 70, 1017–1022.

Mudiam, M.K.R., Ch, R., Saxena, P.N., 2013. Gas chromatography-mass spectrometry based metabolomic approach for optimization and toxicity evaluation of earthworm sub-lethal responses to carbofuran. PLoS One 8, e81077.

Mueller, L.A., Zhang, P., Rhee, S.Y., 2003. AraCyc: a biochemical pathway database for *Arabidopsis*. Plant Physiol. 132, 453–460.

Mülleder, M., Calvani, E., Alam, M.T., Wang, R.K., Eckerstorfer, F., et al., 2016. Functional metabolomics describes the yeast biosynthetic regulome. Cell 167, 553–565.e12.

Nagato, E.G., Simpson, A.J., Simpson, M.J., 2016. Metabolomics reveals energetic impairments in *Daphnia magna* exposed to diazinon, malathion and bisphenol-A. Aquat. Toxicol. 170, 175–186.

Nisa, H., Kamili, A.N., Nawchoo, I.A., Shafi, S., Shameem, N., et al., 2015. Fungal endophytes as prolific source of phytochemicals and other bioactive natural products: a review. Microb. Pathog. 82, 50–59.

OECD, 2004. Test No. 202: *Daphnia* sp. Acute Immobilisation Test, OECD Guidelines for the Testing of Chemicals, Section 2. OECD Publishing, Paris.

OECD, 2006. 221: Lemna sp. Growth Inhibition Test. B: OECD Guidelines for the Testing of Chemicals. OECD Publishing, Paris.

OECD, 2012. Test No. 211: *Daphnia magna* Reproduction Test, OECD Guidelines for the Testing of Chemicals, Section 2. OECD Publishing, Paris.

Oikawa, A., Nakamura, Y., Ogura, T., Kimura, A., Suzuki, H., et al., 2006. Clarification of pathway-specific inhibition by Fourier transform ion cyclotron resonance/mass spectrometry-based metabolic phenotyping studies. Plant Physiol. 142, 398–413.

Olmstead, A.W., Leblanc, G.A., 2003. Insecticidal juvenile hormone analogs stimulate the production of male offspring in the crustacean *Daphnia magna*. Environ. Health Perspect. 111, 919–924.

Ott, K.-H., Aranıbar, N., Singh, B., Stockton, G.W., 2003. Metabonomics classifies pathways affected by bioactive compounds. Artificial neural network classification of NMR spectra of plant extracts. Phytochemistry 62, 971–985.

Petit, A.-N., Fontaine, F., Vatsa, P., Clément, C., Vaillant-Gaveau, N., 2012. Fungicide impacts on photosynthesis in crop plants. Photosynth. Res. 111, 315–326.

Pingali, P.L., 2012. Green revolution: impacts, limits, and the path ahead. Proc. Natl. Acad. Sci. U.S.A. 109, 12302–12308.

Poynton, H.C., Taylor, N.S., Hicks, J., Colson, K., Chan, S., et al., 2011. Metabolomics of microliter hemolymph samples enables an improved understanding of the combined metabolic and transcriptional responses of *Daphnia magna* to cadmium. Environ. Sci. Technol. 45, 3710–3717.

Ramirez-Gaona, M., Marcu, A., Pon, A., Guo, A.C., Sajed, T., et al., 2016. YMDB 2.0: a significantly expanded version of the yeast metabolome database. Nucleic Acids Res. 45, D440–D445.

Ramirez, T., Strigun, A., Verlohner, A., Huener, H.-A., Peter, E., et al., 2018. Prediction of liver toxicity and mode of action using metabolomics in vitro in HepG2 cells. Arch. Toxicol. 92, 893–906.

Rattan, R.S., 2010. Mechanism of action of insecticidal secondary metabolites of plant origin. Crop Protect. 29, 913—920.

Regueiro, J., López-Fernández, O., Rial-Otero, R., Cancho-Grande, B., Simal-Gándara, J., 2015. A review on the fermentation of foods and the residues of pesticides-biotransformation of pesticides and effects on fermentation and food quality. Crit. Rev. Food Sci. Nutr. 55, 839—863.

Robertson, D.G., Watkins, P.B., Reily, M.D., 2010. Metabolomics in toxicology: preclinical and clinical applications. Toxicol. Sci. 120, S146—S170.

Ronchi, A., Farina, G., Gozzo, F., Tonelli, C., 1997. Effects of a triazolic fungicide on maize plant metabolism: modifications of transcript abundance in resistance-related pathways. Plant Sci. 130, 51—62.

Roux, A., Lison, D., Junot, C., Heilier, J.-F., 2011. Applications of liquid chromatography coupled to mass spectrometry-based metabolomics in clinical chemistry and toxicology: a review. Clin. Biochem. 44, 119—135.

Saito, K., Matsuda, F., 2010. Metabolomics for functional genomics, systems biology, and biotechnology. Annu. Rev. Plant Biol. 61, 463—489.

Samuels, G.J., 2006. Trichoderma: systematics, the sexual state, and ecology. Phytopathology 96, 195—206.

Schauer, N., Fernie, A.R., 2006. Plant metabolomics: towards biological function and mechanism. Trends Plant Sci. 11, 508—516.

Schläpfer, P., Zhang, P., Wang, C., Kim, T., Banf, M., et al., 2017. Genome-wide prediction of metabolic enzymes, pathways, and gene clusters in plants. Plant Physiol. 173, 2041—2059.

Scott, M., Rani, M., Samsatly, J., Charron, J.-B., Jabaji, S., 2018. Endophytes of industrial hemp (*Cannabis sativa* L.) cultivars: identification of culturable bacteria and fungi in leaves, petioles, and seeds. Can. J. Microbiol. 64, 664—680.

Sevastos, A., Kalampokis, I., Panagiotopoulou, A., Pelecanou, M., Aliferis, K.A., 2018. Implication of *Fusarium graminearum* primary metabolism in its resistance to benzimidazole fungicides as revealed by ^{1}H NMR metabolomics. Pestic. Biochem. Physiol. 148, 50—61.

Shahzad, R., Waqas, M., Khan, A.L., Asaf, S., Khan, M.A., et al., 2016. Seed-borne endophytic *Bacillus amyloliquefaciens* RWL-1 produces gibberellins and regulates endogenous phytohormones of *Oryza sativa*. Plant Physiol. Biochem. 106, 236—243.

Shannon, P., Markiel, A., Ozier, O., Baliga, N.S., Wang, J.T., et al., 2003. Cytoscape: a software environment for integrated models of biomolecular interaction networks. Genome Res. 13, 2498—2504.

Sharma, H.S., Fleming, C., Selby, C., Rao, J., Martin, T., 2014. Plant biostimulants: a review on the processing of macroalgae and use of extracts for crop management to reduce abiotic and biotic stresses. J. Appl. Phycol. 26, 465—490.

Silva, E., Martins, C., Pereira, A., Loureiro, S., Cerejeira, M., 2018. Toxicity prediction and assessment of an environmentally realistic pesticide mixture to *Daphnia magna* and *Raphidocelis subcapitata*. Ecotoxicology 27, 956—967.

Simonin, M., Colman, B.P., Tang, W., Judy, J.D., Anderson, S.M., et al., 2018. Plant and microbial responses to repeated $Cu(OH)_2$ nanopesticide exposures under different fertilization levels in an agro-ecosystem. Front. Microbiol. 9, 1769.

Sparks, T.C., Dripps, J.E., Watson, G.B., Paroonagian, D., 2012. Resistance and cross-resistance to the spinosyns-A review and analysis. Pestic. Biochem. Physiol. 102, 1—10.

Sparks, T.C., Nauen, R., 2015. IRAC: mode of action classification and insecticide resistance management. Pestic. Biochem. Physiol. 121, 122—128.

Spratlin, J.L., Serkova, N.J., Eckhardt, S.G., 2009. Clinical applications of metabolomics in oncology: a review. Clin. Cancer Res. 15, 431—440.

Strobel, G., Daisy, B., 2003. Bioprospecting for microbial endophytes and their natural products. Microbiol. Mol. Biol. Rev. 67, 491—502.

Sumner, L.W., Lei, Z., Nikolau, B.J., Saito, K., 2015. Modern plant metabolomics: advanced natural product gene discoveries, improved technologies, and future prospects. Nat. Prod. Rep. 32, 212—229.

Tautenhahn, R., Cho, K., Uritboonthai, W., Zhu, Z., Patti, G.J., et al., 2012. An accelerated workflow for untargeted metabolomics using the METLIN database. Nat. Biotechnol. 30, 826.

Tawfike, A.F., Tate, R., Abbott, G., Young, L., Viegelmann, C., et al., 2017. Metabolomic tools to assess the chemistry and bioactivity of endophytic *Aspergillus* strain. Chem. Biodivers. 14, e1700040.

Taylor, N., Gavin, A., Viant, M., 2018. Metabolomics discovers early-response metabolic biomarkers that can predict chronic reproductive fitness in individual *Daphnia magna*. Metabolites 8, 42.

Taylor, N.S., Weber, R.J., Southam, A.D., Payne, T.G., Hrydziuszko, O., et al., 2009. A new approach to toxicity testing in *Daphnia magna*: application of high throughput FT-ICR mass spectrometry metabolomics. Metabolomics 5, 44−58.

Taylor, N.S., Weber, R.J., White, T.A., Viant, M.R., 2010. Discriminating between different acute chemical toxicities via changes in the daphnid metabolome. Toxicol. Sci. 118, 307−317.

Teng, M., Zhu, W., Wang, D., Qi, S., Wang, Y., et al., 2018. Metabolomics and transcriptomics reveal the toxicity of difenoconazole to the early life stages of zebrafish (*Danio rerio*). Aquat. Toxicol. 194, 112−120.

Tilman, D., 1998. The greening of the green revolution. Nature 396, 211−212.

Toyota, K., Gavin, A., Miyagawa, S., Viant, M.R., Iguchi, T., 2016. Metabolomics reveals an involvement of pantothenate for male production responding to the short-day stimulus in the water flea, *Daphnia pulex*. Sci. Rep. 6, 25125.

Trethewey, R.N., 2001. Gene discovery via metabolic profiling. Curr. Opin. Biotechnol. 12, 135−138.

Trouvelot, S., Héloir, M.-C., Poinssot, B., Gauthier, A., Paris, F., et al., 2014. Carbohydrates in plant immunity and plant protection: roles and potential application as foliar sprays. Front. Plant Sci. 5, 592.

Trouvelot, S., Varnier, A.-L., Allegre, M., Mercier, L., Baillieul, F., et al., 2008. A β-1, 3 glucan sulfate induces resistance in grapevine against *Plasmopara viticola* through priming of defense responses, including HR-like cell death. Mol. Plant Microbe Interact. 21, 232−243.

Tufi, S., Leonards, P., Lamoree, M., De Boer, J., Legler, J., et al., 2016a. Changes in neurotransmitter profiles during early zebrafish (*Danio rerio*) development and after pesticide exposure. Environ. Sci. Technol. 50, 3222−3230.

Tufi, S., Wassenaar, P.N., Osorio, V., De Boer, J., Leonards, P.E., et al., 2016b. Pesticide mixture toxicity in surface water extracts in snails (*Lymnaea stagnalis*) by an *in vitro* acetylcholinesterase inhibition assay and metabolomics. Environ. Sci. Technol. 50, 3937−3944.

Urano, K., Kurihara, Y., Seki, M., Shinozaki, K., 2010. 'Omics' analyses of regulatory networks in plant abiotic stress responses. Curr. Opin. Plant Biol. 13, 132−138.

Van Den Berg, R.A., Hoefsloot, H.C., Westerhuis, J.A., Smilde, A.K., Van Der Werf, M.J., 2006. Centering, scaling, and transformations: improving the biological information content of metabolomics data. BMC Genom. 7, 142.

Viant, M.R., Kurland, I.J., Jones, M.R., Dunn, W.B., 2017. How close are we to complete annotation of metabolomes? Curr. Opin. Chem. Biol. 36, 64−69.

Viant, M.R., Pincetich, C.A., Tjeerdema, R.S., 2006. Metabolic effects of dinoseb, diazinon and esfenvalerate in eyed eggs and alevins of Chinook salmon (*Oncorhynchus tshawytscha*) determined by ¹H NMR metabolomics. Aquat. Toxicol. 77, 359−371.

Viant, M.R., Sommer, U., 2013. Mass spectrometry based environmental metabolomics: a primer and review. Metabolomics 9, 144−158.

Vu, T., Sikora, R., Hauschild, R., 2006. *Fusarium oxysporum* endophytes induced systemic resistance against *Radopholus similis* on banana. Nematology 8, 847−852.

Wagner, N.D., Simpson, A.J., Simpson, M.J., 2017. Metabolomic responses to sublethal contaminant exposure in neonate and adult *Daphnia magna*. Environ. Toxicol. Chem. 36, 938−946.

Walters, D.R., Ratsep, J., Havis, N.D., 2013. Controlling crop diseases using induced resistance: challenges for the future. J. Exp. Bot. 64, 1263−1280.

Wang, H.-P., Liang, Y.-J., Zhang, Q., Long, D.-X., Li, W., et al., 2011a. Changes in metabolic profiles of urine from rats following chronic exposure to anticholinesterase pesticides. Pestic. Biochem. Physiol. 101, 232–239.

Wang, H., Wen, K., Zhao, X., Wang, X., Li, A., et al., 2009. The inhibitory activity of endophytic Bacillus sp. strain CHM1 against plant pathogenic fungi and its plant growth-promoting effect. Crop Protect. 28, 634–639.

Wang, Q.-X., Li, S.-F., Zhao, F., Dai, H.-Q., Bao, L., et al., 2011b. Chemical constituents from endophytic fungus *Fusarium oxysporum*. Fitoterapia 82, 777–781.

Wang, Y., Wu, S., Chen, J., Zhang, C., Xu, Z., et al., 2018. Single and joint toxicity assessment of four currently used pesticides to zebrafish (*Danio rerio*) using traditional and molecular endpoints. Chemosphere 192, 14–23.

Wilkowska, A., Biziuk, M., 2011. Determination of pesticide residues in food matrices using the QuEChERS methodology. Food Chem. 125, 803–812.

Wishart, D.S., 2008a. Metabolomics: applications to food science and nutrition research. Trends Food Sci. Technol. 19, 482–493.

Wishart, D.S., 2008b. Quantitative metabolomics using NMR. Trends Anal. Chem. 27, 228–237.

Wishart, D.S., 2016. Emerging applications of metabolomics in drug discovery and precision medicine. Nat. Rev. Drug Discov. 15, 473.

Xia, J., Wishart, D.S., 2016. Using MetaboAnalyst 3.0 for comprehensive metabolomics data analysis. Curr. Protoc. Bioinf. 55, 14.10.1–14.10.91.

Xu, Y.-J., 2017. Foodomics: a novel approach for food microbiology. Trends Anal. Chem. 96, 14–21.

Yi, Z.-B., Yu, Y., Liang, Y.-Z., Zeng, B., 2007. Evaluation of the antimicrobial mode of berberine by LC/ESI-MS combined with principal component analysis. J. Pharmaceut. Biomed. Anal. 44, 301–304.

Yu, Y., Yi, Z.-B., Liang, Y.-Z., 2007. Main antimicrobial components of *Tinospora capillipes*, and their mode of action against *Staphylococcus aureus*. FEBS Lett. 581, 4179–4183.

Yuk, J., Simpson, M.J., Simpson, A.J., 2011. 1-D and 2-D NMR metabolomics of earthworm responses to sublethal trifluralin and endosulfan exposure. Environ. Chem. 8, 281–294.

Zhang, R., Zhou, Z., 2019. Effects of the chiral fungicides metalaxyl and metalaxyl-M on the earthworm *Eisenia fetida* as determined by ^{1}H-NMR-based untargeted Metabolomics. Molecules 24, 1293.

Zhao, L., Huang, Y., Keller, A.A., 2017. Comparative metabolic response between cucumber (*Cucumis sativus*) and corn (*Zea mays*) to a Cu(OH)$_2$ nanopesticide. J. Agric. Food Chem. 66, 6628–6636.

Zhao, L., Ortiz, C., Adeleye, A.S., Hu, Q., Zhou, H., et al., 2016. Metabolomics to detect response of lettuce (*Lactuca sativa*) to Cu(OH)$_2$ nanopesticides: oxidative stress response and detoxification mechanisms. Environ. Sci. Technol. 50, 9697–9707.

Zuluaga, M., Melchor, J.J., Tabares-Villa, F.A., Taborda, G., Sepúlveda-Arias, J.C., 2016. Metabolite profiling to monitor organochlorine pesticide exposure in HepG2 cell culture. Chromatographia 79, 1061–1068.

CHAPTER 7

Metabolomics strategies and analytical techniques for the investigation of contaminants of industrial origin

Òscar Aznar-Alemany, Marta Llorca

Water and Soil Quality Research Group, Department of Environmental Chemistry, IDAEA-CSIC, Barcelona, Spain

Chapter Outline

1. Introduction

During the last years, the use of omics to evaluate the toxic effects in selected organisms has increased. In the specific case of metabolomics, the use of this technique has been applied to the evaluation of toxicity at metabolomics level for chemicals of industrial origin considered persistent organic pollutants (POPs) such as flame retardants (FRs) and perfluoroalkyl substances (PFASs).

FRs are additives to all kind of manufactured materials, such as household appliances, office electronics, textiles, and furniture, to prevent them from ignition. Polybrominated

Environmental Metabolomics. https://doi.org/10.1016/B978-0-12-818196-6.00007-8

diphenyl ethers (PBDEs) and organophosphorus flame retardants (OPFRs), the latter also applied as plasticizers, are the most used FRs in a great variety of indoor and outdoor products (Alaee et al., 2003; van der Veen and de Boer, 2012). PBDEs are considered POPs and have been gradually added to the Stockholm Convention starting with the penta- and octabrominated congeners in 2009. Hexabromocyclododecane (HBCD) is another FR included in the Stockholm Convention. Other brominated as well as chlorinated FRs are thus used as alternatives, e.g., dechloranes (Alaee et al., 2003). PBDEs show toxicity on hormonal regulation and neuronal, thyroid, and liver activity (Mikula and Svobodova, 2006). OPFRs show toxic effects on the reproductive and endocrine systems, as well as systemic and carcinogenic effects (Hou et al., 2016; van der Veen and de Boer, 2012).

PFASs have been extensively used since the 1940s in industrial and commercial applications due to their inertia and resistance to physical, chemical, and biological degradation (Llorca et al., 2012). Because of their persistence in the environment due to their low degradability, it is of high importance to know the toxicological effects that they could have on organisms. For example, it is known that they induce disruption of the thyroid hormones and the metabolism of high-density lipoproteins, cholesterol, and triglycerides (Lau et al., 2007; Peden-Adams et al., 2008). Because of their widespread and low degradability, perfluorooctane sulfonate (PFOS) and its salts are considered POPs and were included in 2009 in the Annex B of Stockholm Convention like some of the FRs.

This chapter is a review of the most commonly used strategies for metabolomics evaluation of FRs and PFAS effects published during the last 5 years.

2. Metabolomics strategies for testing toxicity of chemicals from industrial origin

Metabolomics is based on the evaluation of the response of the metabolomics pathways of biofluids, cells, and tissues exposed to contaminants. In this context, metabolomics is used as a tool for biomarkers discovery (Johnson et al., 2016). In the specific case of metabolomics to assess the effects of the exposure to FRs and PFAS, the available literature since 2015 include a wide variety of target organisms based on in vivo and in vitro experiments (Table 7.1). Regarding in vitro experiments, the most commonly used cells are hepatotoxic cell lines HepG2/C3A from human liver (Ballesteros-Gomez et al., 2015; Gu et al., 2019; Van den Eede et al., 2015; Wang et al., 2016b; Zhang et al., 2016, 2015), lung (Zhang et al., 2015), and mouse preadipocyte T3-L1 (Yang et al., 2018a), all of them for testing the effects after exposition to different types of FRs. Adrenocortical cell line is also used for testing a mixture of chemicals including FRs, PFAS, and endocrine-disrupting compounds (EDCs) (Ahmed et al., 2019). On the other side, a wide set of organisms have been tested in vivo (Table 7.1). They include aquatic organisms like

Table 7.1: Metabolomics strategies for testing the toxicity of FRs and PFAS.

	Metabolomics experiments	Observed effects	References
	Flame retardants		
	In vitro experiments		
RDP	Pooled human liver microsomes (n = 50, mixed gender). Pooled human liver cytosol (mix gender)	Xenometabolomics studies based on transformation products coming from RDP in human liver.	Ballesteros-Gomez et al. (2015)
Nine OPFRs (TDBPP, TMPP, TPhP, TBOEP, TCIPP, TCEP, TEHP, TNBP, and TDCPP)	HepG2 cell line: human liver cancer cell line	Disruptions on the glucocorticoid and mineralocorticoid receptors, metabolism linked with oxidative stress, osmotic pressure equilibrium, and amino acid metabolism.	Gu et al. (2019)
TPhP	Hepatotoxicity testing tool using human HepaRG cell cultures	Identification of biomarkers related to accumulation of phosphoglycerolipids and increase of palmitoyl lysophosphatidylcholine.	Van den Eede et al. (2015)
HBCD	Human hepatoma cells HepG2	The results showed a suppression of the cell uptake of amino acids (inhibition of the activity of membrane transport protein Na^+/K^+-ATPase), downregulated glycolysis, and β-oxidation of long-chain fatty acids (causing a large decrease of ATP production), and the across-membrane transport of amino acids is further inhibited. Significant increase of total phospholipids, mainly through the remodeling of phospholipids from the increased free fatty acids.	Wang et al. (2016b)
HBCD	A549 lung human cells and human hepatoma cells HepG2/C3A	No detected metabolic effects.	Zhang et al. (2015)

Continued

Table 7.1: Metabolomics strategies for testing the toxicity of FRs and PFAS.—cont'd

	Metabolomics experiments	Observed effects	References
Indoor dust standard reference material SRM2585 containing polyaromatic hydrocarbon DMBA and a mixture of four FRs (TCEP, TCIPP, TDCPP, and HBCD)	Human hepatoma cells HepG2/C3A	Indoor dust group: dysregulation of metabolic pathways related to metabolism of amino acids and derivatives (i.e., arginine and proline metabolism, glycine, and serine metabolism) and glutathione conjugation. FRs (HBCD): taurine as biomarker.	Zhang et al. (2016)
BDE-47	Mouse preadipocyte 3 T3-L1 cells	Upregulation of purine metabolism and alterations in glutathione metabolism that promote oxidative stress and uric acid production in adipocytes, elevate mitochondrial respiration, and glycolysis in adipocytes to induce more ATP to combat oxidative stress. In addition, antioxidant treatments (i.e., suppression of xanthine oxidase) have been detected that are used to inhibit the induction of oxidative stress and lipid accumulation. Identification of biomarkers of adipogenesis.	Yang et al. (2018a)
In vivo experiments			
BDE-47	Three subspecies of rice (*Oryza sativa*)—indica, japonica, and indica japonica hybrid: Lianjing-7 (LJ-7), Jinzao-47 (JZ-47), Zhongjiazao-17 (ZJZ-17), Xiushui-134 (XS-134), Yongyou-1540 (YY-1540), Zhongzheyou-1 (ZZY-1), Jiayou-5 (JY-5), Zhejing-88 (ZJ-88), Yongyou-9 (YY-9), and Y-linagyou-1 (Y-LY-1)	YY-9 and LJ-7 were the more sensitive and tolerant varieties. LJ-7: more active in metabolite profiles and adopted more effective antioxidant defense machinery to protect itself against oxidative damages induced by BDE-47—tolerant to BDE-47.YY-9: downregulation of most metabolites indicated—susceptible to BDE-47.	Chen et al. (2018)

Continued

Table 7.1: Metabolomics strategies for testing the toxicity of FRs and PFAS.—cont'd

	Metabolomics experiments	Observed effects	References
α-, β-, and γ-HBCD	Pak choi leaves (*Brassica chinensis* L.)	Thirteen metabolites identified as possible biomarkers related to the lipid, carbohydrate, nucleotide, and amino acid metabolic pathways as well as the induction of the interference of the secondary metabolite pathways. Potential biomarkers were isoorientin, chlorogenic acid, lactose, sn-glycero-3-phosphocholine (GPC), beta-D-glucose, luteolin, sinapic acid, 4-acetamidobutanal, rutin, homovanillic acid, proline, caffeic acid, and 3-methoxy-4-hydroxyphenylglycolaldehyde.	Zhang et al. (2018)
TPhP	Earthworm *Perionyx excavatus*	Biomarkers detected in amino acids, glucose, inosine, and phospholipids metabolism when earthworms were exposed to TPhP by acute experiments.	Wang et al. (2018b)
TPhP	Freshwater microalgae: *Chlorella vulgaris* and *Scenedesmus obliquus*	Decrease of chlorophyll derivatives for both species and concentration- and species-dependent effects for other metabolisms. *C. vulgaris*—enhanced respiration (increase of fumarate and malate) and osmoregulation (increase of sucrose and myo-inositol), synthesis of membrane lipids (accumulation of monogalactosyldiacylglycerol and digalactosyldiacylglycerol, decrease of lysoglycerolipids, fatty acids, and glyceryl-glucoside). *S. obliquus*—enhanced	Wang et al. (2019)

Continued

Table 7.1: Metabolomics strategies for testing the toxicity of FRs and PFAS.—cont'd

	Metabolomics experiments	Observed effects	References
pentaBDE and FM550	Freshwater crustacean *Daphnia magna*	osmoregulation (increase of valine, proline, and raffinose) and lipolysis (decrease of monogalactosyldiacylglycerol, accumulation of fatty acids, lysophospholipids, and glycerol phosphate). The combination of transcriptomics, lipidomics, and metabolomics results induced to hypothesize that pentaPBDE affects transcription and translation, while FM550 elicited significant mRNA changes and impairs nutrient utilization or uptake (changes in histamine, protein biosynthesis, ammonia recycling, nitrogen metabolism, amino acids synthesis, and degradation).	Scanlan et al. (2015)
TCEP, TBOEP, and TPhP with DOM	*D. magna* at sublethal exposure	TCEP: no observed effects. TBOEP: increase in amino acids and decrease in glucose metabolisms. TPhP: decrease in glucose and a significant increase in leucine.	Cui et al. (2008)
TBBPA	Mussels (*Mytilus galloprovincialis*) evaluation of gender-specific responses	Observations of disturbances in energy metabolism and osmotic regulation in females and variation of metabolites related to osmotic regulation in males.	Ji et al. (2016)
TPhP	Zebrafish liver	Metabolomics analysis revealed significant changes in the contents of glucose, UDP-glucose, lactate, succinate, fumarate, choline, acetylcarnitine, and several fatty acids inducing to hypothesize that TPhP disturbs hepatic carbohydrate and lipid metabolism in zebrafish with just 7 days of exposure.	Du et al. (2016)

Continued

Table 7.1: Metabolomics strategies for testing the toxicity of FRs and PFAS.—cont'd

	Metabolomics experiments	Observed effects	References
TBBPA and tetrachlorobisphenol A (TCBPA)	Marine medaka (*Oryzias melastigmas*) embryo model	Biomarkers identified for toxicity related to the presence of TBBPA and TCBPA and related genetic effects are lactate and dopa metabolites. Some of the toxic effects observed include the decrease in the synthesis of nucleosides, amino acids, and lipids, and disruptions in the tricarboxylic acid cycle, glycolysis, and lipid metabolism. These metabolites inhibit the developmental processes of embryos. Other effects are the enhancement of neural activity accompanied by lactate accumulation and accelerated heart rates due to an increase in dopamine pathway and a decrease in inhibitory neurotransmitters. Finally, the disorders observed in the neural system and disruptions in glycolysis, the tricarboxylic acid cycle, nucleoside metabolism, lipid metabolism, glutamate, and aspartate metabolism induced by the toxicants exposure were identified as heritable.	Ye et al. (2016)
TCEP and TDCPP	Mice (*Mus musculus*)	Evaluation of toxicity modulation of OPFRs due to the presence of PE-MPLs and PS-MPLs. Effects detected include oxidative stress, alterations on energy metabolism, and neurotoxic effects.	Deng et al. (2018)

Continued

Table 7.1: Metabolomics strategies for testing the toxicity of FRs and PFAS.—cont'd

	Metabolomics experiments	Observed effects	References
BDE-47	C57BL/6 mice	Fifty-seven differential metabolites that significantly altered in mice serum and striatum. The obtained results suggested that BDE-47 could induce oxidative stress, dysregulate neurotransmitters, and disturb dopaminergic system.	Ji et al. (2017)
α-, γ-, and CM-HBCD	Female C57BL/6 mice (mice serum) 4 days following a single neonatal oral exposure	Differences in endogenous metabolites by treatment and dose groups, including metabolites involved in glycolysis, gluconeogenesis, lipid metabolism, citric acid cycle, and neurodevelopment.	Szabo et al. (2016)
Dec 602	Male mice by oral exposition	Urine: increase in the levels of thymidine and tryptophan as well as a decrease in the levels of tyrosine, 12,13-dihydroxy-9Z-octadecenoic acid, 2-hydroxyhexadecanoic acid, and cuminaldehyde. Sera: decrease in the levels of kynurenic acid, daidzein, adenosine, xanthurenic acid, and hypoxanthine.	Tao et al. (2019)
TPhP and DPP (metabolite of TPhP)	Mice—neonatal exposure from postnatal days 1—10	Sex-specific metabolic disturbance of TPP in male mice: low toxicity dose upregulated lipid-related metabolites and a high toxicity dose downregulated the pyruvate metabolism and citrate cycles.	Wang et al. (2018a)
HBCD	Mice	Biomarkers detected for subchronic exposure to HBCD that caused disturbance in citrate cycle, lipid metabolism, gut microbial metabolism, and homeostasis of amino acids.	Wang et al. (2016a)
BDE-3 (4-bromodiphenyl ether)	Mice exposed intragastrically	Metabolic changes in pathways of tyrosine metabolism, purine metabolism, and riboflavin metabolism.	Wei et al. (2018)

Continued

Table 7.1: Metabolomics strategies for testing the toxicity of FRs and PFAS.—cont'd

	Metabolomics experiments	Observed effects	References
TCEP	Rats exposed through oral gavage	Changes observed in pathways of amino acid and neurotransmitter metabolism, energy metabolism, and cell membrane function integrity.	Yang et al. (2018b)
Perfluoroalkyl substances			
In vivo experiments			
PFOS	*D. magna*	Toxic mode of action is metabolite-specific with some metabolites; disrupts various energy metabolism pathways, and enhances protein degradation. In addition, a nonmonotonic response was observed with higher PFOS exposure concentrations.	Kariuki et al. (2017)
PFOS added in real effluent water from a wastewater treatment plant	*D. magna*	Metabolic profile decreases amino acids and increase sugar metabolites as well as energy molecules especially at the low and high concentrations of PFOS.	Wagner et al. (2019)
PFOS	Neonate and adult microcrustacean *D. magna*	PFOS exposure disrupts various energy metabolic pathways and enhances protein degradation in both neonate and adult exposed to sublethal PFOS concentrations.	Wagner et al. (2017)
PFOS and other 15 xenobiotics individually	Zebrafish (*Danio rerio*) embryo/larvae	Metabolomics perturbations in amino acids, bile acid, fatty acids, sugars, and lipids.	Huang et al. (2016)
PFOS, in parallel experiments with bisphenol A and tributyltin separately	Zebrafish (*D. rerio*) embryo	25 metabolites altered by the presence of PFOS including metabolism of glycerophospholipids, amino acids, purines, and 2-oxocarboxylic acids.	Ortiz-Villanueva et al. (2018)
PFDoA	Male rats	Hepatic lipidosis: elevation in hepatic triglycerides and a decline in serum lipoprotein levels.	Ding et al. (2009)

Continued

Table 7.1: Metabolomics strategies for testing the toxicity of FRs and PFAS.—cont'd

	Metabolomics experiments	Observed effects	References
PFOS	Mouse—in utero exposure and effects on neonatal	PFOS has potential negative impacts on testicular functions.	Lai et al. (2017)
PFOA	Male Balb/c mice	Alterations in metabolic pathways of amino acids, lipids, carbohydrates, and energetics; the mechanisms that affect PFOA-induced hepatotoxicity and neurotoxicity. Brain biomarkers based on neurotransmitters including serotonin, dopamine, norepinephrine, and glutamate. Liver biomarkers mainly involved in amino acid metabolism and lipid metabolism: 5 amino acids, 4 biogenic amines, 6 acylcarnitines, 1 phosphatidylcholine, and 2 polyunsaturated fatty acids.	Yu et al. (2016)
PFOA	Female C57BL/6 mice intraperitoneally injected with PFOA	Metabolites related to Acaca-, Acacb-, and Slc25a20-enriched proteins. PFOA treatment decreased malonyl-CoA in mice and upregulated acetyl-CoA. Therefore, the decrease affects to the downregulation of fatty acid synthesis and upregulated fatty acid oxidation.	Shao et al. (2018)
PFNA and mixture of 14 chemicals	Rats	PFNA affects the steroid hormone synthesis in rats. Corticosterone was affected by PFNA while for androstenedione, testosterone, and dihydrotestosterone the induced effect was caused by the combination with the mixture.	Skov et al. (2015)
PFOS and PFOA	Eggs of *Gallus gallus domesticus* (white leghorn chicken)	PFOS: downregulation of the β-oxidation of fatty and energy metabolism. PFOA: changes in regulated fatty acid metabolism.	Wigh (2017)

Continued

Table 7.1: Metabolomics strategies for testing the toxicity of FRs and PFAS.—cont'd

	Metabolomics experiments	Observed effects	References
PFOA	Blood samples from 120 8-year-old children	Biomarkers identified in keratin sulfate degradation and metabolism of purine, caffeine, vitamin E, linoleate, urea cycle/amino groups, glyoxylate, dicarboxylate, and galactose, consistent with changes to immunological, oxidative stress, and catabolism pathways.	Kingsley et al. (2018)
Mixture of PFAS	Plasma from a cross-sectional study (965 individuals (all aged 70 years, 50% women) sampled in 2001—04	After adjustment with some external factors (sex, smoking, exercise habits, education, energy, and alcohol intake), 15 metabolites, predominantly from lipid pathways, were associated with levels of PFAS. PFNA and perfluoroundecanoic acid (PFUnDA) associated with multiple glycerophosphocholines and fatty acids including docosapentaenoic acid (DPA) and docosahexaenoic acid (DHA). In addition, different PFAS were associated with distinctive metabolic profiles, suggesting potentially different biochemical pathways in humans.	Salihovic et al. (2018)
Mixture of PFAS present in cohort samples	Male cohort serum	Ten potential PFAS biomarkers identified involved in pollutant detoxification, antioxidation, and nitric oxide signal pathways. These suggested adverse impacts on glutathione cycle, Krebs cycle, nitric oxide generation, and purine oxidation in humans at low environmental levels of PFAS.	Wang et al. (2017)

Continued

Table 7.1: Metabolomics strategies for testing the toxicity of FRs and PFAS.—cont'd

	Metabolomics experiments	Observed effects	References
Mixture of industrial chemicals			
In vitro experiments			
Mixture of POPs and endocrine-disrupting compounds including PFAS—PFOA, PFOS, PFDA, PFNA, PFHxS, and PFUnDA; Brominated FRs—BDE209, BDE-47, BDE-99, BDE-100, BDE-153, BDE-154, and HBCD; Chlorinated FRs—PCB-138, PCB-153, PCB-101, PCB-180, PCB-52, PCB-28, PCB-118; and other EDCs	Human H295R adrenocortical cell line	The results showed • Downregulation in steroid production when cells were exposed to the highest concentration of a mixture of brominated and fluorinated compounds; • Upregulation with estrone and some other steroids when cells were exposed to a PFAS mixture (1000-times human blood values); • Downregulation of the mixture of chlorinated and PFAS at low concentration. • PFAS mixture alone produced albeit small of pregnenolone. • The total mixture a similar effect on 17-hydroxypregnenolone.	Ahmed et al. (2019)

microalgae (Wang et al., 2019), microcrustacean *Daphnia magna* (Cui et al., 2008; Kariuki et al., 2017; Scanlan et al., 2015; Wagner et al., 2017, 2019), filterers like bivalves (Ji et al., 2016), fishes, including the two most commonly tested species zebrafish (Du et al., 2016; Huang et al., 2016) and marine medaka (Ye et al., 2016), terrestrial organisms like earthworm—soil ecosystem (Wang et al., 2018b), chickens (Wigh, 2017), mice (Deng et al., 2018; Ji et al., 2017; Shao et al., 2018; Szabo et al., 2016; Tao et al., 2019; Wang et al., 2016a, 2017, 2018a; Wei et al., 2018; Yu et al., 2016), rats (Skov et al., 2015; Yang et al., 2018b), and humans in cohort studies (Kingsley et al., 2018; Salihovic et al., 2018; Wang et al., 2017). In the available bibliography related to FRs and PFAS, two types of metabolomics exposure can be identified: (i) exposition to a single contaminant and (ii) exposition to a mixture of contaminants.

2.1 Exposition to a single contaminant

In the case of a single contaminant, the metabolomics pathways affected are different depending on the contaminant and the species exposed. For example, the study of BDE-47 in rice showed that 13 amino acids and 24 organic acids (i.e., L-glutamic acid, β-alanine, glycolic acid, and glyceric acid) were upregulated significantly, which contributed to scavenging reactive oxygen species (ROS) (Chen et al., 2018). In contrast, mouse cell lines exposed to the same contaminant (BDE-47) allowed the identification of biomarkers of adipogenesis through the acid production in adipocytes, elevated mitochondrial respiration and glycolysis in adipocytes which induced more adenosine triphosphate to combat oxidative stress (Yang et al., 2018a). On the other hand, the exposure of the same species to different isomers of contaminants could induce stereospecific effects. For example, the metabolomics effects observed in neonatal female mice exposed to 4-day oral administration of a commercial mixture of hexabromocyclododecane (CM-HBCD) containing 10% of α-, 10% of γ-, and 80% of β-HBCD resulted in different serum metabolomics profiles (Szabo et al., 2016). These profiles indicated stereospecific and mixture-specific effects (Szabo et al., 2016). A similar pattern of stereospecific effects of α-, β-, and γ-HBCD was observed by Zhang et al. (2018) in pak choi leaves (*Brassica chinensis* L.). The common isomer effects include the system activation of the stress defenses (signaling pathway, antioxidant defense system, shikimate, and phenylpropanoid metabolism), the carbohydrate and amino acid metabolisms were also disturbed, and the lipid, amino acid, and secondary metabolite metabolisms were regulated for HBCD stress prevention (Zhang et al., 2018).

2.2 Exposition to a mixture of contaminants

The exposition to a mixture of contaminants combines the mixture of common organic environmental contaminants only and also the effects of the coexposure of some FRs or PFAS with microplastics (MPLs) (Deng et al., 2018) or real environmental waters with different contents of dissolved organic material (DOM) (Kovacevic et al., 2018). These works evaluate not only the metabolomics response to selected contaminants but also the modulation of toxicity due to the presence of other co-contaminants. For example, the coexposure of some FRs with MPLs to mice showed that MPLs are able to modulate the toxicity of FRs showing a synergism compared with the results of FRs and MPLs when exposed alone (Deng et al., 2018). In the same line, there are few studies evaluating the metabolomics response to real samples containing industrial chemicals or by spiking real environmental samples with those industrial compounds. An example is given by the interaction of OPFRs (tris(2-chloroethyl) phosphate (TCEP), tris(2-butoxyethyl) phosphate (TBOEP), and triphenyl phosphate (TPhP)) with DOM and how it affects the exposition of *D. magna* at metabolomics level (Kovacevic et al., 2018). The results showed that TCEP did not produce significant metabolic changes and DOM did not alter the metabolic response. In contrast, TBOEP showed significant increase in amino acids and a decrease in glucose. In the case of TPhP, DOM seems to be an additional stressor for daphnids (Kovacevic et al., 2018). In this other example, daphnids are exposed to real effluent from a wastewater treatment plant spiked with different concentrations of PFOS (Wagner et al., 2019). Different metabolomics profiles for daphnids were detected depending on whether the exposition was just to effluent or to effluent spiked with PFOS. The majority of biomarkers identified belong to amino acid metabolites, sugar metabolites, and energy molecules (Wagner et al., 2019). Another example is the assessment of the effects of the exposure to specific and environmentally relevant mixtures of chlorinated and brominated FRs, PFAS, and other EDCs on steroidogenesis carried out by Ahmed et al. (2019). In the same context, Skov et al. (2015) evaluated the metabolomics response of rats exposed to perfluorononanoic acid (PFNA) alone and to a mixture of 14 relevant environmental contaminants including one grapefruit constituent, one liquorice constituent, one plastic additive, one preservative, two plasticizers, two sun filters, and six pesticides. The authors observed that the combination of contaminants of emerging concern with PFNA induced effects on androstenedione, testosterone, and dihydrotestosterone in rat metabolism (Skov et al., 2015).

2.3 Identification of biomarkers in real samples

The identification of environmental relevant biomarkers has also been carried out in real samples taken from fish and humans. For instance, Huang et al. (2016) characterized the metabolomics perturbations in zebrafish larvae after exposure to 16 xenobiotics individually including PFOS, one of the most recalcitrant PFASs. The authors identified different biomarkers as a signal of zebrafish contamination by selected xenobiotics and, after that, the exposition of the zebrafish to real effluent allowed to detect those biomarkers denoting the presence of the tested contaminants in such waters and their related toxicity to zebrafish (Huang et al., 2016). In humans, there are different studies that investigate any possible biomarker related to the exposure to PFAS. For example, the cross-sectional study developed by Salihovic et al. (2018) in Sweden for the identification of metabolic profiles in human plasma associated with human exposure to PFAS, where the authors identified 15 metabolites and the results suggested potentially different biochemical pathways in humans. In the same line, Wang et al. (2017) developed a cohort study in serum of Chinese males trying to identify some biomarkers for PFAS toxic effects. The main biomarkers included pollutant detoxification, antioxidation, and nitric oxide signal pathways (Wang et al., 2017).

2.4 Metabolomics link to other techniques for the assessment of potential effects

2.4.1 Combination with other omics techniques

In order to have a wide spectrum of the effects that FRs and PFAS have on organisms, the metabolomics results can be combined with other omics providing knowledge at different levels of cellular organization. For example, Du et al. (2016) exposed zebrafish for 7 days to TPhP at different concentrations. The omics analyses of fish liver showed effects in the metabolism as well as in cell cycles through the observations of deoxyribonucleic acid replication (transcriptomics) (Du et al., 2016). In the same line of omics combinations, Zhang et al. (2015) evaluated the in vitro exposition to HBCD of lung and hepatocyte human cell lines with proteomics and metabolomics strategies. The results showed that the dosage of exposition did not have toxic risk for the selected cell lines. Other authors have investigated also combined effects like, for example, Ji et al. (2016) exposed mussels to tetrabromobisphenol A and evaluated the combination of metabolomics and proteomics studies . Scanlan et al. exposed the microcrustacean *D. magna* to seven novel FRs and evaluated their effects through the combination of messenger ribonucleic acid expression, metabolomics, and lipidomics profiling (Scanlan et al., 2015). Another example is the study that combined proteomics metabolomics to determine the protein targets of PFOA in female mice injected intraperitoneally (Shao et al., 2018). Zhang et al. (2016) studied through transcriptomics, metabolomics, and lipidomics the effects of indoor dust spiked with a mixture of four FRs on human hepatoma cells. The results showed changes in gene expression associated with the metabolism of xenobiotics in the dust extract group but not in the FR mixture group.

2.4.2 Combination with other toxicological techniques

There are works that combine the changes detected on endogenous and exogenous metabolites with the study of external factors, like changes in morphology and physiology (Chen et al., 2018; Yang et al., 2018b) (i.e., chlorophylls integrity in algae (Wang et al., 2019)), or the study of other toxicological effects, like endocrine disruption (Wang et al., 2018a, 2019) and cellular effects on ROS. As an example, Yang et al. (2018b) investigated the effects on rats exposed to TCEP by combining metabolomics with neuropathology and neurobehavioral analyses. The combination of the results showed that the FR produced neurotoxic effects in female rats, while the Morris water maze (test of spatial learning for rodents) results revealed a dose-dependent decline in spatial learning and memory functions of the exposed rats. Through the pathological examination of the rats' brain, the authors observed apoptotic and necrotic lesions in cells of the hippocampus as well as inflammatory cells and calcified/ossified foci in the cortex areas (Yang et al., 2018b). All these observations were combined with the metabolomics results where TCEP interfered with amino acid and neurotransmitter metabolism, energy metabolism, and cell membrane function integrity by changing the concentrations of related metabolites and metabolites linked to antioxidant physiological processes (Yang et al., 2018b). Another example is the study carried out by Yu et al. (2016) on mice exposed to PFOA. The combination of the study of other factors with the metabolomics studies in mice gave novel insights into mechanisms of PFOA-induced neurobehavioral effects and PFOA-induced hepatotoxicity as well as alterations in metabolism of arachidonic acid, which suggests potential of PFOA to cause an inflammation response in liver (Yu et al., 2016).

It is noteworthy to mention that the metabolomics effects need to be evaluated not only through the exposition of organisms to an industrial toxicant but also through the presence of their environmental degradation products because different pathways could be affected. For example, the exposure of mice to BDE-47 could induce oxidative stress, dysregulate neurotransmitters, and disturb dopaminergic system (Ji et al., 2017), while its photodegradation product, 4-bromodiphenyl ether (BDE-3), via intragastrical exposure showed metabolic disturbances in the tyrosine metabolism, the purine metabolism, and the riboflavin metabolism (Wei et al., 2018).

3. Analytical techniques

The main analytical processes for metabolomics evaluation include sample preparation followed by liquid chromatography (LC) or gas chromatography (GC) separation coupled to mass spectrometry (MS) analysis or to proton nuclear magnetic resonance (^{1}H-NMR). Table 7.2 summarizes the state of the art since 2015 regarding the analysis of environmental metabolomics with chemicals of industrial origin.

3.1 Sample preparation

The preparation of the samples depends on the biological matrix (serum, liver, cells, urine, brain, etc.) as well as on whether a targeted or untargeted analysis is performed. In this latter, the process for screening purposes is usually based on ultrasonic assisted extraction with a mixture of chloroform:water:methanol (e.g., 1:1:1) if the analysis is performed by LC-MS (Huang et al., 2016; Ortiz-Villanueva et al., 2018; Zhang et al., 2016) or by NMR (Du et al., 2016; Gu et al., 2019; Ji et al., 2016; Scanlan et al., 2015), while the chloroform fraction is derivatized if the analysis is carried out by GC-MS (Chen et al., 2018; Yang et al., 2018b; Yu et al., 2016). Other methodologies employ protein precipitation and cellular disruption (Wang et al., 2016a,b, 2018b, 2019; Wei et al., 2018), with an organic solvent extraction like methanol (Van den Eede et al., 2015) or a mixture of methanol and acetonitrile (Tao et al., 2019). The sample extraction for NMR analysis is based mainly in phosphate buffer extraction followed by homogenization and centrifugation (Deng et al., 2018; Ding et al., 2009; Kovacevic et al., 2018; Wagner et al., 2017, 2019; Wang et al., 2016a, 2018a) or saline solution extraction following homogenization and centrifugation (Szabo et al., 2016; Wang et al., 2018a).

3.2 Instrumental analysis

The analysis of the samples for metabolomics evaluation in the case of experiments with PFAS and FRs are based on MS or NMR. In the case of NMR, samples are directly injected and the analysis is based on ^{1}H-NMR at 298−300K, working between 500 and

Table 7.2: Sample preparation and instrumental analysis.

Continued

Flame retardants

Sample type	Extract preparation	Instrumental technique	Flame retardants	References
HepG2 cell line: human liver cancer cell line	1. UAE with chloroform: water: methanol. 2. Centrifugation. 3. Lyophilization of methanol fraction with water. 4. Dilution again in optimum solvent for NMR.	^{1}H-NMR: 600 MHz 298K Analyzed by Chenomx NMR Suite (version 7.0, Chenomx Inc., Canada) and the HMDB database	Metabolites linked to glucocorticoid and mineralocorticoid receptors, oxidative stress, osmotic pressure equilibrium, and amino acid metabolism. Potential biomarker trimethylamine N-oxide (TMAO) as a differential metabolite for six of the tested OPFRs.	Gu et al. (2019)
Human HepaRG cell cultures	1. Four hundred µL of a precooled methanol:MilliQ water (80%, v/v) transferred to a precooled vial containing 480 µL of chloroform and 285 µL MilliQ water. 2. Rinsed well with another 400 µL of the methanol:MilliQ water mixture (80%, v/v) and transferred to a vial. 3. Spiked with cholesterol-d4, TEHP and lauric acid-d3 (internal standards for the nonpolar fraction) and tryptophane-d5 and succinic acid-d4 (internal standards for the polar fraction).	UHPLC-ESI-Q-TOF-MS: Ionization source: ESI (positive and negative) Column: Kinetex XB-C18 100 A column (150 mm × 2.10 mm, 1.7 µm) Mobile phase: 10 mM ammonium acetate in MilliQ water and methanol Acquisition: full scan Resolution: 10,000	Alterations observed for metabolites: phosphatidylethanolamine, 1-acyl phosphatidylcholine, diacylglycerophosphate, diacylglycerol, phosphatidylcholine, acetylcholine, trimethylaminobutyraat, 5-D-(5/6)-5-C-(hydroxymethyl)-2,6-dihydroxycyclohex-2-en-1-one, acetylcarnitine, lysylvaline, isovalerylcarnitine, adenosine, methyladenosine, cortisolsulfate.	Van den Eede et al. (2015)

Nine OPFRs (TDBPP, TMPP, TPhP, TBOEP, TCIPP, TCEP, TEHP, TNBP, and TDCPP)

TPhP

Table 7.2: Sample preparation and instrumental analysis.—cont'd

Sample type	Extract preparation	Instrumental technique	Flame retardants	References
	4. Vortexed and centrifuged for 7 min at 2200 g. 5. Layer separation of polar and nonpolar fractions. 6. Polar fraction spiked with TPHP-d15 and diphenyl phosphate-d10 (external standard) and evaporated to dryness using a nitrogen flow at room temperature. 7. Nonpolar fraction spiked with triamyl phosphate and α-HBCD-d18 (external standard solution) and evaporated to dryness using a nitrogen flow. 8. Reconstitution in optimal LC solvent.			
Human cell lines	1. UAE with chloroform:water:methanol (1:0.9:1). 2. Collection of both phases for MS analyses and dried. 3. Polar and nonpolar extracts for positive ion mode analysis reconstituted in methanol:water (80:20, v/v) with 0.25% formic acid. 4. Polar and nonpolar extracts for negative ion mode analysis reconstituted in methanol:water (80:20, v/v) with 20 mM ammonium acetate.	LTQ FT: Ionization source: nanoelectrospray ion source (positive and negative) Column: direct infusion Acquisition: full scan	Identification of metabolic changes in taurine.	Zhang et al. (2016)
Indoor dust and TCEP, TCIPP, TDCPP, and HBCD				

Continued

HBCD	HepG2 cells	1. Addition of 1 mL of ultra-pure water. 2. Ultrasonic disruption in an ice–water bath for 3 min. 3. Suspension containing cell fragments transferred into an Eppendorf tube. 4. Freeze-drying simultaneously for the disrupted cells containing intracellular metabolites and collected culture medium containing extracellular metabolites. 5. Dissolved in 0.5 mL of 80% methanol, vortexed for 20 min, and centrifuged for 20 min at 13,000 g and 8°C. 6. Supernatant filtered by an organic-phase filter and transferred to a vial for metabolite analysis.	UHPLC-ESI-QTrap and UHPLC-ESI-Q-TOF: Ionization source: ESI (positive and negative) Column: UPLC BEH C8 column (2.1 mm × 100 mm, 1.7 μm, Waters, USA), positive ionization Acquity UPLC HSS T3 column (2.1 mm × 100 mm, 1.8 μm, Waters, USA), negative ionization Acquisition: full scan and multiple reaction monitoring	The main metabolic pathways perturbed are amino acid metabolism, protein biosynthesis, fatty acid metabolism, and phospholipid metabolism.	Wang et al. (2016b)
BDE-47	Mouse preadipocyte 3 T3-L1 cells	1. Cells washed with phosphate buffered saline solution twice. 2. Cells quenched with chilled methanol:water (80:20, v/v). 3. Centrifugation and the supernatant evaporated to dryness. 4. Reconstitution with 100 μL methanol:water (50:50, v/v) and spiked with 4-chloro-	LC-Q Orbitrap: Ionization source: ESI (positive and negative) Column: ACQUITY UPLC HSS T3 column (100 mm × 2.1 mm i.d., 1.7 μm particle size) Mobile phase: water (0.1% formic acid) and acetonitrile (0.1% formic acid)	Metabolites related to purine metabolism, pyrimidine metabolism, glutathione metabolism, citrate cycle, riboflavin metabolism, pentose and glucuronate interconversions, starch and sucrose metabolism, and glycolysis or gluconeogenesis process.	Yang et al. (2018a)

Table 7.2: Sample preparation and instrumental analysis.—cont'd

	Sample type	Extract preparation	Instrumental technique	Flame retardants	References
RDP	Human liver cells	phenylalanine as internal standard. Salting-out extraction in mixtures with acetonitrile:water with ammonium acetate.	Full scan acquisition mode LC-QTOF Ionization source: ESI Column: Agilent Extend Zorbax C18 column (50 mm × 2.1 mm, 3.5 μm) Mobile phase: 2 mM ammonium acetate (water) and acetonitrile	Transformation products coming from RDP in human liver.	Ballesteros-Gomez et al. (2015)
BDE-47	Rice	1. UAE with chloroform:water:methanol (2 mL, 2:2:5, v/v/v) for 40 min at room temperature. 2. Centrifugation. 3. Derivatization (silylation).	• Morphological analysis • GC-Q(MS) Ionization source: EI (70 eV) Column: DB-5 MS (30 m, 0.25 mm, 0.25 μm) Carrier gas: He Flow rate of 1.0 mL/min Transfer line: 280°C Ion source: 230°C	Metabolites including amino acids, saccharides, tricarboxylic acid cycle, and glyoxylate and dicarboxylate metabolism—related metabolites among others. Combination of the results with morphological analyses.	Chen et al. (2018)
α-, β-, and γ-HBCD	Pak choi leaves	1. Leaves crushed with a mortar in liquid nitrogen, freeze-dried, and transferred to an Eppendorf tube. 2. Addition of four nonendogenous metabolites and analyzed 24 h after for traceability purposes.	LC-QqTOF: Ionization source: ESI (positive and negative) Column: ZORBAX SB-Aq column (2.1 mm × 100 mm, 1.8 μm) Mobile phase: water	Metabolites were identified through open access databases METLIN and HMDB databases. Metabolites that change for the exposition to HBCD isomers include lipid metabolism	Zhang et al. (2018)

Continued

TPhP	Earthworms	3. Before analysis, addition of 200 μL of methanol, 540 μL of MTBE, and 360 μL of water sequentially. 4. Vortexed for 30 s before and after the addition of these compounds. 5. Mixture placed at room temperature for 10 min to allow equilibration. 6. The two phases transferred into a new tube and reduced to incipient dryness. 7. Reconstitution in methanol:water (1:4, v/v). 8. Centrifugation at 13,000 rpm for 5 min and the supernatant transferred into an LC-vial. Simultaneous extraction of exogenous and endogenous metabolites: 1. Flash-frozen and ground in liquid nitrogen using mortar and pestle. 2. 50 mg worm tissue extracted with 600 μL ice cold methanol:water (4:1, v/v) with N-9-Fmoc-L-glycine as internal standard. 3. Homogenization with three zirconium oxide beads (medium speed for 60 s)	(0.1% formic acid) and acetonitrile Acquisition: full scan and multiple reaction monitoring for structure confirmation GC-MS/MS: Ionization source: EI (70 eV) Column: DB-5 MS (30 m, 0.25 mm, 0.25 μm) Carrier gas: He Flow rate of 1.0 mL/min Transfer line: 290°C Ion source: 230°C Acquisition: full scan LC-QqTOF: Ionization source: ESI (positive and negative)	(i.e., eicosatrienoic acid), carbohydrate metabolism (i.e., L-galactonate, lactose), nucleotide synthesis (i.e., adenylic acid, adenine, guanine), secondary metabolites (luteolin, sinapic acid, coniferol), and amino acids (proline, glutamic acid, glutamine, isoleucine). Exogenous metabolites: phase I metabolites (i.e., x diphenyl phosphate, monohydroxylated, and di-hydroxylated TPhP) and phase II metabolites (i.e., thiol conjugates including mercaptolactic acid, cysteine, cysteinylglycine, and mercaptoethanol conjugates; and glucoside conjugates including glucoside, glucoside-phosphate, and $C_{14}H_{19}O_{10}P$ conjugates).	Wang et al. (2018b)

Table 7.2: Sample preparation and instrumental analysis.—cont'd

Sample type	Extract preparation	Instrumental technique	Flame retardants	References
	and two times in a Minilys homogenizer. 4. Mixture vortexed for 5 min and kept on dry ice for 15 min. 5. Centrifugation at 14,000 rpm for 10 min at 4°C. 6. Supernatant transferred to an LC-vial for LC-QTOF analysis. 7. 50 µL aliquot dried in a vacuum concentrator and derivatized for GC-MS analysis by silylation.	Column: Zorbax Eclipse Plus C18 column (2.1 mm × 100 mm, 3.5 µm) Mobile phase: 0.1% formic acid in water and acetonitrile for positive mode and 10 mM ammonium formate in water and acetonitrile in negative mode Acquisition: full scan and information-dependent acquisition in parallel	Endogenous metabolites: asparagine, leucine, lysine, phenylalanine, proline, serine, threonine, valine, glucose, inosine, phospholipids PE, and PC.	
Freshwater microalgae TPhP	1. Centrifugation of algal suspension (10,621 g, 5 min, 4°C). 2. Cell pellet mixed with 1 mL 80% cold methanol containing internal standard N-9-Fmoc-L-glycine. 3. Mixture freeze-thawed for three cycles in liquid nitrogen. 4. Centrifuged (10,621 g, 5 min, 4°C) and pellet sonicated for 20 min with 0.5 mL of the same extraction solvent as in 2. Supernatants of the two	GC-MS/MS: Ionization source: EI (70 eV) Column: DB-5 MS (30 m, 0.25 mm, 0.25 µm) Carrier gas: He Flow rate of 1.0 mL/min. Transfer line: 290°C Ion source: 230°C Acquisition: full scan LC-QqTOF: Ionization source: ESI (positive and negative) Column: Zorbax Eclipse Plus C18	Changes identified in the following metabolites: organic acids (fumarate malate), amino acids (valine, proline), sugars (sucrose, myo-inositol, trehalose, glyceryl-glucoside), phosphates (phosphate, glycerol-phosphate, adenosine monophosphate, phosphatidylglycerol), N-acyl-taurines, fatty acids (linolenic acid, eicosatrienoic acid, octadecadienoic acid, margaric acid), methyl esters (octadecenoic acid methyl ester, hexadecenoic acid	Wang et al. (2019)

Continued

Compound	Organism	Method	Instrument	Results	Reference
		extraction processes combined. 5. Direct analysis by LC-QTOF. 6. Aliquot of 100 μL supernatant dried in a vacuum concentrator and derivatized by silylation for GC-MS analysis.	column (2.1 mm × 100 mm, 3.5 μm) Mobile phase: 0.1% formic acid in water and acetonitrile for positive mode and 10 mM ammonium formate in water and acetonitrile in negative mode Acquisition: full scan and information-dependent acquisition in parallel	methyl ester), chlorophyll derivatives (pheophorbide a, chlorophyllone a, pyropheophorbide a, $C_{33}H_{32}N_4O_4$, $C_{33}H_{32}N_4O_5$, $C_{34}H_{34}N_4O_5$, $C_{32}N_4O_8$), diacylglycoglycerolipids (MGDG and DGDG), and monoacylglycerolipids (MGMG, lysoPC, lysoPI, lysoPG and lysoPE).	
TCEP, TBOEP, and TPhP with the presence of DOM	*Daphnia magna*	1. Daphnids flash-frozen in liquid nitrogen and lyophilized. 2. Metabolites extraction with 40 μL of a D2O 0.2 M phosphate buffer with NaN3 as a preservation agent and an internal calibrant. 3. Samples vortexed 30 s and 15 min sonication. 4. Centrifugation at 12,000 rpm for 20 min. 5. Supernatant pipetted and introduced into 1.7 mm capillary NMR tubes.	^{1}H NMR: 500 MHz Direct injection	Metabolites that changed within the treatments include amino acids.	Kovacevic et al. (2018)
pentaBDE and FM550	Hemolymph of *D. magna*	1. UAE with chloroform:water:methanol (4:2.85:4, v/v/v).	^{1}H-NMR: 600 MHz 298K Direct injection	Changes observed in histidine, glucose, phosphocholine, glutamine, creatine, acetate, alanine,	Scanlan et al. (2015)

Table 7.2: Sample preparation and instrumental analysis.—cont'd

Sample type	Extract preparation	Instrumental technique	Flame retardants	References
	2. Aqueous-phase reconstituted in 220 μL of 0.1M sodium phosphate buffered, deuterium oxide (pH 7.4) containing 20 μM sodium 2,2-dimethyl-2-silapentane-5-sulfonate (DSS).		valine, glutamate, choline, leucine, isoleucine, malate, and lysine metabolites.	
TPhP	Zebrafish liver	1. UAE with chloroform:water:methanol (4:3.6:4, v/v/v) at room temperature for 60 s in a vortex. 2. Centrifugation. 3. Lyophilization of methanol fraction with water. 4. Vacuum drying of the chloroform fraction. 5. Dilution again in optimum solvent for NMR.	^{1}H-NMR: 600.17 MHz 298.5K Analyzed by GSD, Global Spectral Deconvolution using MestReNova v6.1.0 −6224 software	Metabolomics analysis revealed significant changes in the contents of glucose, UDP-glucose, lactate, succinate, fumarate, choline, acetylcarnitine, and several fatty acids inducing to hypothesize that TPhP disturbs hepatic carbohydrate and lipid metabolism in zebrafish with just 7 days of exposure.	Du et al. (2016)
TBBPA and TCBPA	Marine medaka embryos	1. Offspring embryos in an Eppendorf tube. 2. Addition of steel balls and 600 μL methanol:water solution (4:1, v/v) and internal standard. 3. Homogenization at 33 times/s for 1.5 min and centrifuged at 12,000 rpm at 4°C for 15 min. 4. Four hundred an 80 μL supernatant vacuum-dried and addition of 50 μL methoxyamine solution.	GC-MS/MS: Ionization source: EI (70 eV) Column: DB-5 MS (30 m, 0.25 mm, 0.25 μm) Carrier gas: He Flow rate of 1.0 mL/min. Transfer line: 280°C Ion source: 230°C Acquisition: full scan	Metabolite pathways that were affected include glycolysis (glucose, sorbitol, fructose, mannose, glucose-6-phosphate, fructose-6-phosphate, pyruvic acid, lactic acid), in tricarboxylic acid cycle, and alanine, aspartate, and glutamate metabolism (malic acid, citrate, fumaric acid, succinic acid, alanine glutamic acid, asparagine, aspartic acid, N-acetyl-	Ye et al. (2016)

TBBPA	Mussel gills	5. Vortexed for 30 s and then placed in a 37°C water bath for 1.5 h of oximation reaction. 6. Silylation for subsequent GC-MS analysis. 1. UAE with chloroform:water:methanol. 2. Centrifugation. 3. Lyophilization of methanol fraction with water. 4. Dilution again in optimum solvent for NMR.	^{1}H-NMR: 500.18 MHz 298K Analyzed by custom-written ProMetab software in MATLAB version 7.0 (The MathsWorks, Natick, MA, USA) ^{1}H-NMR	aspartic acid), in nucleoside metabolism (threonine, glycine, inosine, ribose, serine, adenine hypoxanthine, uric acid), and in neural system (dopa). 9 metabolites and 67 proteins were altered in mussel gills. Combination of metabolomics results with proteomics analyses.	Ji et al. (2016)
OPFRs	Mice serum	1. Phosphate buffer with serum (1:1) 300 μL serum. 2. Homogenization and centrifugation.		The metabolites related to pathways of amino acid and energy metabolism showed more toxic effects when OPFRs were coexposed with MPLs than when the OPFRs were exposed alone. Combination of the results with histological analyses (liver and guts) and biochemical analyses with commercial kits for liver with superoxide dismutase (SOD), catalase (CAT), and malonaldehyde (MDA).	Deng et al. (2018)
BDE-47	Mice serum and striatum	1. Serum and striatum mixed with methanol (0.1% formic acid) for protein precipitation. 2. Centrifugation.	LC-Orbitrap MS Ionization source: ESI (positive and negative) Column: Waters ACQUITY UPLC HSS	Identification of metabolites through databases searching and authentic standards confirmation. Identification of metabolites	Ji et al. (2017)

Continued

Table 7.2: Sample preparation and instrumental analysis.—cont'd

Sample type	Extract preparation	Instrumental technique	Flame retardants	References
α-, γ-, and CM-HBCD	3. Supernatant dried under nitrogen and reconstituted in LC mobile phase.	T3 (1.8 μm, 2.1 mm × 100 mm) Mobile phase: 0.1% formic acid (water) and acetonitrile Resolution: 60,000 Full scan acquisition mode	related to purine metabolism, alanine, aspartate and glutamate, tryptophan, phenylalanine, and glutathione metabolisms.	Szabo et al. (2016)
Mice serum	1. 60 μL of serum mixed with 80 μL of solution containing 5 mM formate, 0.2% NaN3 in D2O, and 260 μL of saline (0.9% NaCl in D2O). 2. 400 μL of solution transferred into 5 mm NMR tube.	^{1}H NMR: 600 MHz 298K Direct injection	Changes observed in glucose, pyruvate, and alanine metabolites. Mice exposed to α-HBCD showed a decrease in amino acids glutamate (excitatory neurotransmitter in learning and memory), while mice exposed to γ-HBCD showed decrease in phenylalanine (neurotransmitter precursor).	
Dec 602	Sera and urine samples: 1. Thawed and vortexed at room temperature 100 mL sera or urine with 600 mL of ice-cold methanol:acetonitrile (1:9, v/v). 2. Vortexed for 2 min and sonicated for 1 min. 3. Centrifuged at 12,000 rpm for 15 min at 4°C. 4. Filtration of the supernatant through a 0.22 mm membrane filter then transferred to an LC vial.	UHPLC-ESI-IT-TOF-MS: Ionization source: ESI (positive and negative) Column: Eclipse plus C18 column (1.8 mm, 2.1 mm × 100 mm, Agilent) Mobile phase: 0.1% formic acid (water) and 0.1% formic acid (methanol) Acquisition: full scan	Endogenous metabolites including ubiquinone and other terpenoid quinone metabolites, phenylalanine, tyrosine, tryptophan, purine, amnoacyl, pyrimidine, among others.	Tao et al. (2019)
Mice sera and urine				

Continued

TPhP and DPP (metabolite of TPhP)	Mice serum	1. 200 µL serum + 300 µL saline solution + 100 µL D2O (5 mM formate). 2. Centrifugation at 14,000 g for 10 min. 3. 550 µL of the supernatant transferred into a 5-mm NMR tube.	¹H NMR: 600 MHz 300K Direct injection	Alterations observed in lipid metabolism (lipid, UFA, PUFA, choline, phosphocholine, taurine metabolites), pyruvate metabolism (lactate, acetate, pyruvate metabolites), citrate cycle (succinate, citrate, fumarate metabolites), and branch-chain amino acid metabolism (valine, leucine, isoleucine metabolites).	Wang et al. (2018a)
HBCD	Mice urine	Metabolic analysis by ¹H NMR: 1. 400 µL urine mixed with 400 µL phosphate buffer (pH 7.4). 2. Centrifugation at 12,000 rpm for 5 min. 3. Supernatant mixed with 50 µL of TSP-d4/D2O (29.02 mM) solution. Targeted amino acids by LC-MS/MS: 1. 50 µL urine + 20 µL internal standard. 2. Dilution to 200 µL with 0.1% formic acid in water. 3. Derivatization with propylchloroformate. 4. Extraction of derivates with 250 µL of ethyl acetate. 5. Supernatant evaporated to dryness under a nitrogen stream. 6. Redissolved in 200 µL 0.1% formic acid in water.	¹H NMR: 600 MHz 298K Direct injection LC-MS/MS: Ionization source: ESI positive Column: EZ:faast 4u AAA-MS Mobile phase: acetonitrile (0.1% formic acid) and water (0.1% formic acid) Acquisition: multiple reaction monitoring	Changes detected in some metabolites including some amino acids (alanine, lysine, and phenylalanine), malonic acid, trimethylamine, citric acid, 2-ketoglutarate, acetate, formate, trimethylamine (TMA), and 3-hydroxybutyrate metabolites.	Wang et al. (2016a)

Table 7.2: Sample preparation and instrumental analysis.—cont'd

Sample type	Extract preparation	Instrumental technique	Flame retardants	References
Mice urine and serum	1. 100 μL urine or serum + 300 μL methanol with 2-chloro-L-phenylalanine internal standard. 2. Vortexed for 5 min and centrifuged at 13,000 rpm, 4 C for 15 min. 3. 150 μL of supernatant added into the LC vials.	LC-QqTOF: Ionization source: ESI (positive and negative) Column: Waters XSelect HSS T3 (2.1 mm × 100 mm, 2.5 μm) Mobile phase: water (0.1% formic acid) and acetonitrile (0.1% formic acid) Acquisition: full scan and multiple reaction monitoring for structure confirmation	2-Aminoadenosine, 5-methylcytidine, adenosine, alexine, deoxythymidylate (dTMP), L-glutamate. Urine—alterations in L-homocysteine, L-prolinamide, lysoPC, malonylcarnitine, mevalonic acid, N-acetylserotonin, N-formylmethionine, nicotinamide N-oxide, phosphatidylglycerol, phosphocreatine, prenalterol, pyroglutamic acid, S-(formylmethyl) glutathione, suberylglycine, ubiquinone, 3-hydroxy-L-proline, acetylpyruvate, alanopine, β-hydroxyarginine, creatine, D-biotin, L-leucine, L-thyronine, L-tyrosine, lyso-PAF, PG, PGE2, reduced riboflavin (vitamin B2), succinic anhydride, and xanthurenic acid metabolites. Serum—alterations in N-acetyl-L-phenylalanine, L-proline, L-valine, ketoleucine, L-leucine, 2-phenylacetamide, L-	Wei et al. (2018)
BDE-3				

Continued

| TCEP | Brain rats | 1. Brain tissues dissected and snap frozen in liquid nitrogen.
2. 200 mg of the frozen cerebrum suspended in methanol and double distilled water and vortexed.
3. Addition of chloroform and 50% chloroform.
4. Suspension kept on ice for 30 min and centrifuged at 10,009 g for 30 min at 4°C.
5. Upper phase (aqueous phase) collected and evaporated to dryness under a stream of nitrogen.
6. Residue reconstituted with 580 mL D2O with 3-trimethylsilyl 1-[2,2,3,3,-2H4]propionate.
7. Centrifugation at 12,000 g for 5 min.
8. Supernatant transferred to a 5-mm NMR tube. | ¹H NMR | methionine, 4-imidazolone-5-propionic acid, ʟ-carnitine, ʟ-phenylalanine, ʟ-tyrosine, phenyl sulfate, ʟ-tyrosine, 2-hendecenoic acid, ʟ-tryptophan, pyrocatechol sulfate, indoxylsulfuric acid, dodecanedioic acid, inosine, lysoPE, lysoPC, and cholic acid glucuronide metabolites.
Observed metabolites that can be biomarkers of exposition include glutamate, γ-aminobutyric acid, N-acetyl-ᴅ-aspartate, creatine and lactic acid metabolites, taurine, myo-inositol, and choline metabolites. | Yang et al. (2018b) |

Table 7.2: Sample preparation and instrumental analysis.—cont'd

Sample type	Extract preparation	Instrumental technique	Flame retardants	References
		Perfluoroalkyl substances		
D. magna PFOS in parallel with propranolol, and atrazine separately	1. Sixty to 100 individuals of neonates (approximately 1 mg). 2. Daphnids homogenized into a fine powder with a 5-mm stainless steel spatula. 3. Addition of 45 μL of 0.2M phosphate-buffered in D2O containing 0.1 w/v sodium azide and 2,2-dimethyl-2-silapentane-5-sulfonate sodium salt (internal calibrate). 4. Vortexed for 30 s and sonicated for 15 min. 5. Centrifugation at 14,000 rpm at 48°C for 20 min. 6. Supernatant removed and transferred into a 1.7-mm NMR tube.	^{1}H NMR: 500 MHz Direct injection	Changes observed in energy metabolic pathways and protein degradation pathways.	Wagner et al. (2017)
D. magna PFOS	1. Daphnids flash-frozen in liquid nitrogen and lyophilized to limit enzyme activity. 2. Metabolites extraction with 40 μL of a D2O 0.2 M phosphate buffer with NaN3 as a preservation agent and an internal calibrant. 3. Samples vortexed 45 s and 15 min sonication.	^{1}H NMR: 500 MHz Direct injection	Metabolites that are affected include amino acids (alanine, arginine, asparagine, isoleucine, glutamate, glutamine, glycine, leucine, lysine, phenylalanine, serine, threonine, tryptophan, tyrosine, valine), sugar glucose, and maltose, uracil, and the energy molecule ATP.	Ding et al. (2009)

Analyte/condition	Organism	Procedure	Analytical technique	Findings	Reference
PFOS spiked in real effluent wastewater	*D. magna*	1. Daphnids flash-frozen in liquid nitrogen and lyophilized. 2. Metabolites extraction with 45 µL of a D2O 0.2 M phosphate buffer with NaN3 as a preservation agent and an internal calibrant. 3. 15 min sonication. 4. Centrifugation at 14,000 rpm for 20 min. 5. Supernatant pipetted and introduced into 1.7 mm capillary NMR tubes.	^{1}H NMR: 500 MHz Direct injection	Metabolites that changed within the treatments include amino acids, sugar metabolites, and energy molecules.	Wagner et al. (2019)
PFOS and other 15 xenobiotics	Larvae zebrafish	1. Microcentrifuge tube containing 80 animals + zirconium oxide beads + 200 µL methanol. 2. Samples blended for 2 min and centrifugation for 30 s at 10,000 rpm. 3. Methanol extract removed and process repeated with 200 µL of chloroform. 4. Aliquot 1 dried under nitrogen stream and derivatized for amino acids and biogenic amines analyses previous addition of corresponding internal standards.	LC-MS/MS: (amino acids and biogenic amines) Ionization source: ESI (positive) Column: Agilent Zorbax Eclipse XDB-C18 Acquisition: multiple reaction monitoring ($\sum$Hexose, bile acids and fatty acids) Ionization source: ESI (negative) Column: Waters XTerra MS C18	21 amino acids, 21 biogenic amines, 4 bile acids, $\sum$Hexose, 17 fatty acids, 40 acylcarnitines, 89 phosphatidylcholine, and 15 sphingomyelins.	Huang et al. (2016)

Continued

Table 7.2: Sample preparation and instrumental analysis.—cont'd

Sample type	Extract preparation	Instrumental technique	Flame retardants	References
	5. Aliquot 2 added to a well plate with corresponding internal standards for analyses of the rest of metabolites. 6. Dilution of aliquot 2 with 5 mM ammonium acetate (methanol) up to 250 μL. 7. Shaked at 300 rpm for 10 min and 80 μL of aliquot 2 removed and diluted with equal amount of methanol for lipid analysis. 8. Remaining portion of aliquot 2 diluted with an equal amount of water for bile acids, fatty acids, and hexose analyses.	Acquisition: multiple reaction monitoring (Acylcarnitines, phosphatidylcholine, and sphingomyelins Ionization source: ESI (positive) Column: no column/flow injection Acquisition: multiple reaction monitoring		
PFOS, in parallel experiments with bisphenol A and tributyltin separately	Zebrafish (*Danio rerio*) embryos	1. Embryos thawed in water bath at room temperature. 2. Extraction with 900 μL of methanol containing methionine sulfone (surrogate). 3. Vortexed 15 s, sonicated for 15 min and vortexed again 15 s. Centrifugation at 23,500 g for 10 min at 4°C. 4. Supernatant + 500 μL of water + 300 μL of chloroform. 5. Vortexed 15 s, placed on ice for 10 min and centrifuged. 6. Aqueous fractions evaporated to dryness under nitrogen gas. 7. Reconstituted with acetonitrile:water (1:1, v/v) containing internal standard.	LC-HRMS QExative: Ionization source: heated ESI (positive and negative) Column: HILIC—TSK Gel Amide-80 (250 mm × 2.1 mm; 5 μm) Mobile phase: acetonitrile and 5 mM of ammonium acetate in water Acquisition: full scan	Changes detected on 25 metabolites. Metabolites related to metabolism of glycerophospholipids, amino acids, purines, and 2-oxocarboxylic acids.	Ortiz-Villanueva et al. (2018)

PFOA	Mice brain and liver	8. Extracts filtrated with 0.22 μm filters at 11,000 g for 4 min and stored at −80°C until the analysis. 1. Frozen tissue homogenized in extraction buffer using Precellys beads. 2. Centrifugation and supernatant collected for metabolites purification. 3. Concentration of metabolites with AbsoluteIDQTM p180 kits (Biocrates life sciences, Innsbruck, Austria), amino acids, acylcarnitines, and dicarboxylacylcarnitines, sphingomyelins, phosphatidylcholines, and biogenic amines by LC-MS. 4. Sample silylation for fatty acids determination by GC-MS.	GC-MS: Ionization source: EI (70 eV) LC-QTrap: Ionization source: ESI	Metabolites evaluation including amino acids, acylcarnitines, and dicarboxylacylcarnitines, sphingomyelins, phosphatidylcholines, biogenic amines, and fatty acids.	Yu et al. (2016)
PFOS and PFOA	*Gallus gallus domesticus* embryos liver and hearts	1. Addition of PBS buffer to samples (1:1) and homogenization. 2. Aliquot + internal standard + control + methanol and vortexed. 3. Addition of MTBE and kept at room temperature for 60 min. 4. Water addition, 10 min incubation, and centrifugation at 1000 rcf for 10 min.	GC-MS: Ionization source: EI (70 eV) Column: DB-5 MS (30 m, 0.25 mm, 0.25 μm) Carrier gas: He Acquisition: selected ion monitoring	Fatty acid metabolites.	Wigh (2017)

Continued

Table 7.2: Sample preparation and instrumental analysis.—cont'd

Sample type	Extract preparation	Instrumental technique	Flame retardants	References
Human serum in a cohort study	5. Organic phase evaporated under nitrogen. 6. Derivatization of the extract by fatty acid methyl esters. 1. 600 μL cold methanol added to 200 μL serum and shaken. 2. Mixture stored for 10 min and then centrifuged at 12,000 g for 10 min. 3. Supernatant filtered through a 0.22 mm syringe filter.	HPLC-ESI-Q-Orbitap: Ionization source: ESI (negative) Column: Kinetex XB-C18 100 A column (150 mm × 2.10 mm, 2.6 μm) Mobile phase: 0.1% formic acid in water and 0.1% formic acid in methanol Acquisition: full scan	Metabolites that changed in compared groups include D-glucurono-6,3-lactone, α-carboxyethyl hydroxychromanol, arachidonic acid, hypoxanthine, oxoglutaric acid, pyroglutamic acid, tetrahydrobiopterin, and xanthine; deoxyarabinohexonic acid, and hydroxybutyric acid.	Wang et al. (2017)

Mixtures

Sample type	Extract preparation	Instrumental technique	Flame retardants	References
Human cell lines	1. Cell medium extracted using liquid—liquid extraction. 2. 85 μL of sample were used for analysis in addition to 10 μL of internal standard.	LC-MS/MS: Ionization source: ESI (positive and negative) Acquisition: multiple reaction monitoring	Metabolites related to endocrine disruption including pregnenolone, progesterone, 11-deoxycorticosterone, corticosterone, aldosterone, 17-hydroxypregnenolone, 17-hydroxyprogesterone, cortisol, cortisone, estrone, estroneS, estradiol, estriol, 7-ketodehydroepiandrosterone (DHEA), testosterone, and 5α-dihydrotestosterone (DHT).	Ahmed et al. (2019)

(Row labels at far left: "Mixture of PFAS present in human serum samples" and "Mix of POPs and EDCs")

600 MHz. Afterward, the data must be deconvoluted using specific software such as MestReNova (Du et al., 2016), Chenomx NMR Suite (Gu et al., 2019), or a custom-written deconvolution software in MATLAB (Ji et al., 2016).

Regarding MS, different types of strategies are adopted according to the endpoint. Three groups can be identified including untargeted metabolomics, targeted metabolomics, and imaging metabolomics. According to Johnson et al. (2016), untargeted metabolomics allow metabolites assessment extracted from the sample and can reveal novel and unanticipated perturbations. Their main advantage is that they can provide a detailed assessment of the metabolites. Untargeted methodologies are most effective working with high-resolution mass spectrometry (HRMS). However, there are many unknown metabolites that remain unannotated in metabolite databases which increase the complexity of their identification. In addition, those databases are dependent on pH, solvent, chromatographic column, and ionization technique, among others (Johnson et al., 2016). There are some examples of this strategy, which that has been applied to metabolomics characterization of experiments performed with FRs (Ji et al., 2017; Wang et al., 2018b, 2019) and PFAS (Kingsley et al., 2018; Ortiz-Villanueva et al., 2018). Conversely, targeted metabolomics measure the concentration of a predefined set of metabolites (Ahmed et al., 2019; Johnson et al., 2016; Shao et al., 2018; Wang et al., 2016a; Wigh, 2017). Finally, the localization of selected metabolites within some tissue can be performed by imaging MS techniques such as matrix-assisted laser desorption ionization, nanostructure-imaging mass spectrometry, desorption electrospray ionization mass spectrometry, and secondary ion mass spectrometry, among others (Johnson et al., 2016).

MS is nowadays considered the best option for metabolomics studies due to its higher sensitivity compared to NMR (Bloszies and Fiehn, 2018). MS is used during the identification of metabolites of interest involved in the metabolomics pathway of different organisms trying to identify the biomarker related to the presence of an exogenous compound such as FRs or PFAS. In addition, these techniques are also applied for the tentative identification of exogenous compound transformation products generated during the metabolomics processes. In general, low-resolution mass spectrometry is used when the work is focused on targeted metabolites (Wang et al., 2016b). However, researchers tend to work with HRMS because of the high sensitivity of the new-generation instruments when working with exact mass. This last instrumentation allows analyzing samples by performing a full scan mass-to-charge ratio (m/z) screening that can later be used for the tentative identification of known and unknown (suspect) metabolites within an error below 5 ppm. The most commonly used analyzers are based on time of flight (TOF) and Orbitrap as well as hybrid quadrupole-TOF (QToF) (Ballesteros-Gomez et al., 2015), hybrid quadrupole-Orbitrap (QExactive), hybrid ion trap-Orbitrap (IT-Orbitrap), or by direct infusion without chromatographic separation in a linear ion trap with a Fourier-transform ion cyclotron resonance (LTQ FT-ICR) MS detector (Zhang et al., 2016).

HRMS is usually coupled to LC or GC providing the necessary separation for isomers in some cases (Zhang et al., 2018) or minimizing the coelution of some metabolites that occurs when working by direct infusion.

In the case of the identification of metabolites generated due to the presence of an exogenous substance like FRs and PFAS, it is necessary to assume that the target compound in the metabolome complex system could be chemically or enzymatically modified (Bloszies and Fiehn, 2018). For this reason, the use of in silico compound libraries or databases that would help during metabolite identification is indispensable. These tools propose transformation products based on solid chemical or biochemical foundations that can be generated during different metabolism pathways by cytochrome P (CYP), phosphatases, alcohol dehydrogenases (ADH) enzymes, sulfotransferase (SULT) enzymes, and glucuronosyltransferase (UGT) enzymes, among others. One example is the prediction software Meteor Nexus program (Lhasa limited, UK) used during the evaluation of in vitro human metabolism of the resorcinol bis(diphenyl phosphate) (RDP) (Ballesteros-Gomez et al., 2015). The other option is to use open access metabolomics tools such as KEGG database (KEGG_PATHWAY) used to identify the biochemical pathways, Metabolic In Silico Network Expansions (MINEs) (ChemAxon, 2014) that predicts the formation of metabolites from known enzymatic processes (Bloszies and Fiehn, 2018), as well as the METLIN Metabolomics Database (METLIN), MassBank (MassBank), and the Human Metabolome Database (HMDB). Other specific software for data interpretation include the open access Madison Metabolomics Consortium Database (Cui et al., 2008; Wagner et al., 2017) (specific for NMR analyses) and nonopen access SIEVE software (from Thermo Fisher) (Wang et al., 2017) for HRMS data.

On the other side, the data generated must be statistically analyzed with advanced chemometrics data analysis methods in order to elucidate the different patterns showed by the exposure organisms versus the nonexposed organisms. Most of the works related to metabolomics effects of the exposition to FRs and PFAS are chemometrically processed using MATLAB (Ji et al., 2016; MathWorks; Ortiz-Villanueva et al., 2018), although other open access software is available such as R analysis software. However, the use of this software needs specialized researchers.

4. Future trends

The metabolomics strategies for the investigation of contaminants of industrial origin are nowadays mainly based on in vivo experiments and, in most cases, on the evaluation of just one chemical. Currently, researchers are starting to focus in the development of non-aggressive methods for those investigations. An example is the increasing number of in vitro experiments. In addition, the interaction of chemicals among them in the environment should be addressed more consistently. In this context, the efforts are focused on the assessment of metabolomics effects of a mixture of chemicals or of a real environmental sample that, for instance, contains different amounts of dissolved organic matter or other co-contaminants. The interaction of those chemicals with the other co-contaminants, such as MPLs or DOM, must be evaluated since metabolomics effects could be modulated. Furthermore, it is important to combine metabolomics studies not only with other omics but also with other type of toxicological studies as mentioned in this chapter. Finally, the use of new-generation HRMS for the study of metabolites offers a good alternative to conventional 1H-RMN.

References

Ahmed, K.E.M., Frøysa, H.G., Karlsen, O.A., Blaser, N., Zimmer, K.E., Berntsen, H.F., Verhaegen, S., Ropstad, E., Kellmann, R., Goksøyr, A., 2019. Effects of defined mixtures of POPs and endocrine disruptors on the steroid metabolome of the human H295R adrenocortical cell line. Chemosphere 218, 328–339.

Alaee, M., Arias, P., Sjödin, A., Bergman, Å., 2003. An overview of commercially used brominated flame retardants, their applications, their use patterns in different countries/regions and possible modes of release. Environ. Int. 29, 683–689.

Ballesteros-Gomez, A., Van den Eede, N., Covaci, A., 2015. In vitro human metabolism of the flame retardant resorcinol bis (diphenylphosphate)(RDP). Environ. Sci. Technol. 49, 3897–3904.

Bloszies, C.S., Fiehn, O., 2018. Using untargeted metabolomics for detecting exposome compounds. Curr. Opin. Toxicol. 8, 87–92.

ChemAxon, 2014. Metabolic In Silico Network Expansion Databases (MINE).

Chen, J., Li, K., Le, X.C., Zhu, L., 2018. Metabolomic analysis of two rice (Oryza sativa) varieties exposed to 2, 2′, 4, 4′-tetrabromodiphenyl ether. Environ. Pollut. 237, 308–317.

Cui, Q., Lewis, I.A., Hegeman, A.D., Anderson, M.E., Li, J., Schulte, C.F., Westler, W.M., Eghbalnia, H.R., Sussman, M.R., Markley, J.L., 2008. Metabolite identification via the madison metabolomics consortium database. Nat. Biotechnol. 26, 162–164.

Deng, Y., Zhang, Y., Qiao, R., Bonilla, M.M., Yang, X., Ren, H., Lemos, B., 2018. Evidence that microplastics aggravate the toxicity of organophosphorus flame retardants in mice (*Mus musculus*). J. Hazard Mater. 357, 348–354.

Ding, L., Hao, F., Shi, Z., Wang, Y., Zhang, H., Tang, H., Dai, J., 2009. Systems biological responses to chronic perfluorododecanoic acid exposure by integrated metabolomic and transcriptomic studies. J. Proteome Res. 8, 2882–2891.

Du, Z., Zhang, Y., Wang, G., Peng, J., Wang, Z., Gao, S., 2016. TPhP exposure disturbs carbohydrate metabolism, lipid metabolism, and the DNA damage repair system in zebrafish liver. Sci. Rep. 6, 21827.

Gu, J., Su, F., Hong, P., Zhang, Q., Zhao, M., 2019. 1H NMR-based metabolomic analysis of nine organophosphate flame retardants metabolic disturbance in Hep G2 cell line. Sci. Total Environ. 665, 162–170.

HMDB, The Human Metabolome Database.

Hou, R., Xu, Y., Wang, Z., 2016. Review of OPFRs in animals and humans: absorption, bioaccumulation, metabolism, and internal exposure research. Chemosphere 153, 78–90.

Huang, S.S.Y., Benskin, J.P., Chandramouli, B., Butler, H., Helbing, C.C., Cosgrove, J.R., 2016. Xenobiotics produce distinct metabolomic responses in Zebrafish larvae (*Danio rerio*). Environ. Sci. Technol. 50, 6526.

Ji, C., Li, F., Wang, Q., Zhao, J., Sun, Z., Wu, H., 2016. An integrated proteomic and metabolomic study on the gender-specific responses of mussels to tetrabromobisphenol A (TBBPA). Chemosphere 144, 527–539.

Ji, F., Luan, H., Huang, Y., Cai, Z., Li, M., 2017. MS-based metabolomics for the investigation of neuro-metabolic changes associated with BDE-47 exposure in C57bl/6 mice. J. Anal. Test. 1, 233–244.

Johnson, C.H., Ivanisevic, J., Siuzdak, G., 2016. Metabolomics: beyond biomarkers and towards mechanisms. Nat. Rev. Mol. Cell Biol. 17, 451.

Kariuki, M., Nagato, E., Lankadurai, B., Simpson, A., Simpson, M., 2017. Analysis of sub-lethal toxicity of perfluorooctane sulfonate (PFOS) to daphnia magna Using 1H nuclear magnetic resonance-based metabolomics. Metabolites 7, 15.

KEGG_PATHWAY, KEGG PATHWAY: Metabolic Pathways.

Kingsley, S.L., Walker, D.I., Chen, A., Calafat, A., Lanphear, B., Yolton, K., Pennell, K., Braun, J.M., 2018. Metabolomics of childhood exposure to perfluorooctanoic acid: a cross-sectional study. In: ISEE Conference Abstracts.

Kovacevic, V., Simpson, A., Simpson, M., 2018. Investigation of *Daphnia magna* sub-lethal exposure to organophosphate esters in the presence of dissolved organic matter using 1H NMR-based metabolomics. Metabolites 8, 34.

Lai, K.P., Lee, J.C., Wan, H.T., Li, J.W., Wong, A.Y., Chan, T.F., Oger, C., Galano, J.M., Durand, T., Leung, K.S., Leung, C.C., Li, R., Wong, C.K., 2017. Effects of in utero PFOS exposure on transcirptome, lipidome, and function of mouse testis. Environ. Sci. Technol. 51, 8782.

Lau, C., Anitole, K., Hodes, C., Lai, D., Pfahles-Hutchens, A., Seed, J., 2007. Perfluoroalkyl acids: a review of monitoring and toxicological findings. Toxicol. Sci. 99, 366−394.

Llorca, M., Farré, M., Picó, Y., Müller, J., Knepper, T.P., Barceló, D., 2012. Analysis of perfluoroalkyl substances in waters from Germany and Spain. Sci. Total Environ. 431, 139−150.

MassBank, NORMAN MassBank.

MathWorks, MATLAB (Matrix Laboratory): Maths, Graphs, Computational.

METLIN, METLIN Metabolomics Database.

Mikula, P., Svobodova, Z., 2006. Brominated flame retardants in the environment: their sources and effects (a review). Acta Vet. BRNO 75, 587−599.

Ortiz-Villanueva, E., Jaumot, J., Martínez, R., Navarro-Martín, L., Piña, B., Tauler, R., 2018. Assessment of endocrine disruptors effects on zebrafish (*Danio rerio*) embryos by untargeted LC-HRMS metabolomic analysis. Sci. Total Environ. 635.

Peden-Adams, M., Keller, J., Eudaly, J., Berger, J., Gilkeson, G., Keil, D., 2008. Suppression of humoral immunity in mice following exposure to perfluorooctane sulfonate. Toxicol. Sci. 104, 144−154.

Salihovic, S., Fall, T., Ganna, A., Broeckling, C.D., Prenni, J.E., Hyötyläinen, T., Kärrman, A., Lind, P.M., Ingelsson, E., Lind, L., 2018. Identification of metabolic profiles associated with human exposure to perfluoroalkyl substances. J. Expo. Sci. Environ. Epidemiol. 1.

Scanlan, L.D., Loguinov, A.V., Teng, Q., Antczak, P., Dailey, K.P., Nowinski, D.T., Kornbluh, J., Lin, X.X., Lachenauer, E., Arai, A., 2015. Gene transcription, metabolite and lipid profiling in eco-indicator Daphnia magna indicate diverse mechanisms of toxicity by legacy and emerging flame-retardants. Environ. Sci. Technol. 49, 7400−7410.

Shao, X., Ji, F., Wang, Y., Zhu, L., Zhang, Z., Du, X., Chung, A.C.K., Hong, Y., Zhao, Q., Cai, Z., 2018. Integrative chemical proteomics-metabolomics approach reveals Acaca/Acacb as direct molecular targets of PFOA. Anal. Chem. 90, 11092−11098.

Skov, K., Hadrup, N., Smedsgaard, J., Frandsen, H.L., 2015. Metabolomics−an Analytical Strategy for Identification of Toxic Mechanism of Action. National Food Institute, Technical University of Denmark.

Szabo, D.T., Pathmasiri, W., Sumner, S., Birnbaum, L.S., 2016. Serum metabolomic profiles in neonatal mice following oral brominated flame retardant exposures to hexabromocyclododecane (HBCD) alpha, gamma, and commercial mixture. Environ. Health Perspect. 125, 651−659.

Tao, W., Tian, J., Xu, T., Xu, L., Xie, H.Q., Zhou, Z., Guo, Z., Fu, H., Yin, X., Chen, Y., 2019. Metabolic profiling study on potential toxicity in male mice treated with Dechlorane 602 using UHPLC-ESI-IT-TOF-MS. Environ. Pollut. 246, 141−147.

Van den Eede, N., Cuykx, M., Rodrigues, R.M., Laukens, K., Neels, H., Covaci, A., Vanhaecke, T., 2015. Metabolomics analysis of the toxicity pathways of triphenyl phosphate in HepaRG cells and comparison to oxidative stress mechanisms caused by acetaminophen. Toxicol. In Vitro 29, 2045−2054.

van der Veen, I., de Boer, J., 2012. Phosphorus flame retardants: properties, production, environmental occurrence, toxicity and analysis. Chemosphere 88, 1119–1153.

Wagner, N.D., Helm, P.A., Simpson, A.J., Simpson, M.J., 2019. Metabolomic responses to pre-chlorinated and final effluent wastewater with the addition of a sub-lethal persistent contaminant in *Daphnia magna*. Environ. Sci. Pollut. Control Ser. 1–13.

Wagner, N.D., Simpson, A.J., Simpson, M.J., 2017. Metabolomic responses to sublethal contaminant exposure in neonate and adult *Daphnia magna*. Environ. Toxicol. Chem. 36.

Wang, D., Zhang, P., Wang, X., Wang, Y., Zhou, Z., Zhu, W., 2016a. NMR-and LC–MS/MS-based urine metabolomic investigation of the subacute effects of hexabromocyclododecane in mice. Environ. Sci. Pollut. Control Ser. 23, 8500–8507.

Wang, D., Zhu, W., Chen, L., Yan, J., Teng, M., Zhou, Z., 2018a. Neonatal triphenyl phosphate and its metabolite diphenyl phosphate exposure induce sex-and dose-dependent metabolic disruptions in adult mice. Environ. Pollut. 237, 10–17.

Wang, F., Zhang, H., Geng, N., Zhang, B., Ren, X., Chen, J., 2016b. New insights into the cytotoxic mechanism of hexabromocyclododecane from a metabolomic approach. Environ. Sci. Technol. 50, 3145–3153.

Wang, L., Huang, X., Laserna, A.K.C., Li, S.F.Y., 2018b. Untargeted metabolomics reveals transformation pathways and metabolic response of the earthworm *Perionyx excavatus* after exposure to triphenyl phosphate. Sci. Rep. 8, 16440.

Wang, L., Huang, X., Lim, D.J., Laserna, A.K.C., Li, S.F.Y., 2019. Uptake and toxic effects of triphenyl phosphate on freshwater microalgae *Chlorella vulgaris* and *Scenedesmus obliquus*: insights from untargeted metabolomics. Sci. Total Environ. 650, 1239–1249.

Wang, X., Liu, L., Zhang, W., Zhang, J., Du, X., Huang, Q., Tian, M., Shen, H., 2017. Serum metabolome biomarkers associate low-level environmental perfluorinated compound exposure with oxidative/nitrosative stress in humans. Environ. Pollut. 229, 168–176.

Wei, Z., Xi, J., Gao, S., You, X., Li, N., Cao, Y., Wang, L., Luan, Y., Dong, X., 2018. Metabolomics coupled with pathway analysis characterizes metabolic changes in response to BDE-3 induced reproductive toxicity in mice. Sci. Rep. 8, 5423.

Wigh, V., 2017. Metabolomic Study of the Effects of Perfluorinated Compounds on the Fatty Acid Metabolism during the Development of Gallus gallus Domesticus.

Yang, C., Wong, C.-M., Wei, J., Chung, A.C., Cai, Z., 2018a. The brominated flame retardant BDE 47 upregulates purine metabolism and mitochondrial respiration to promote adipocyte differentiation. Sci. Total Environ. 644, 1312–1322.

Yang, W., Zhao, F., Fang, Y., Li, L., Li, C., Ta, N., 2018b. 1H-nuclear magnetic resonance metabolomics revealing the intrinsic relationships between neurochemical alterations and neurobehavioral and neuropathological abnormalities in rats exposed to tris (2-chloroethyl) phosphate. Chemosphere 200, 649–659.

Ye, G., Chen, Y., Wang, H.-o., Ye, T., Lin, Y., Huang, Q., Chi, Y., Dong, S., 2016. Metabolomics approach reveals metabolic disorders and potential biomarkers associated with the developmental toxicity of tetrabromobisphenol A and tetrachlorobisphenol A. Sci. Rep. 6, 35257.

Yu, N., Wei, S., Li, M., Yang, J., Li, K., Jin, L., Xie, Y., Giesy, J.P., Zhang, X., Yu, H., 2016. Effects of perfluorooctanoic acid on metabolic profiles in brain and liver of mouse revealed by a high-throughput targeted metabolomics approach. Sci. Rep. 6, 23963.

Zhang, J., Abdallah, M.A.-E., Williams, T.D., Harrad, S., Chipman, J.K., Viant, M.R., 2016. Gene expression and metabolic responses of HepG2/C3A cells exposed to flame retardants and dust extracts at concentrations relevant to indoor environmental exposures. Chemosphere 144, 1996–2003.

Zhang, J., Williams, T.D., Abdallah, M.A.-E., Harrad, S., Chipman, J.K., Viant, M.R., 2015. Transcriptomic and metabolomic approaches to investigate the molecular responses of human cell lines exposed to the flame retardant hexabromocyclododecane (HBCD). Toxicol. In Vitro 29, 2116–2123.

Zhang, Y., Guo, Q., Tan, D., He, Z., Wang, Y., Liu, X., 2018. Effects of low-levels of three hexabromocyclododecane diastereomers on the metabolic profiles of pak choi leaves using high-throughput untargeted metabolomics approach. Environ. Pollut. 242, 1961–1969.

CHAPTER 8

Mass spectrometry to explore exposome and metabolome of organisms exposed to pharmaceuticals and personal care products

Frédérique Courant, Hélène Fenet, Bénilde Bonnefille, Thibaut Dumas, Elena Gomez

HydroSciences, Univ Montpellier, CNRS, IRD, Montpellier, France

Chapter Outline

1. Pharmaceuticals and personal care products are widespread in the environment

Pharmaceuticals and personal care products (PPCPs) have been detected in the environment worldwide. The first studies on the presence of drugs in the environment were conducted in the 1970s, whereas personal care products (PCPs) have been the focus of more recent attention. Research on PPCPs in the environment has been increasing exponentially since the late 1990s (Daughton and Ternes, 1999), bolstered by novel state-of-the-art analytical methods. Pharmaceutical products (PPs) and PCPs are now recognized

Environmental Metabolomics. https://doi.org/10.1016/B978-0-12-818196-6.00008-X

environmental contaminants that are often introduced by human and livestock breeding activities (Boxall et al., 2012; Daughton, 2016).

Treated or untreated wastewaters are the main sources of water contamination by human PPs and some PCPs. Human therapeutic uses account for the release of PPs, as parent compounds or metabolites, into wastewater via urine and feces. PCPs include chemical ingredients in hygienic and cosmetic products, such as soaps, detergents, sun creams, moisturizers, lipsticks, hair dyes, deodorants, lotions, creams, shampoos, toothpastes, and repellents (Hopkins and Blaney, 2016). These compounds are also emitted into the environment via domestic or industrial wastewater (Fig. 8.1). Over a 100 different PPCP molecules have thus been screened for and detected in wastewater, as mentioned in a recent review (Tran et al., 2018). As wastewater treatment plants are not designed to remove these contaminants, PPCP residues are continuously discharged into the receiving waterbodies, as well as in soils when treated or untreated wastewater is used for crop irrigation. Moreover, some PPCPs are present in sewage sludge or in manure from treated animals applied as fertilizer on soils (Verlicchi and Zambello, 2015). In addition, some PPs administered to animals could directly reach the soil and some PPCPs are also directly released in water via

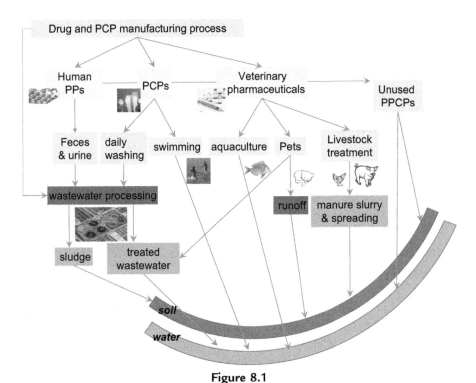

Figure 8.1

Pharmaceutical and personal care product (PPCP) fate and transport in the environment. *PCP, personal care product; PP, pharmaceutical product.*

PPs from aquaculture sources and via PCPs during recreational activities, as shown for ultraviolet (UV) filters and musks detected in lakes and marine water (Langford et al., 2015). Consequently, PPCPs are being detected worldwide in surface water (Brausch and Rand, 2011; Pal et al., 2010; Balakrishna et al., 2017; Madikizela et al., 2017), groundwater (Lapworth et al., 2012), and coastal zones (Arpin-Pont et al., 2014; Gaw et al., 2014). They are also present in soil, biota, and foodstuffs (Álvarez-Muñoz et al., 2015; Bourdat-Deschamps et al., 2017; Christou et al., 2019; Li, 2014; Madikizela et al., 2018).

PPCPs detected and quantified in the aquatic environment worldwide in previously reported monitoring studies belong to different classes. For PPs, classes such as analgesics, antipyretic and antiinflammatory drugs (acetaminophen, diclofenac, ibuprofen, etc.), antibiotics (e.g., ciprofloxacin), anticonvulsants/antidepressants (carbamazepine, venlafaxine, etc.), beta-blocking agents (metoprolol, etc.), and lipid-regulating drugs (fibrates) have been investigated (Vandermeersch et al., 2015). For PCPs, research data have been reported on UV filters and stabilizers, antimicrobials, and preservatives such as parabens and musks (Vandermeersch et al., 2015). The concentrations measured in different matrices vary markedly of course depending on the human and veterinary prescriptions and uses, household PCP usage, the efficacy of wastewater treatment systems (when present), the physicochemical properties of the molecules, along with their biodegradation and sorption profiles, the characteristics of the receiving environments, and the climatic conditions. However, in a very schematic way, pharmaceutical concentrations are generally higher in wastewater (reported concentrations in the hundreds of ng/L to μg/L range) than in groundwater (from the limit of quantification to a few ng/L). It should also be noted that levels found in the marine environment were in general lower than those found in the freshwater systems (Rodriguez-Mozaz et al., 2017).

One specificity of PPs (unlike other contaminants) is that they are administered to humans and animals, after which they are often absorbed and metabolized before excretion. The majority of studies conducted on PPs in the environment have focused on the parent molecules, but in recent years the importance of also taking excreted metabolites into account has been stressed. Recent research has revealed the presence of metabolites along with the parent compound, e.g., for carbamazepine (Zhang et al., 2008). In addition to these human or animal metabolites, transformation products (TPs) resulting from biodegradation and physical processes, such as oxidation and photolysis, in the receiving environment must be considered (Boix et al., 2016). Thus, the nature of the molecules present is very diverse. Note also that these PPCP residues in the environment may be found alongside other micropollutants, such as pesticides and surfactants (Kuzmanovic et al., 2015).

Because of their continuous use and release into the environment, PPCPs have a pseudopersistence behavior, suggesting that the organisms present may be chronically exposed. Various compounds have been detected in wild freshwater and marine species (mollusks, crustaceans, and fishes) (Arpin-Pont et al., 2014; Martínez Bueno et al., 2013; Miller et al., 2018; Picot Groz et al., 2014), as well as in soil organisms and plants (Klampfl et al., 2019). Furthermore, mollusks, crustaceans, and fishes have been shown to accumulate some PPCPs under controlled laboratory conditions (Gago-Ferrero et al., 2012; Gomez et al., 2012; Valdès et al., 2016). These findings highlight the need to characterize bioaccumulation and biological effects in nontarget organisms.

2. PPCPs can disturb the normal functioning of organisms

PPCPs present in biotopes and biota elicit adverse impacts on wild organisms. PPs have therapeutic effects by targeting specific metabolic, enzymatic, or cell signaling mediators. They also likely have concomitant effects on nontarget organisms, with highly variable manifestations depending on the exposure level, time, and species. Effects at environmentally realistic exposure have been highlighted by measuring a great diversity of biomarkers of both exposure (e.g., carboxylesterase activity) and effect (e.g., lipid peroxidation or vitellogenin quantification) (Chandra et al., 2012; Solé and Sanchez-Hernandez, 2018; Maulvault et al., 2019).

In recent decades, commonly measured biomarkers have also revealed that a wide variety of PPs impair most biological functions in living organisms at different biological organization levels (i.e., molecular, population, or community). For instance, Sehonova et al. (2018) reviewed the literature on the effects of waterborne antidepressants in aquatic organisms, including invertebrates and vertebrates (amphibians and fish), and concluded that fluoxetine, sertraline, or citalopram exposure results in behavioral, reproductive, and developmental alterations. Other impacts of antidepressants identified in laboratory studies on nontarget organisms included swimming alterations, decreased camouflage efficiency, increased aggressiveness, and impairment of nest building and defense behavior (Sehonova et al., 2018; Weinberger and Klaper, 2014). PPs also modulate early development, e.g., carbamazepine has been shown to disturb the normal growth of exposed zebrafish embryos and larvae (Qiang et al., 2016). At the molecular level, exposure to an environmentally relevant concentration (1 µg/L) of carbamazepine also altered the expression pattern of neural-related genes of zebrafish embryos and larvae (Fraz et al., 2018), while modulating the visual motor response in zebrafish larvae (Huang et al., 2019). The review of Fabbri and Franzelitti (2016) illustrates that a single PP is potentially able to alter various crucial homeostasis mechanisms of organisms. Thus, based on published findings, there is evidence that PPs disrupt key biological functions in aquatic organisms, such as metabolic activity, stress response, behavior, reproduction, and development.

When key biological functions are disrupted by exposure to PPs, thus impairing the normal reproduction or survival of organisms, the entire population is then affected. From the 1990s, a striking >97% Gyps vulture population decline on the Indian subcontinent was found to be caused by the consumption of diclofenac-treated livestock carcasses (Oaks et al., 2004; Taggart et al., 2007). As nonsteroidal antiinflammatory drugs (NSAIDs), diclofenac is known to cause renal failure and visceral gout in several vertebrates, thus explaining this alarming case of high vulture mortality. This also revealed the extent of ecotoxicological effects that can be reached by PPCPs exposure.

PPCPs belonging to different classes have proven to be able to alter the reproductive function of organisms by acting as endocrine disrupters in both field and laboratory studies (Caliman and Gavrilescu, 2009). The molecule present in contraceptive pills, 17-alpha-ethynylestradiol (EE2), is the most widely studied PP. EE2 induces estrogenic effects in fish at concentrations as low as ng/L (Aris et al., 2014). A remarkable study revealed the near extinction of fathead minnows chronically exposed to low concentrations (5–6 ng/L) of potent EE2 residues in an experimental lake in Canada (Kidd et al., 2007). The population collapse was related to the feminization of males through the production of vitellogenin and the impact on gonadal development, as shown by intersex in males and altered oogenesis in females. Some other pharmaceuticals are suspected to produce endocrine disruption, e.g., gemfibrozil and carbamazepine, which decrease steroid production in zebrafish testes (Fraz et al., 2018). Common exposure to both PPs and PCPs leads to questions about the effects of multiple exposure since some PCPs are also considered to be endocrine disruptors (Fent et al., 2008).

Among PCPs, the significant increase in sunscreen use has become a serious environmental concern. In addition to their endocrine disruption properties demonstrated in fish (Fent et al., 2008), sunscreen agents also have negative impacts on marine ecosystems (Gago-Ferrero et al., 2012). Organic UV filters—along with other contaminants and global warming—may be involved in the decline of coral reefs. Danovaro et al. (2008) demonstrated that UV filters participated in the rapid complete bleaching of hard corals in laboratory and in situ experiments.

PPCPs are degraded in the environment by soil and water microorganisms. Among PPs, antibiotics affect microbial community structure and functioning through their bactericide and bacteriostatic actions. In soils, PPs are partly responsible for the development of antibiotic-resistant bacteria, and in return bacterial strains degrade antibiotics through metabolic or cometabolic processes (Barra Caracciolo et al., 2015; Grenni et al., 2018). Even if PPs elicit acute toxicity to soil organisms at higher concentrations than in water, as demonstrated with the earthworm *Eisenia fetida*, little is known regarding chronic effects on these organisms (Pino et al., 2015). When this earthworm was exposed to ciprofloxacin-polluted soil, it exhibited reactive oxygen species overproduction that enhanced an active detoxification and defense system (Wang et al., 2018). Moreover, soil-borne PPCPs are transferred to plants and there is substantial evidence on the uptake, metabolism, and binding of PPs/PPCPs to plant cell walls (Klampfl et al., 2019). It is also commonly recognized that drugs in soils are responsible for adverse effects on plants, as shown by studies on the effects of veterinary antibiotics (Bártíková et al., 2016). However, so far few studies have investigated adverse outcomes on PPCP exposure for this compartment beyond the transfer to foodstuffs.

These examples, although not exhaustive, demonstrate that the environmental impact of PPCPs is a major concern. The mechanisms of action of PPCPs in nontarget organisms are still not totally understood. Research is needed to acknowledge the relationship between PPCP exposure and key molecular events leading to adverse effects at higher biological levels. For that purpose, mass spectrometry (MS) technology can be effective in the detection of both contaminants for characterizing the exposure and naturally occurring metabolites for highlighting molecular events.

3. Contribution of high-resolution mass spectrometry—based approaches to exposome characterization

MS is used increasingly in environmental science research. It offers specificity and sensitivity when the aim is to monitor organic contaminants at trace levels in very complex matrices. MS can also be combined with gas chromatography (GC) and liquid chromatography (LC), with the former being best suited for apolar and volatile compounds (PCPs such as musks and UV filters), and the latter better adapted to polar and semipolar molecules along with ionic compounds (most pharmaceuticals). There is increased interest in using high-resolution mass spectrometry (HRMS) in environmental science research as it is suitable for both targeted (with reference standards) and nontargeted analyses (without reference standards, suspect screening, or nontarget screening) (Krauss et al., 2010).

Targeted analysis is very powerful for the extraction, identification, and quantification of PCPPs in environmental matrices (water, soil, sediment, wastewater effluent) (Hernandez et al., 2012, 2018) and biota (Miller et al., 2018). A quite clear picture of the exposome of organisms emerges provided that a broad range of molecules is investigated. However, targeted analyses are usually focused on just one or two different classes of organic pollutants. Moreover, if one class of contaminants is monitored, the analysis is focused on the best-known substances in this class. A comprehensive approach (nontargeted analysis) can therefore be more effective to characterize the exposome of organisms (totality of environmental exposure) or nonpriority substances that are not covered by targeted method (but which have bioaccumulation and persistence potential). Specifically regarding PPCPs, the proof-of-concept has already been successfully applied on various environmental matrices: surface water (Park et al., 2018), wastewater effluent (Hernandez et al., 2018; Singer et al., 2016), groundwater (Soulier et al., 2016), and sediment (Terzic and Ahel, 2011). Nontargeted approaches have the advantage of detecting known pharmaceuticals that have not been previously reported in these matrices or are not on priority lists. Another advantage is that they may also detect PPCP TPs and/or metabolites in environmental matrices which may be active exposome components (Bletsou et al., 2015; Chen et al., 2018). Indeed, TPs and/or PPCP impurities from commercial formulations can be present at relevant concentrations and may be as toxic as the parent compound, or even more. Organisms may then be exposed to TPs since their abiotic and biotic formation has been observed in the laboratory in all environmental compartments, including river water (Jimenez et al., 2017, 2018), sediments (Li et al., 2014; Su et al., 2016), activated sludge (Beretsou et al., 2016), soil (Koba et al., 2016, 2017), as well as in the field (Kosma et al., 2017; Hernandez et al., 2011).

As HRMS can detect (in a nontarget way) organic contaminants (and their TPs) in environmental matrices, it can be used to quantify internal concentrations or body burdens of these contaminants in biota, thus enhancing exposome characterization. Most current approaches for quantifying multiple classes of organic contaminants in biota are targeted methods that benefit from the potential of HRMS (full scan acquisition). For instance, Jia et al. (2017) developed and validated an analytical method for simultaneous analysis of 137 veterinary drug residues and metabolites from 16 different classes in tilapia using HRMS data acquisition with fragmentation. Thanks to this nontargeted data acquisition, they were also able to detect pharmaceuticals that were not initially included in the method. With a similar strategy, Inostroza et al. (2016) developed a method that allows extraction and quantification of organic micropollutants of diverse chemical classes and physicochemical properties in gammarids. They used a suspect and nontarget screening tool based on HRMS full scan analysis to detect brominated and chlorinated compounds that were not initially included in the method. However, regarding the exposome, so far few studies have focused in a real nontargeted analysis of PPCPs in biota. Indeed, PPCPs are mostly present at trace levels in organisms, which are very complex matrices, and currently available analytical techniques are not sufficiently sensitive for these investigations when conducted on a large scale. In recent years, the contribution of ultraperformance (UP) or nano-LC-HRMS has significantly improved this potential. Based on a metabolomics strategy, a differential UPLC-HRMS profiling approach was used to profile a range of chemical xenobiotics and their metabolites in rainbow trout exposed to WWTP effluents. The chemical profiles of exposed or vehicle-exposed (control) rainbow trout were compared through dedicated bioinformatics and statistical tools. Accumulation of surfactants, naphthols, chlorinated xylenols, phenoxyphenols, chlorophenes, resin acids, steroidal alkaloids, and PPCPs was demonstrated in the bile or plasma of exposed fish (Al-Salhi et al., 2012). A few years later, the same group developed nano-LC-HRMS (usually used for proteomics analyses) to circumvent the lack of sensitivity of conventional MS (Chetwynd et al., 2014). Compared to conventional LC-HRMS, this latter approach enhanced the limits of detection of a variety of (xeno-) metabolites by 2- to 2000-fold. The methodology was found to be highly repeatable and reproducible for the analysis of fish urine and plasma samples. Moreover, phospholipid filtration plates combined with polymeric or mixed mode exchange solid-phase extraction were developed for removing highly abundant compounds (proteins, lipids) that may interfere during nano-LC-HRMS measurement. This sample preparation method combined with nano-LC-HRMS was very efficient for untargeted profiling of xenobiotics in blood plasma from fish (David et al., 2014). This approach, when applied to roaches exposed to WWTP effluents, allowed characterization of the pharmaceutical exposome. Thirty-one pharmaceuticals belonging to 11 classes were detected (including 26 that were confirmed by pure standards) in exposed roach plasma and tissues, thus confirming the high sensitivity of this approach and the promise of next-generation analytical methods for future biota exposome characterization (David et al., 2017).

Beyond exposome studies, HRMS can also be used to study PPCP biotransformation products in organisms in order to characterize the detoxification pathways of exposed organisms. The comparison through a differential approach of organisms exposed to PP and control animals identified the formation of metabolites specific to the drug-exposed group. This approach has demonstrated its efficacy in several studies focused on pharmaceutical metabolism. The metabolisms of carbamazepine and ibuprofen in *Solea senegalensis* (Acena et al., 2017), clofibric acid in *Danio rerio* embryos (Brox et al., 2016), and antihelmintics in *Campanula rotundifolia* (Stuchlíková et al., 2016) were successfully documented. The metabolism of fenhexamid produced by *Lactobacillus casei* was revealed using a data mining strategy to screen for chlorine-containing metabolites on the basis of the MS chlorine isotope ratio (Lénárt et al., 2013). The same strategy was also successfully implemented by Bonnefille et al. (2017), and new undescribed metabolites were detected when investigating diclofenac metabolism in mussels. Regarding PCPs, octocrylene biotransformation has been studied in coral. Among the different metabolites identified, the formation of fatty acid conjugates via oxidation of the ethylhexyl chain was demonstrated, yielding very lipophilic octocrylene analogues able to accumulate in coral tissues (Stien et al., 2019). These different studies have highlighted the substantial potential of these approaches, which were found to be able to detect metabolites that were previously unknown or not predicted by dedicated software.

Finally, HRMS-based chemical profiles collected on fluids or biological tissues of organisms are likely to reflect exposure by the direct nontargeted detection of parent compounds and their metabolites, as demonstrated in this section. They may also reflect effects by detection and characterization of the modulation of endogenous metabolites (endometabolome) in exposed organisms that may lead to physiological disturbances induced by the chemical exposure.

4. Metabolomics provides information on expected and unexpected molecular effects in organisms

The metabolomics approach can be applied in different fields. It has recently been found to be of interest in ecotoxicology research, especially for assessing the effects of organic micropollutants on organisms. The first studies focused on determining the effects of PPCPs on organisms were published in the late 2000s. These studies mainly involved the use of H^1 NMR, in accordance with the practices and techniques available at the time (Ekman et al., 2008a,b; Samuelsson et al., 2006). Since then, a growing number of studies involving GC and LC-(HR)MS have been published, especially since 2015. In this section, we will focus particularly on the results obtained in studies using MS.

The metabolomics approach is useful for investigating the mode of action (MoA) of PPs in organisms. The effectiveness of this nontargeted approach for revealing the MoA of PPs, identical or different from the human one, has been demonstrated in several studies. For example, the effects of various fluoxetine exposure concentrations (0.012−700 µg/L) in 2 h postfertilization zebrafish embryos (*D. rerio*) were studied over a 94 h period by GC-MS (Mishra et al., 2017). A clear separation was revealed between the metabolomics fingerprints of control individuals and those of embryos exposed to concentrations higher than 12 µg/L. At these exposure concentrations, disruptions in the metabolic activity of amino acids such as alanine, aspartate, glutamate, phenylalanine, tyrosine, and tryptophan were generally observed. Some of these disruptions were in accordance with the fluoxetine MoA, i.e., inhibition of serotonin reuptake leading to disruption of the cyclic adenosine monophosphate signaling pathway, which also involves various other amino acid pathways. Other metabolic pathways appeared to be modulated, leading to possible neuroendocrine and energy metabolism disruption, as well as homeostasis and immune system disruption—effects that appeared to be beyond the MoA of fluoxetine (Mishra et al., 2017). These results are in accordance with those obtained following 24 h fluoxetine exposure of *D. rerio* larvae studied by LC-HRMS, which revealed the modulation of serotonin and tryptophan concentrations, in addition to the disruption of other amino acid and energy metabolisms (Huang et al., 2017).

The metabolomics approach provides relevant information on the biological function that can be impaired by the exposure. The effects of NSAIDs (ibuprofen, diclofenac, nimesulide) were investigated through metabolomics in different aquatic species. Metabolic disruptions in amino acids, neurotransmitter modulation, or energy metabolism impairment were highlighted (Bonnefille et al., 2018; Currie et al., 2016; Gómez-Canela et al., 2016; Prud'homme et al., 2018; Song et al., 2018). As specific metabolites are known to govern biological functions within species, the implications at the organism/population level of NSAIDs exposure can be hypothesized and may lead to immunoregulation disruption, cytotoxicity, and DNA damage in adult zebrafish, for instance (Song et al., 2018), whereas osmoregulation and reproduction may be of concern in mussels (Bonnefille et al., 2018). However, more functional studies are needed to pinpoint the metabolites that regulate biological and physiological functions in many species. To this end, metabolomics can help to gain further insight into the regulation of vital functions in different species throughout their entire life cycle. It would help in the interpretation of the results obtained at the molecular scale in order to better hypothesize the possible adverse outcomes at the individual level. Moreover, the same metabolites can be modulated as a result of the exposure of different species to a PP. Nevertheless, they may not necessarily be involved in the same biological functions, and their modulation may therefore not lead to the same outcome in different species. This aspect would be interesting to study in future investigations.

Metabolomics is sensitive enough to identify differences in effects of PPCP enantiomer exposure. Enantiomers are known to be present in the environment, and studies have already reported effects depending on the isomeric form (De Andrés et al., 2009; Sanganyado et al., 2017; Stanley et al., 2007; Ye et al., 2015). Among them, ibuprofen is a chiral compound whose (S)-enantiomer is more active than (R)-enantiomer in domestic animals (Landoni et al., 1997). This compound has been detected in water and sediments (Ali et al., 2009). Metabolomics has proven to be sensitive enough to exhibit different effects when organisms are exposed to (S) of (R) ibuprofen. Indeed, the effects of 5 µg/L of ibuprofen (R(−), S(+), or a racemic mixture) were studied in adult zebrafish brain (*D. rerio*) after 28 days exposure (Song et al., 2018). The exposure affected various metabolic pathways, including effects on amino acids, neurotransmitters, oligopeptides, and oligosaccharides. Differences in metabolite modulation were noted depending on the ibuprofen enantiomer. For example, phosphoribosyl formamido carboxamide was highly impacted by exposure to R(−)-ibuprofen and racemic mixture, but was not affected by exposure to S(+)-ibuprofen. Conversely, glutathione was sensitive to both S(+) and racemic ibuprofen, but not to R(−)-ibuprofen. The overall evaluation of the results showed that R(−)-ibuprofen was less disruptive in terms of the number of metabolites modulated by exposure than S(+)-ibuprofen or racemic mixture, in accordance with previously published findings (Landoni et al., 1997). Finally, ibuprofen exposure, regardless of enantiomer, led to oxidative stress, energy metabolism disturbance, and alteration of various physiological functions in fish, such as immunity (Song et al., 2018). This issue could also be explored with regard to PCPs, some of which have also been shown to have different enantiomers (Liang et al., 2017).

Metabolomics may generate new insight into the impact of chronic and transgenerational exposure. A recent study conducted on two generations of a mosquito (*Aedes aegypti*) revealed that metabolomics, combined with other approaches (targeted and nontargeted transcriptomics, targeted analysis of ecdysteroids, and phenotypic observations at the macroscopic scale), provided a clearer picture of PPCP transgenerational effects. The effects of ibuprofen (1 µg/L) on a chronically exposed mosquito F0 population (from eggs to mature adults) and on an F1 population (produced by F0 mature females) were studied. No metabolomics effects were observed on the F0 population for both developmental stages studied (fourth larval instar and mature adults). However, downmodulation of 20 metabolites in the 0–4 h F1 egg population was highlighted (Prud'homme et al., 2018). Among them, amino acids, carbohydrates, polyols, phosphoric acid, and ornithine were lower compared to the control egg findings. Moreover, the metabolic resource internalization in F1 eggs was impacted by parental ibuprofen exposure, in accordance with the higher survival rate under starvation conditions of exposed eggs compared to control eggs (Prud'homme et al., 2018). Despite the fact that only one study was conducted using metabolomics to assess the transgenerational effects of PPCP exposure, the authors highlighted the advantages of this approach combined with other approaches for gaining insight into the mechanisms involved.

Metabolomics is useful to generate information on the effects of PCPs and their TPs. Only a few metabolomics studies have been conducted on PCPs and, to our knowledge, only one study has assessed the effects of UV filters on a coral species. The exposure of adult *Pocillopora damicornis* coral to octocrylene concentrations ranging from 5 to 1000 µg/L over a 7-day period revealed major modulation of acylcarnitine concentrations after LC-HRMS metabolomic profiling (Stien et al., 2019). The authors explained that acylcarnitine modulation might be due to an abnormal fatty acid metabolism in relation to mitochondrial dysfunction. They also pointed out that increased acylcarnitine concentrations are often related to cell toxicity, in accordance with macroobservations in an exposed aquarium (polyps closed). Triclosan, an antimicrobial agent that is used in many PCPs, seems to be the most widely studied compound. Although triclosan appears to be of low metabolic toxicity in both zebrafish (Huang et al., 2016) and earthworms (Gillis et al., 2017), the presence of its degradation products in the environment would likely lead to ecotoxicological risks. A GC-MS metabolomics study was conducted to assess the effects of various methyl-triclosan concentrations (0.001–400 µg/L) on zebrafish embryos (*D. rerio*) after 94 h of exposure (Fu et al., 2019). The authors documented impacts on various metabolic pathways, i.e., glycerophosphate, amino acid, energy, carbohydrate, and lipid metabolism. The authors hypothesized that the binding of methyl-triclosan to proteins prevents them from fulfilling their role, leading to increased amino acid production to ensure higher protein production. Moreover, they revealed that amino acid pathway modulation and other metabolic perturbations caused nitrogenous and ammonia waste production, which activated respiratory processes, in turn resulting in increased energy demand (Fu et al., 2019). This study illustrates the advantage of focusing on TPs even if the parent compound has no metabolic effects.

Is metabolomics sensitive enough to identify effects at environmentally realistic concentrations? To date, very few metabolomics studies have investigated PPCP effects at very low exposure levels (concentrations) in organisms, i.e., under 1 µg/L in water and 1 µg/g in soil. Some authors pointed out difficulties in highlighting metabolic modulation triggered by low dose exposure (Davis et al., 2017; Fu et al., 2019; Mishra et al., 2017). This may be due to (i) a lack of sensitivity/reproducibility of the analytical method used (GC-(HR)MS versus LC-(HR)MS), or of the statistical power of the experimental design, (ii) possible insufficient maturity in data processing methods to be able to reveal the discrete signature of low dose exposure, or (iii) a lack of sensitivity of the organism studied to the target molecule. The challenge of highlighting low dose exposure may also be associated with the nontargeted nature of metabolomics, i.e., broader scope but lower accuracy than targeted methods. Indeed, it has been found that targeted ecotoxicological studies are able to detect effects at very low exposure concentrations, as demonstrated for fluoxetine (Ford and Fong, 2016; Franzellitti et al., 2013; Franzellitti and Fabbri, 2013). Further methodological improvements in metabolomics may thus be required to be able to highlight consistent effects at low environmental concentrations. Nevertheless, some researchers have succeeded in documenting low dose effects through metabolomics analyses. Exposure of juvenile gilt-head bream (*Sparus aurata*) to 0.2 µg/L amitriptyline (antidepressant) for 2, 4, and 7 h revealed an alteration of this species' amino acid metabolism, in accordance with the amitriptyline MoA (nonselective inhibition of monoamine recapture), as well as energy metabolism disruption (Ziarrusta et al., 2019). Moreover, Huang et al. (2017) revealed that exposure of *D. rerio* larvae to diphenhydramine generated a metabolic profile that diverged more from that of the controls when exposed to the lower dose (0.51 µg/L; 0.002 µM) compared to the higher concentrations (51.7 and 510.7 µg/L; 0.2 and 2 µM). This metabolomics study indicated that the dose–response relationship of pharmaceuticals might not be monotonic, as already suggested for acetaminophen, carbamazepine, fluoxetine, triclosan, and triiodothyronine (Huang et al., 2016, 2017).

In summary, the studies reported in this section highlight that metabolomics is a relevant approach for acquiring additional information on PPCP effects on organisms. The approach helps reveal metabolite and metabolic pathway modulations in accordance with their known MoA in humans, as well as beyond their known MoA, while also sometimes highlighting biomarker of effect. It provides information on PPCP effects regarding current ecotoxicological issues, such as chronic and transgenerational effects, effects induced by different enantiomers of the same compound, and effects at environmentally realistic concentrations. Some of the limitations of metabolomics may also be revealed: the approach can lack sensitivity under certain conditions, and technological and/or methodological adjustments may be necessary to meet this challenge. In addition, metabolomics studies conducted to evaluate PPCP effects have mainly been conducted in aquatic organisms, especially in fish. This could be explained by the fact that water is the main gateway by which organic contaminants reach the environment. However, PPCPs are able to adsorb to suspended matter and may consequently be present in sewage sludge, depending on their physicochemical properties (Li et al., 2016; McClellan and Halden, 2010; Verlicchi and Zambello, 2015). Terrestrial organisms are also of concern since sewage sludge from wastewater treatment plants is used to enrich soils. Further studies on terrestrial organisms (animals or plants) are therefore needed to assess the ecotoxicological impact of such practices. Finally, metabolomics has so far been used very little to investigate the biotransformation products and effects of PCP exposure, while evidence gathered on PPs suggests that this approach is useful for elucidating the onset of adverse environmental effects.

5. Environmental metabolomics may contribute to the definition and application of adverse outcome pathways

An adverse outcome pathway (AOP) is a conceptual framework that pools existing knowledge on the linkage between a direct molecular initiating event (e.g., a molecular interaction between a xenobiotic and a specific biomolecule) and an adverse outcome at a biological organization level relevant to risk assessment (Fig. 8.2).

As such, AOPs are generally sequential series of events that, by definition, span multiple biological organization levels (Ankley et al., 2010). AOP provides a coherent framework for organizing (eco-)toxicological knowledge on effects at different biological organization levels and thus facilitate (i) more effective application and integration of diverse information and (ii) identification of uncertainties and research needs. An AOP may be developed to establish linkages between biological changes that are relevant to risk assessors and molecular/cellular alterations that might be detectable at earlier stages of chemical exposure. As such, environmental metabolomics may contribute to the definition and application of

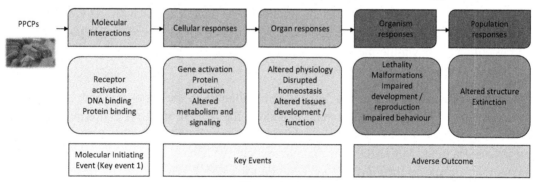

Figure 8.2

Schematic representation of the adverse outcome pathway illustrated with examples of key events/outcomes. *PPCP*, Pharmaceutical and personal care product.

AOPs along with other omics approaches (e.g., transcriptomics and proteomics). Omics analyses may identify molecular initiating events and provide supportive evidence of key events at different biological organization levels and across taxonomic groups. They may reveal mechanistic evidence to support chemical read-across, weight of evidence information for MoA assignment, and allow the development of extrapolations of species sensitivity and the assessment of target conservation (Brockmeier et al., 2017). To date, metabolomics has been applied in very few studies to shed light on AOPs of pharmaceuticals. Davis et al. (2017) examined metabolomics responses along with changes in fecundity, VTG gene expression and protein concentrations, and plasma sex steroid levels of fathead minnows (*Pimephales promelas*) exposed to different spironolactone concentrations for 21 days. Their main goal was to gather AOP data relevant to the activation of the androgen receptor, while also exploring other biological impacts possibly unrelated to this receptor. The authors confirmed that spironolactone exposure altered profiles of liver endogenous metabolites. Interestingly, plasma VTG concentrations and female fecundity (lower at higher exposure levels) were closely correlated with metabolite profiles at the individual fish level. Metabolites that covaried the most with VTG and fecundity could be considered as key events in the androgen receptor activation AOP. Furthermore, untargeted analysis of metabolite changes demonstrated additional altered biochemical pathways (involved in osmoregulation and membrane transport) that may reflect additional AOPs initiated by spironolactone exposure, but potentially unrelated to androgen receptor activity. The results of this study exemplify how HRMS and metabolomics can complement existing AOP knowledge and generate hypotheses on further biological impacts requiring the development of specific AOPs.

6. Summary and future research

The rise in the number of metabolomics studies focused on PPCPs in recent years highlights the growing interest in this approach. Supported by highly sensitive MS techniques, the great potential of metabolomics can be beneficial for both: (i) identifying chemicals in biofluids or tissues of nontarget organisms in order to monitor exposure and (ii) targeting effects via the measurement of endogenous compounds which may be disrupted by chemical exposure (Fig. 8.3).

The metabolomics approach—when applied to highlight both the exposome and metabolome—offers significant potential to address the issue of multiple exposures when studying organisms in a complex real-world environment. David et al. (2017) applied a sensitive nontargeted MS profiling technique to identify changes in the exposome and metabolome of roaches (*Rutilus rutilus*) exposed for 15 days to a wastewater treatment work effluent. They revealed the accumulation of 31 pharmaceuticals and metabolites in plasma and tissues. Meanwhile, they demonstrated that effluent exposure resulted in a reduction in

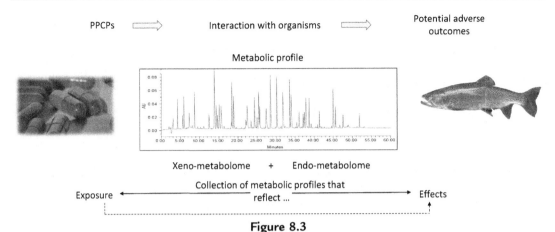

Figure 8.3

Metabolomics can generate information on both exposure and effects. *PPCP*, pharmaceutical and personal care product.

prostaglandin profiles, as well as tryptophan/serotonin, bile acid, and lipid metabolism disruptions in fish. Future developments will likely enhance the exposome and metabolome coverage, while advancing our understanding of the link between exposure to complex mixtures of contaminants and their impacts on multiple metabolite pathways. Metabolomics findings are thus of high interest, especially for environmental risk assessment. However, to gain greater insight into the link between disruptions in multiple interacting pathways and adverse outcomes on wildlife, metabolomics findings should be combined with other results obtained at molecular (genomic, transcriptomic, proteomic results obtained via targeted or nontargeted approaches), organism, and population levels (Fig. 8.4).

Indeed, by combining metabolomics results with those obtained through other molecular approaches (that are earlier in the biochemical cascade, e.g., transcriptomics or proteomics) the metabolite modulations observed could be confirmed. Furthermore, metabolomics can help more precisely define molecular initiating events, along with key molecular events triggered by the chemicals, to gain insight into the MoA in nontarget organisms.

Metabolomics may also be used to pinpoint metabolites that regulate biological and physiological functions, which may be very helpful in ecotoxicology research since metabolomics can readily help generate hypotheses. Metabolomics is considered as the molecular approach the closest to the phenotype. Knowledge on metabolite disruption following exposure may give rise to hypotheses regarding biological functions that could be further impaired with prolonged exposure. The design of additional experiments based on higher biological organization levels (individual or population) can thus be facilitated while possibly confirming assumptions derived from metabolomics and omics approaches. Such experiments have sometimes even been carried out previously without any knowledge of the MoA associated with the adverse outcomes observed. The AOP concept thus has great

AOP construction

Figure 8.4

Metabolomics at the crossroads between molecular and phenotypic effects. *AOP*, adverse outcome pathway.

potential and should make it possible to link the different molecular or phenotypic events. This may help scientists highlight consistent adverse outcomes following exposure, while also assisting risk assessors in drawing up new environmental policies.

Expectations regarding the potential of metabolomics for characterizing the exposome and metabolome of wild organisms are now being met, as demonstrated by the successful investigations on PPCP exposure and effects reported in this chapter. Beyond elucidating the AOPs in which PPCP substances could be involved, the possibility of measuring molecules (within the same organism/organ/tissue), their metabolites and TPs, and endogenous metabolites modulated simultaneously (or over a time course), is very promising with regard to determining the causal relationships. This approach and its potential is a new asset in the ecotoxicology research field as it lays new foundations for generating knowledge via both fundamental and applied research. In addition to the approaches discussed in this chapter, metabolomics has the potential to strengthen and facilitate decision-making for sustainable environmental management. The metabolomics approach has paved the way for novel fields of investigation likely with a bright future ahead of them.

Acknowledgment

Authors would like to thank the Agence Nationale de la Recherche (IMAP ANR-16-CE34-0006-01) for funding.

References

Acena, J., Perez, S., Eichhorn, P., Sole, M., Barcelo, D., 2017. Metabolite profiling of carbamazepine and ibuprofen in *Solea senegalensis* bile using high-resolution mass spectrometry. Anal. Bioanal. Chem. 409, 5441–5450.

Ali, I., Singh, P., Aboul-Enein, H.Y., Sharma, B., 2009. Chiral analysis of ibuprofen residues in water and sediment. Anal. Lett. 42, 1747–1760.

Al-Salhi, R., Abdul-Sada, A., Lange, A., Tyler, C.R., Hill, E.M., 2012. The xenometabolome and novel contaminant markers in fish exposed to a wastewater treatment works effluent. Environ. Sci. Technol. 46, 9080–9088.

Alvarez-Munoz, D., Rodríguez-Mozaz, S., Maulvault, A.L., Tediosi, A., Fernández-Tejedor, M., Van den Heuvel, F., Kotterman, M., Marques, A., Barceló, D., 2015. Occurrence of pharmaceuticals and endocrine disrupting compounds in macroalgaes, bivalves, and fish from coastal areas in Europe. Environ. Res. 143, 56–64.

Ankley, G.T., Bennett, R.S., Erickson, R.J., Hoff, D.J., Hornung, M.W., Johnson, R.D., Mount, D.R., Nichols, J.W., Russom, C.L., Schmieder, P.K., Serrrano, J.A., Tietge, J.E., Villeneuve, D.L., 2010. Adverse outcome pathways: a conceptual framework to support ecotoxicology research and risk assessment. Environ. Toxicol. Chem. 29, 730–741.

Aris, A.Z., Shamsuddin, A.S., Praveena, S.M., 2014. Occurrence of 17α-ethynylestradiol (EE2) in the environment and effect on exposed biota: a review. Environ. Int. 69, 104–119.

Arpin-Pont, L., Bueno, M.J., Gomez, E., Fenet, H., 2014. Occurrence of PPCPs in the marine environment: a review. Environ. Sci. Pollut. Res. 23, 4978–4991.

Balakrishna, K., Rath, A., Praveenkumarreddy, Y., Guruge, K.S., Subedi, B., 2017. A review of the occurrence of pharmaceuticals and personal care products in Indian water bodies. Ecotoxicol. Environ. Saf. 137, 113−120.

Barra Caracciolo, A., Topp, E., Grennia, P., 2015. Pharmaceuticals in the environment: biodegradation and effects on natural microbial communities. J. Pharmaceut. Biomed. Anal. 106, 25−36.

Bártíková, H., Podlipná, R., Skálová, L., 2016. Veterinary drugs in the environment and their toxicity to plants. Chemosphere 144, 2290−2301.

Beretsou, V.G., Psoma, A.K., Gago-Ferrero, P., Aalizadeh, R., Fenner, K., Thomaidis, N.S., 2016. Identification of biotransformation products of citalopram formed in activated sludge. Water Res. 103, 205−214.

Bletsou, A.A., Jeon, J., Hollender, J., Archontaki, E., Thomaidis, N.S., 2015. Targeted and non-targeted liquid chromatography-mass spectrometric workflows for identification of transformation products of emerging pollutants in the aquatic environment. Trends Anal. Chem. 66, 32−44.

Boix, C., Ibanez, M., Sancho, J.V., Parsons, J.R., Hernandez, F., 2016. Biotransformation of pharmaceuticals in surface water and during wastewater treatment: identification and occurrence of transformation products. J. Hazard. Mater. 302, 175−187.

Bonnefille, B., Arpin-Pont, L., Gomez, E., Fenet, H., Courant, F., 2017. Metabolic profiling identification of metabolites formed in Mediterranean mussels (*Mytilus galloprovincialis*) after diclofenac exposure. Sci. Total Environ. 583, 257−268.

Bonnefille, B., Gomez, E., Alali, M., Rosain, D., Fenet, H., Courant, F., 2018. Metabolomics assessment of the effects of diclofenac exposure on *Mytilus galloprovincialis*: potential effects on osmoregulation and reproduction. Sci. Total Environ. 613−614, 611−618.

Bourdat-Deschamps, M., Ferhi, S., Bernet, N., Feder, F., Crouzet, O., Patureau, D., Montenach, D., Moussard, G.D., Mercier, V., Benoit, P., Houot, S., 2017. Fate and impacts of pharmaceuticals and personal care products after repeated applications of organic waste products in long-term field experiments. Sci. Total Environ. 607−608, 271−280.

Boxall, A.B.A., Rudd, M.A., Brooks, B.W., Caldwell, D.J., Choi, K., Hickmann, S., Innes, E., Ostapyk, K., Staveley, J.P., Verslycke, T., Ankley, G.T., Beazley, K.F., Belanger, S.E., Berninger, J.P., Carriquiriborde, P., Coors, A., Deleo, P.C., Dryer, S.D., Ericson, J.F., Gagne, F., Giesy, J.P., Gouin, T., Hallstrom, L., Karlsson, M.V., Joakim Larsson, D.G., Lazorchak, J.M., Mastrocco, F., Mclaughlin, A., Mcmaster, M.E., Meyerhoff, R.D., Moore, R., Parrott, J.L., Snape, J.R., Murray-Smith, R., Servos, M.R., Sibley, P.K., Oliver Straub, J., Szabo, N.D., Topp, E., Tetreault, G.R., Trudeau, V.L., Der Kraak, G.V., 2012. Pharmaceuticals and personal care products in the environment: what are the big questions? Environ. Health Perspect. 120, 1221−1229.

Brausch, J.M., Rand, G.M., 2011. A review of personal care products in the aquatic environment: environmental concentrations and toxicity. Chemosphere 82, 1518−1532.

Brockmeier, E.K., Hodges, G., Hutchinson, T.H., Butler, E., Hecker, M., Tollefsen, K.E., Garcia-Reyero, N., Kille, P., Becker, D., Chipman, K., Colbourne, J., Collette, T.W., Cossins, A., Cronin, M., Graystock, P., Gutsell, S., Knapen, D., Katsiadaki, I., Lange, A., Marshall, S., Owen, S.F., Perkins, E.J., Plaistow, S., Schroeder, A., Taylor, D., Viant, M., Ankley, G., Falciani, F., 2017. The role of omics in the application of adverse outcome pathways for chemical risk assessment. Toxicol. Sci. 158 (2), 252−262.

Brox, S., Seiwert, B., Haase, N., Küster, E., Reemtsma, T., 2016. Metabolism of clofibric acid in zebrafish embryos (*Danio rerio*) as determined by liquid chromatography−high resolution−mass spectrometry. Comp. Biochem. Physiol. C Toxicol. Pharmacol. 185−186, 20−28.

Caliman, F.A., Gavrilescu, M., 2009. Pharmaceuticals, personal care products and endocrine disrupting agents in the environment − a review. Clean 37, 277−303.

Chandra, K., Bosker, T., Hogan, N., Lister, A., MacLatchy, D., Currie, S., 2012. Sustained high temperature increases the vitellogenin response to 17α-ethynylestradiol in mummichog (*Fundulus heteroclitus*). Aquat. Toxicol. 118−119, 130−140.

Chen, W.L., Cheng, J.Y., Lin, X.Q., 2018. Systematic screening and identification of the chlorinated transformation products of aromatic pharmaceuticals and personal care products using high-resolution mass spectrometry. Sci. Total Environ. 637−638, 253−263.

Chetwynd, A.J., David, A., Hill, E.M., Abdul-Sada, A., 2014. Evaluation of analytical performance and reliability of direct nanoLC-nanoESI-high resolution mass spectrometry for profiling the (xeno) metabolome. J. Mass Spectrom. 49, 1063−1069.

Christou, A., Kyriacou, M.C., Georgiadou, E.C., Papamarkou, R., Fatta-Kassinos, D., 2019. Uptake and bioaccumulation of three widely prescribed pharmaceutically active compounds in tomato fruits and mediated effects on fruit quality attributes. Sci. Total Environ. 647, 1169−1178.

Currie, F., Broadhurst, D.I., Dunn, W.B., Sellick, C.A., Goodacre, R., 2016. Metabolomics reveals the physiological response of *Pseudomonas putida* KT2440 (UWC1) after pharmaceutical exposure. Mol. BioSyst. 12, 1367−1377.

Danovaro, R., Bongiorni, L., Corinaldesi, C., Giovannelli, D., Damiani, E., Astolfi, P., Greci, L., Pusceddu, A., 2008. Sunscreens cause coral bleaching by promoting viral infections. Environ. Health Perspect. 116, 441−447.

Daughton, C.G., Ternes, T.A., 1999. Pharmaceuticals and personal care products in the environment: agents of subtle change? Environ. Health Perspect. 107 (Suppl. 6), 907−938.

Daughton, C.G., 2016. Pharmaceuticals and the environment (PiE): evolution and impact of the published literature revealed by bibliometric analysis. Sci. Total Environ. 562, 391−426.

David, A., Abdul-Sada, A., Lange, A., Tyler, C.R., Hill, E.M., 2014. A new approach for plasma (xeno) metabolomics based on solid-phase extraction and nanoflow liquid chromatography-nanoelectrospray ionisation mass spectrometry. J. Chromatogr. A 1365, 72−85.

David, A., Lange, A., Abdul-Sada, A., Tyler, C.R., Hill, E.M., 2017. Disruption of the prostaglandin metabolome and characterization of the pharmaceutical exposome in fish exposed to wastewater treatment works effluent as revealed by nanoflow-nanospray mass spectrometry-based metabolomics. Environ. Sci. Technol. 51, 616−624.

Davis, J.M., Ekman, D.R., Skelton, D.M., LaLone, C.A., Ankley, G.T., Cavallin, J.E., Villeneuve, D.L., Collette, T.W., 2017. Metabolomics for informing adverse outcome pathways: androgen receptor activation and the pharmaceutical spironolactone. Aquat. Toxicol. 184, 103−115.

De Andrés, F., Castañeda, G., Ríos, Á., 2009. Use of toxicity assays for enantiomeric discrimination of pharmaceutical substances. Chirality 21, 751−759.

Ekman, D.R., Teng, Q., Villeneuve, D.L., Kahl, M.D., Jensen, K.M., Durhan, E.J., Ankley, G.T., Collette, T.W., 2008a. Investigating compensation and recovery of fathead minnow (*Pimephales promelas*) exposed to 17α-ethynylestradiol with metabolite profiling. Environ. Sci. Technol. 42, 4188−4194.

Ekman, D.R., Teng, Q., Villeneuve, D.L., Kahl, M.D., Jensen, K.M., Durhan, E.J., Ankley, G.T., Collette, T.W., 2008b. Profiling lipid metabolites yields unique information on sex- and time-dependent responses of fathead minnows (*Pimephales promelas*) exposed to 17α-ethynylestradiol. Metabolomics 5, 22.

Fabbri, E., Franzelitti, S., 2016. Human pharmaceuticals in the marine environment: focus on exposure and biological effects in animal species. Environ. Toxicol. Chem. 35 (4), 799−812.

Fent, K., Kunz, P., Gomez, E., 2008. UV filters in the aquatic environment induce hormonal effects and affect fertility and reproduction in fish. Chimia 62, 1−8.

Ford, A.T., Fong, P.P., 2016. The effects of antidepressants appear to be rapid and at environmentally relevant concentrations. Environ. Toxicol. Chem. 35, 794−798.

Franzellitti, S., Buratti, S., Valbonesi, P., Fabbri, E., 2013. The mode of action (MOA) approach reveals interactive effects of environmental pharmaceuticals on *Mytilus galloprovincialis*. Aquat. Toxicol. 140−141, 249−256.

Franzellitti, S., Fabbri, E., 2013. Cyclic-AMP mediated regulation of ABCB mRNA expression in mussel haemocytes. PLoS One 8, e61634.

Fraz, S., Lee, A.H., Wilson, J.Y., 2018. Gemfibrozil and carbamazepine decrease steroid production in zebrafish testes (*Danio rerio*). Aquat. Toxicol. 198, 1–9.

Fu, J., Gong, Z., Bae, S., 2019. Assessment of the effect of methyl-triclosan and its mixture with triclosan on developing zebrafish (*Danio rerio*) embryos using mass spectrometry-based metabolomics. J. Hazard. Mater. 368, 186–196.

Gago-Ferrero, P., Díaz-Cruz, M.S., Barceló, D., 2012. An overview of UV-absorbing compounds (organic UV filters) in aquatic biota. Anal. Bioanal. Chem. 404, 2597–2610.

Gaw, S., Thomas, K.V., Hutchinson, T.H., 2014. Sources, impacts and trends of pharmaceuticals in the marine and coastal environment. Phil. Trans. R. Soc. B 369, 2013586.

Gillis, J.D., Price, G.W., Prasher, S., 2017. Lethal and sub-lethal effects of triclosan toxicity to the earthworm *Eisenia fetida* assessed through GC–MS metabolomics. Special Issue on Emerging Contaminants in engineered and natural environment. J. Hazard. Mater. 323, 203–211.

Gomez, E., Bachelot, M., Boillot, C., Munaron, D., Chiron, S., Casellas, C., Fenet, H., 2012. Bioconcentration of two pharmaceuticals (benzodiazepines) and two personal care products (UV filters) in marine mussels (*Mytilus galloprovincialis*) under controlled laboratory conditions. Environ. Sci. Pollut. Res. 19, 2561–2569.

Gómez-Canela, C., Miller, T.H., Bury, N.R., Tauler, R., Barron, L.P., 2016. Targeted metabolomics of *Gammarus pulex* following controlled exposures to selected pharmaceuticals in water. Sci. Total Environ. 562, 777–788.

Grenni, P., Ancona, V., Barra Caracciolo, A., 2018. Ecological effects of antibiotics on natural ecosystems: a review. Microchem. J. 136, 25–39.

Hernandez, F., Castiglioni, S., Covaci, A., de Voogt, P., Emke, E., Kasprzyk-Hordern, B., Ort, C., Reid, M., Sancho, J.V., Thomas, K.V., van Nuijs, A.L.N., Zuccato, E., Bijlsma, L., 2018. Mass spectrometric strategies for the investigation of biomarkers of illicit drug use in wastewater. Mass Spectrom. Rev. 37, 258–280.

Hernandez, F., Ibanez, M., Gracia-Lor, E., Sancho, J.V., 2011. Retrospective LC-QTOF-MS analysis searching for pharmaceutical metabolites in urban wastewater. J. Separ. Sci. 34, 3517–3526.

Hernandez, F., Sancho, J.V., Ibanez, M., Abad, E., Portoles, T., Mattioli, L., 2012. Current use of high-resolution mass spectrometry in the environmental sciences. Anal. Bioanal. Chem. 403, 1251–1264.

Hopkins, Z.R., Blaney, L., 2016. An aggregate analysis of personal care products in the environment: identifying the distribution of environmentally-relevant concentrations. Environ. Int. 92–93, 301–316.

Huang, I.J., Sirotkin, H.I., McElroy, A.E., 2019. Varying the exposure period and duration of neuroactive pharmaceuticals and their metabolites modulates effects on the visual motor response in zebrafish (*Danio rerio*) larvae. Neurotoxicol. Teratol. 72, 39–48.

Huang, S.S.Y., Benskin, J.P., Chandramouli, B., Butler, H., Helbing, C.C., Cosgrove, J.R., 2016. Xenobiotics produce distinct metabolomic responses in zebrafish larvae (*Danio rerio*). Environ. Sci. Technol. 50, 6526–6535.

Huang, S.S.Y., Benskin, J.P., Veldhoen, N., Chandramouli, B., Butler, H., Helbing, C.C., Cosgrove, J.R., 2017. A multi-omic approach to elucidate low-dose effects of xenobiotics in zebrafish (*Danio rerio*) larvae. Aquat. Toxicol. 182, 102–112.

Inostroza, P.A., Wicht, A.J., Huber, T., Nagy, C., Brack, W., Krauss, M., 2016. Body burden of pesticides and wastewater-derived pollutants on freshwater invertebrates: method development and application in the Danube River. Environ. Pollut. 214, 77–85.

Jia, W., Chu, X., Chang, J., Wang, P.G., Chen, Y., Zhang, F., 2017. High-throughput untargeted screening of veterinary drug residues and metabolites in tilapia using high resolution orbitrap mass spectrometry. Anal. Chim. Acta 957, 29–39.

Jimenez, J.J., Munoz, B.E., Sanchez, M.I., Pardo, R., 2018. Forced and long-term degradation assays of tenoxicam, piroxicam and meloxicam in river water. Degradation products and adsorption to sediment. Chemosphere 191, 903–910.

Jimenez, J.J., Sanchez, M.I., Pardo, R., Munoz, B.E., 2017. Degradation of indomethacin in river water under stress and non-stress laboratory conditions: degradation products, long-term evolution and adsorption to sediment. J. Environ. Sci. 51, 13–20.

Kidd, K.A., Blanchfield, P.J., Mills, K.H., Palace, V.P., Evans, R.E., Lazorchak, J.M., Flick, R.W., 2007. Collapse of a fish population after exposure to a synthetic estrogen. Proc. Natl. Acad. Sci. U.S.A. 104, 8897–8901.

Klampfl, C.W., Mzukisi Madikizela, L., Ncube, S., Chimuka, L., 2019. Metabolization of pharmaceuticals by plants after uptake from water and soil: a review. Trends Anal. Chem. 111, 13–26.

Koba, O., Golovko, O., Kodešová, R., Fér, M., Grabic, R., 2017. Antibiotics degradation in soil: a case of clindamycin, trimethoprim, sulfamethoxazole and their transformation products. Environ. Pollut. 220, 1251–1263.

Koba, O., Golovko, O., Kodešová, R., Klement, A., Grabic, R., 2016. Transformation of atenolol, metoprolol, and carbamazepine in soils: the identification, quantification, and stability of the transformation products and further implications for the environment. Environ. Pollut. 218, 574–585.

Kosma, C.I., Lambropoulou, D.A., Albanis, T.A., 2017. Photochemical transformation and wastewater fate and occurrence of omeprazole: HRMS for elucidation of transformation products and target and suspect screening analysis in wastewaters. Sci. Total Environ. 590–591, 592–601.

Krauss, M., Singer, H., Hollender, J., 2010. LC–high resolution MS in environmental analysis: from target screening to the identification of unknowns. Anal. Bioanal. Chem. 397, 943–951.

Kuzmanovic, M., Ginebreda, A., Petrovic, M., Barcelo, D., 2015. Risk assessment based prioritization of 200 organic micropollutants in 4 Iberian rivers. Sci. Total Environ. 503–504, 289–299.

Landoni, M.F., Soraci, A.l., Delatour, P., Lees, P., 1997. Enantioselective behaviour of drugs used in domestic animals: a review. J. Vet. Pharmacol. Ther. 20, 1–16.

Langford, K.H., Reid, M.J., Fjeld, E., Oxnevad, S., Thomas, K.V., 2015. Environmental occurrence and risk of organic UV filters and stabilizers in multiple matrices in Norway. Environ. Int. 80, 1–7.

Lapworth, D.J., Baran, N., Stuart, M.E., Ward, R.S., 2012. Emerging organic contaminants in groundwater: a review of sources, fate and occurrence – review article. Environ. Pollut. 163, 287–303.

Lénárt, J., Bujna, E., Kovács, B., Békefi, E., Száraz, L., Dernovics, M., 2013. Metabolomic approach assisted high resolution LC–ESI-MS based identification of a xenobiotic derivative of fenhexamid produced by *Lactobacillus casei*. J. Agric. Food Chem. 61, 8969–8975.

Li, Z., Maier, M.P., Radke, M., 2014. Screening for pharmaceutical transformation products formed in river sediment by combining ultrahigh performance liquid chromatography/high resolution mass spectrometry with a rapid data-processing method. Anal. Chim. Acta 810, 61–70.

Li, M., Sun, Q., Li, Y., Lv, M., Lin, L., Wu, Y., Ashfaq, M., Yu, C., 2016. Simultaneous analysis of 45 pharmaceuticals and personal care products in sludge by matrix solid-phase dispersion and liquid chromatography tandem mass spectrometry. Anal. Bioanal. Chem. 408, 4953–4964.

Li, W.C., 2014. Occurrence, sources, and fate of pharmaceuticals in aquatic environment and soil. Environ. Pollut. 187, 193–201.

Liang, Y., Zhan, J., Liu, X., Zhou, Z., Zhu, W., Liu, D., Wang, P., 2017. Stereoselective metabolism of the UV-filter 2-ethylhexyl 4-dimethylaminobenzoate and its metabolites in rabbits in vivo and vitro. RSC Adv. 7, 16991–16996.

Madikizela, L.M., Ncube, S., Chimuka, L., 2018. Uptake of pharmaceuticals by plants grown under hydroponic conditions and natural occurring plant species: a review. Sci. Total Environ. 636, 477–486.

Madikizela, L.M., Tavengwa, N.T., Chimuka, L., 2017. Status of pharmaceuticals in African water bodies: occurrence, removal and analytical methods. J. Environ. Manag. 193, 211–220.

Martínez Bueno, M.J., Boillot, C., Fenet, H., Chiron, S., Casellas, C., Gómez, E., 2013. Fast and easy extraction combined with high resolution-mass spectrometry for residue analysis of two anticonvulsants and their transformation products in marine mussels. J. Chromatogr. A 1305, 27–34.

Maulvault, A.L., Camacho, C., Barbosa, V., Alves, R., Anacleto, P., Pousão-Ferreira, P., Rui, R., Marques, A., Souza Diniz, M., 2019. Living in a multi-stressors environment: an integrated biomarker approach to assess the ecotoxicological response of meagre (*Argyrosomus regius*) to venlafaxine, warming and acidification. Environ. Res. 169, 7–25.

McClellan, K., Halden, R.U., 2010. Pharmaceuticals and personal care products in archived U.S. biosolids from the 2001 EPA national sewage sludge survey. Water Res. 44, 658–668.

Miller, T.H., Bury, N.R., Owen, S.F., MacRae, J.I., Barron, L.P., 2018. A review of the pharmaceutical exposome in aquatic fauna. Environ. Pollut. 239, 129–146.

Mishra, P., Gong, Z., Kelly, B.C., 2017. Assessing biological effects of fluoxetine in developing zebrafish embryos using gas chromatography-mass spectrometry based metabolomics. Chemosphere 188, 157–167.

Oaks, J.L., Gilbert, M., Virani, M.Z., Watson, R.T., Meteyer, C.U., Rideout, B.A., Shivaprasad, H.L., Ahmed, S., Iqbal Chaudhry, M.J., Arshad, M., Mahmood, S., Ali, A., Ahmed Khan, A., 2004. Diclofenac residues as the cause of vulture population decline in Pakistan. Nature 427, 630–633.

Park, N., Choi, Y., Kim, D., Kim, K., Jeon, J., 2018. Prioritization of highly exposable pharmaceuticals via a suspect/non-target screening approach: a case study for Yeongsan River, Korea. Sci. Total Environ. 639, 570–579.

Pal, A., Gin, K.Y., Lin, A.Y., Reinhard, M., 2010. Impacts of emerging organic contaminants on freshwater resources: review of recent occurrences, sources, fate and effects. Sci. Total Environ. 408, 6062–6069.

Picot Groz, M., Martínez Bueno, M.J., Rosain, D., Fenet, H., Casellas, C., Pereira, C., Gomez, E., 2014. Detection of emerging contaminants (UV filters, UV stabilizers and musks) in marine mussels from Portuguese coast by QuEChERS extraction and GC-MS/MS. Sci. Total Environ. 493, 162–169.

Pino, R.P., Val, J., Mainar, A.M., Zuriaga, E., Español, C., Langa, E., 2015. Acute toxicological effects on the earthworm *Eisenia fetida* of 18 common pharmaceuticals in artificial soil. Sci. Total Environ. 518–519, 225–237.

Prud'homme, S.M., Renault, D., David, J.-P., Reynaud, S., 2018. Multiscale approach to deciphering the molecular mechanisms involved in the direct and intergenerational effect of ibuprofen on mosquito *Aedes aegypti*. Environ. Sci. Technol. 52, 7937–7950.

Qiang, L., Cheng, J., Yi, J., Rotchell, J.M., Zhu, X., Zhou, J., 2016. Environmental concentration of carbamazepine accelerates fish embryonic development and disturbs larvae behavior. Ecotoxicology 25 (7), 1426–1437.

Rodriguez-Mozaz, S., Alvarez-Muñoz, D., Barceló, D., 2017. Pharmaceuticals in marine environment: analytical techniques and applications. In: Environmental Problems in Marine Biology: Methodological Aspects and Applications. Taylor & Francis Publisher, pp. 268–316.

Samuelsson, L.M., Förlin, L., Karlsson, G., Adolfsson-Erici, M., Larsson, D.G.J., 2006. Using NMR metabolomics to identify responses of an environmental estrogen in blood plasma of fish. Aquat. Toxicol. 78, 341–349.

Sanganyado, E., Lu, Z., Fu, Q., Schlenk, D., Gan, J., 2017. Chiral pharmaceuticals: a review on their environmental occurrence and fate processes. Water Res. 124, 527–542.

Sehonova, P., Svobodova, Z., Dolezelova, P., Vosmerova, P., Faggio, C., 2018. Effects of waterborne antidepressants on non-target animals living in the aquatic environment: a review. Sci. Total Environ. 631–632, 789–794.

Singer, H.P., Wössner, A.E., McArdell, C.S., Fenner, K., 2016. Rapid screening for exposure to "non-target" pharmaceuticals from wastewater effluents by combining HRMS-based suspect screening and exposure modeling. Environ. Sci. Technol. 50, 6698–6707.

Solé, M., Sanchez-Hernandez, J.C., 2018. Elucidating the importance of mussel carboxylesterase activity as exposure biomarker of environmental contaminants of current concern: an in vitro study. Ecol. Indicat. 85, 432–439.

Song, Y., Chai, T., Yin, Z., Zhang, X., Zhang, W., Qian, Y., Qiu, J., 2018. Stereoselective effects of ibuprofen in adult zebrafish (*Danio rerio*) using UPLC-TOF/MS-based metabolomics. Environ. Pollut. 241, 730–739.

Soulier, C., Coureau, C., Togola, A., 2016. Environmental forensics in groundwater coupling passive sampling and high resolution mass spectrometry for screening. Sci. Total Environ. 563, 845–854.

Stanley, J.K., Ramirez, A.J., Chambliss, C.K., Brooks, B.W., 2007. Enantiospecific sublethal effects of the antidepressant fluoxetine to a model aquatic vertebrate and invertebrate. Chemosphere 69, 9–16.

Stien, D., Clergeaud, F., Rodrigues, A.M.S., Lebaron, K., Pillot, R., Romans, P., Fagervold, S., Lebaron, P., 2019. Metabolomics reveal that octocrylene accumulates in *Pocillopora damicornis* tissues as fatty acid conjugates and triggers coral cell mitochondrial dysfunction. Anal. Chem. 91, 990–995.

Stuchlíková, L., Jirásko, R., Skálová, L., Pavlík, F., Szotáková, B., Holčapek, M., Vaněk, T., Podlipná, R., 2016. Metabolic pathways of benzimidazole anthelmintics in harebell (*Campanula rotundifolia*). Chemosphere 157, 10–17.

Su, T., Deng, H., Benskin, J.P., Radke, M., 2016. Biodegradation of sulfamethoxazole photo-transformation products in a water/sediment test. Chemosphere 148, 518–525.

Taggart, M.A., Senacha, K.R., Green, R.E., Jhala, Y.V., Raghavan, B., Rahmani, A.R., Cuthbert, R., Pain, D.J., Meharg, A.A., 2007. Diclofenac residues in carcasses of domestic ungulates available to vultures in India. Environ. Int. 33, 759–765.

Terzic, S., Ahel, M., 2011. Nontarget analysis of polar contaminants in freshwater sediments influenced by pharmaceutical industry using ultra-high-pressure liquid chromatography-quadrupole time-of-flight mass spectrometry. Environ. Pollut. 159, 557–566.

Tran, N.H., Reinhard, M., Yew-Hoong, G.K., 2018. Occurrence and fate of emerging contaminants in municipal wastewater treatment plants from different geographical regions-a review. Water Res. 133, 182–207.

Valdès, M.E., Huerta, B., Wunderlin, D.A., Bistoni, M.A., Barcelo, D., Rodriguez-Mozaz, S., 2016. Bioaccumulation and bioconcentration of carbamazepine and other pharmaceuticals in fish under field and controlled laboratory experiments. Evidences of carbamazepine metabolization by fish. Sci. Total Environ. 557–558, 58–67.

Vandermeersch, G., Lourenço, H.M., Alvarez-Muñoz, D., Cunha, S., Diogène, J., Cano-Sancho, G., Sloth, J.J., Kwadijk, C., Barcelo, D., Allegaert, W., Bekaert, K., Fernandes, J.O., Marques, A., Robbens, J., 2015. Environmental contaminants of emerging concern in seafood–European database on contaminant levels. Environ. Res. 143, 29–45.

Verlicchi, P., Zambello, E., 2015. Pharmaceuticals and personal care products in untreated and treated sewage sludge: occurrence and environmental risk in the case of application on soil — a critical review. Sci. Total Environ. 538, 750–767.

Wang, C., Rong, H., Liu, H., Wang, X., Gao, Y., Deng, R., Liu, R., Liu, Y., Zang, D., 2018. Detoxification mechanisms, defense responses, and toxicity threshold in the earthworm *Eisenia foetida* exposed to ciprofloxacin-polluted soils. Sci. Total Environ. 612, 442–449.

Weinberger, J., Klaper, R., 2014. Environmental concentrations of the selective serotonin reuptake inhibitor fluoxetine impact specific behaviors involved in reproduction, feeding and predator avoidance in the fish *Pimephales promelas* (fathead minnow). Aquat. Toxicol. 151, 77–83.

Ye, J., Zhao, M., Niu, L., Liu, W., 2015. Enantioselective environmental toxicology of chiral pesticides. Chem. Res. Toxicol. 28, 325–338.

Zhang, Y., Geißen, S., Gal, C., 2008. Carbamazepine and diclofenac: removal in wastewater treatment plants and occurrence in water bodies. Chemosphere 73, 1151–1161.

Ziarrusta, H., Ribbenstedt, A., Mijangos, L., Picart-Armada, S., Perera-Lluna, A., Prieto, A., Izagirre, U., Benskin, J.P., Olivares, M., Zuloaga, O., Etxebarria, N., 2019. Amitriptyline at an environmentally relevant concentration alters the profile of metabolites beyond monoamines in gilt-head bream. Environ. Toxicol. Chem. https://doi.org/10.1002/etc.4381.

CHAPTER 9

Metabolomics effects of nanomaterials: an ecotoxicological perspective

Marinella Farré[1], Awadhesh N. Jha[2]
[1]Water and Soil Quality Research Group, Department of Environmental Chemistry, IDAEA-CSIC, Barcelona, Spain; [2]University of Plymouth, Plymouth, Devon, United Kingdom

Chapter Outline

1. Introduction

Nanomaterials (NMs) are defined as structures with at least one dimension of less than 100 nm (Potocnik, 2011). However, a more specific definition is provided by the European Commission (EU) that recommends defining a NM as "*a natural, incidental or manufactured material containing particles, in an unbound state or as an aggregate or as an agglomerate and where, for 50% or more of the particles in the number size distribution, one or more external dimensions is in the size range 1 nm-100 nm*." Additionally, the EU recommends considering micron-size aggregates of fullerenes, single-wall carbon nanotubes (CNTs), and graphene flakes as NMs regardless of their global size.

Environmental Metabolomics. https://doi.org/10.1016/B978-0-12-818196-6.00009-1

Cero valent NPs	E.g., Au-NPs	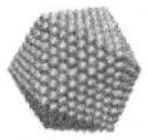
Metal oxide NPs	E.g., CeO$_2$-NPs	
Quantum-Dots	E.g., CdSe/ZnS core shell	
Carbon-based NMs	E.g. C$_{60}$ fullerenes	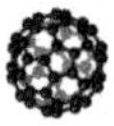
	Single wall carbon nanotubes (SWCNTs)	
	Multi wall carbon nanotubes (MWCNTs)	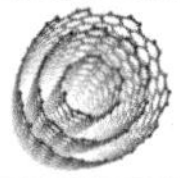
	Graphene	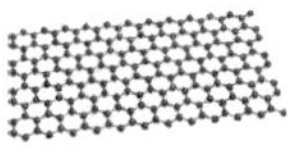
Dendrimers	E.g. Poli(amidoamine) G3	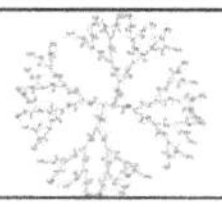
Liposomes	E.g. unsaturated phosphatidylcholine layers	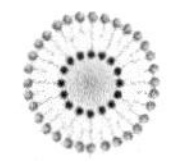
Nanoplastics	Polystyrene	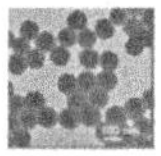

Figure 9.1

Summary of main classes of nanomaterials.

In Fig. 9.1, the main classes of NMs are summarized.

The unique physicochemical characteristics of NMs have propelled nanotechnology as a rapidly developing field. Two main factors cause the properties of NMs to differ significantly from other materials: increased relative surface area and quantum effects. These factors can change or enhance properties (e.g., reactivity, strength and electrical properties, and optical characteristics). Nanotechnology is already having an impact on products as diverse as medicine, pharmaceuticals, microelectronics, personal care products, pesticides, chemical coatings, and food. The NMs success has resulted in a potential increasing occurrence in the environment and the potential of human exposure. However, due to the unique properties of these new materials, they can exert an unknown impact and different from other contaminants. The investigations of their potential toxicological effects on the environment and biological systems have yet to catch up with the rapid nanotechnology development. Currently, it is known that nanoparticles (NPs) are prone to cross cell barriers, enter cells, and interact with subcellular structures. A large number of studies have reported the induction of oxidative stress and inflammation as a response to NPs exposure in a number of in vitro and in vivo models. It should, however, be highlighted that the large number of different NMs present in the environment could induce various potential effects. Besides, most of them have not shown acute effects but can be associated with chronic toxicological damages or even modulate the toxicity of other contaminants present in the same mixtures, as happens with fullerenes. In addition, most of these NMs have not shown acute toxicological effects but could induce biological effects under chronic conditions or even modulate the toxic effects of other contaminants. For these reasons, high-throughput methods are required to assess the potential risk of NPs. The advances during recent years on methodologies and techniques of toxicity testing changed and these developments have also influenced the testing of NPs, raising our expectations to determine potential biological responses, particularly at low, chronic level exposures. One of these changes was the introduction of quantitative analysis of molecular and functional changes in multiple levels of biological organization. Systems toxicology, "*omics*" techniques (genomics, transcriptomics, proteomics, and metabolomics), changed the current toxicology. One of the main advantages might be the identification of new targets and markers for NP toxicity. *Omics* techniques are useful because exposure to NPs occurs at low levels and at these levels of exposure conventional in vitro testing, in general, do not show phenotypic changes because exposure duration is too short. Moreover, the omics techniques are well suited to assess toxicity in both in vitro and in vivo experiments.

In the 1990s, metabolomics was defined for the first time as the study of the metabolites presents within a cell, tissue, or organism during a genetic alteration or physiological stimulus (Nicholson et al., 1999; Oliver et al., 1998). But it was long before as the 1950s when the study of the entire set of biochemical reactions in the organisms to extract the meaningful biological and clinical characteristics started.

Compared to other *omics* techniques such as transcriptomics and proteomics, which provide information about potential hazards, metabolomics identifies phenotypic changes by measuring profiling variations in the entire metabolome (carbohydrate, lipid, and amino acid patterns). Moreover, metabolomics differs from the former techniques in the sense that it is not organism-specific and does not have a fixed code (De La Luz-Hdez, 2012).

It should be highlighted that metabolomics presents a significant advantage compared to other *omics* approaches for the assessment of the toxicity of NPs. Metabolomics describes the net results of genomic, transcriptomic, and proteomic changes and provides general information since a given metabolite, unlike a gene or a protein, is the same for every organism that possesses it. Therefore, metabolomics is an ideal choice for investigating NP—organism interaction at different molecular levels.

In this chapter, an overview of recent metabolomics studies is summarized as well as the opportunities and challenges faced in nanotoxicology.

2. Metabolomics strategies

Metabolomics is based on the systematic characterization of the changes in small cellular molecules in response to xenobiotic stimuli. Environmental metabolomics can be carried out according to different types of exposure experiments:

- Laboratory exposure under controlled conditions
- Field studies

In both cases, organisms are sampled to assess biological effects by a specific kind of stressors, and the terms of exposure are predefined and controlled (laboratory studies) or are determined (field studies). In any case, the number of samples should be enough to obtain statistically representative results.

In the case of environmental metabolomics studies of NMs, due to the difficulties of analyzing certain NMs in real complex samples, the studies have been performed under controlled laboratory conditions. In these studies, the concentrations of NMs to be tested, the number of samples, the number of replicates, types of samples, and exposure conditions are previously defined.

Besides, in regard to the type of analysis performed, metabolomics studies can be

- Untargeted
- Targeted

Untargeted metabolomics is based on comparison of metabolite profiles pre- and postexposure and identify and quantify molecules which concentrations are significantly different. On the other hand, targeted metabolomics is based on the analysis of a specified list of metabolites related to particular pathways of interest.

Nuclear magnetic resonance (NMR) techniques and gas chromatography (GC) or liquid chromatography coupled to high-resolution mass spectrometry (LC-HRMS) based on Orbitrap technology are the main techniques employed for untargeted metabolomics. These techniques can be considered complementary, presenting both advantages and limitations.

3. Common analytical tools for metabolomics

Advancements in analytical tool facilitated the increased of metabolite coverage. The most common analytical techniques are NMR spectroscopy and chromatography to mass spectrometry (MS).

3.1 NMR-based metabolomics

NMR is a highly reproducible and nondestructive technique that offers good quantification power and ease of sample handling without requirements of extraction (Larive et al., 2015; Markley et al., 2017). One of the main advantages is that NMR can be used in in vivo metabolomics studies using stable isotope tracing. However, this approach lacks sensitivity and requires high concentrations, typically in the millimolar range. Other limitations of this technique are the difficulties in automation and for high-throughput analyses (Larive et al., 2015; Markley et al., 2017).

Different types of NMR experiments can be performed, including different nuclei such as ^{1}H, ^{13}C, ^{5}N, and ^{31}P. Moreover, experiments with various levels of correlation can be carried out using one- and two-dimensional NMR as well as correlated spectroscopy, total correlation spectroscopy, and heteronuclear single-quantum spectroscopy. The two-dimensional techniques improve sensitivity, reduce acquisition times, and metabolite identification (Markley et al., 2017). Besides, NMR is not limited to liquid samples, but intact tissue samples can also be analyzed through high-resolution magic-angle spinning NMR (Emwas et al., 2013) with similar results (Huang et al., 2019).

3.2 MS-based metabolomics

Nowadays, the most common analytical technique for metabolomics analysis is MS due to its high sensitivity and suitability for high-throughput analysis (Aretz and Meierhofer, 2016). MS-based methods are based on the monitoring mass-to-charge ratios (m/z) of all ionizable molecules present in a sample. Moreover, these methods can offer quantification by the use of standards or other reference compounds such as surrogates (De Hoffmann et al., 1996). However, MS is a destructive technique (Viant and Sommer, 2013).

In addition, different separation techniques can be coupled to MS such as LC, GC, and capillary electrophoresis, but due to the versatility and robustness, LC-MS and GC-MS are the most common ones. Main advantages of GC-MS over LC-MS are the high chromatographic resolution and reproducible retention times. GC is usually coupled to hard ionization sources such as electron impact ionization, which allows for in-source fragmentation and identification of the molecular ion with the extensive databases available for GC-MS (Gross, 2011). However, the use of GC-MS is limited to volatile analytes; in order to detect nonvolatile compounds such as amino acids, sugars, and organic acids, derivatization is required before they can be analyzed by GC-MS.

Alternatively, LC-MS provides high sensitivity and selectivity for the analysis of nonvolatile compounds. However, separation by LC is susceptible to retention time shifts (Kuehnbaum and Britz-Mckibbin, 2013). The usage of two-dimensional LC and GC has also been implemented for metabolomics in order to minimize interferences from complex matrices to increase the chromatographic resolution and the peak capacity (Navarro-Reig et al., 2017).

Analysis by LC-MS is commonly achieved using HRMS that provides accurate metabolite identification by highly accurate mass measurements of the precursor and fragment ions (Gross, 2011). HRMS instrumentation allows for the identification of additional metabolites because of the increased selectivity and the ability to determine the exact mass of molecular and fragment ions. This allows the determination of the elemental composition of unknown metabolites. For this reason, HRMS is suited for untargeted metabolomics. Some examples of HRMS instrumentation include time-of-flight (ToF), quadrupole time-of-flight (QToF), Fourier transform ion cyclotron resonance, and Orbitrap mass spectrometers.

4. Metabolomics in nanotoxicology

4.1 Metal nanoparticles

These NPs are typically obtained by reducing solutions of metal salts, and their physical properties can be tuned by varying the reduction conditions (Farré et al., 2011). Most consumer-product applications using zero-valence NPs have involved silver NPs (AgNPs) because of their bactericidal properties (Sharma et al., 2009). These applications include socks and other textiles; toothpaste; air filters; vacuum cleaners; and washing machines (Dastjerdi and Montazer, 2010). There is a variety of active silver nanostructures (Meng et al., 2010) (e.g., metallic AgNPs). The antimicrobial activity of AgNPs is most often attributed to the dissolved cation rather than to their large surface area. Another important group of metal-NPs are gold-NPs (Au-NPs) because their applications have been exploited due to their catalytic activity (Shiju and Guliants, 2009). In addition, a large part of Au-NP investigation is devoted to their possible use in medical applications and in bioanalytics (Tan et al., 2009).

Metabolomics provides valuable information regarding the overall changes of metabolites and biochemical pathways that might be altered in response to NP and particular metal-NPs. However, only a limited number of studies have been carried out until now.

Pietrzak et al. (2016) studied the effects of AgNP on molds. In this study, the metabolomics changes on *Aspergillus niger* and *Penicillium chrysogenum* were conducted by laser desorption/ionization time-of-flight mass spectrometry (LDI-ToF-MS) analysis. The metabolome analysis revealed that AgNPs (62 ppm) caused the downregulation of 446 (*A. niger*) and 48 (*P. chrysogenum*) compounds, which were below m/z 600. These results suggest that AgNPs cause the inhibition of metabolic pathways that regulate small molecules, and, like in bacteria, the mechanism of AgNPs on molds is multidirectional. In another study working with microorganisms, the change of the metabolic profiling of AgNPs in *Microcystis aeruginosa* was investigated by Zhang et al. (2018b). The results reveal that a total of 97 metabolites and 16 specific metabolic pathways were significantly altered by exposure to AgNPs, many of them associated with the cellular stress response. The metabolic analysis revealed that some metabolic pathways were regulated explicitly by AgNPs, such as the arginine and proline metabolism, indole alkaloid biosynthesis, and phospholipid metabolism. The study also showed that exposure increased the superoxide dismutase activity and malondialdehyde generation and interrupts the endocytosis process in algae cell. In more recent work (Wang et al., 2019b), the biological effects of AgNPs on *Scenedesmus obliquus*, a freshwater microalgae, was evaluated by nontargeted metabolomics. After 48 h exposure to AgNPs, nine metabolites had significant changes compared to the control group, including D-galactose, sucrose, and D-fructose. The carbohydrates that significantly changed are involved in the repair of cell walls. Besides, glycine increased with AgNP exposure concentration increasing, likely to counteract an increased intracellular oxidative stress (Wang et al., 2019b).

To study the subacute aquatic toxicology, different metabolomics studies have been performed with daphnids. For example, *Daphnia magna* exposed to AgNP and Ag^+ exhibited significant changes in their metabolic profile related to oxidative stress and disturbances of energy metabolism (Li et al., 2015a). In both cases, the effects are due to the release of Ag^+, produce most of same metabolic changes, suggesting that the dominant effect is due to the released Ag^+. However, also some effects are NP-dependent, for example, the levels of lactate were elevated in all AgNP-treated groups but not in the Ag^+-treated groups (Li et al., 2015a).

The same group of authors (Li et al., 2015b) studied the contribution of ionic Ag^+ and nanoparticulate Ag to the overall toxicity of AgNPs to the earthworm, *Eisenia fetida*. A similar alteration was seen in groups exposed to both smaller AgNPs (10 nm) and larger AgNPs (40 nm), indicating that these effects in *E. fetida* were induced by exposure to released Ag^+. Also, different metabolic responses, including decreased malate and glucose levels in 10 nm AgNP-exposed earthworms, could be associated with exposure to NPs, while the increase of leucine and arginine and decreased ATP and inosine levels were observed only in groups exposed to 40 nm AgNP exposures, which clearly demonstrated size-dependent effects of AgNPs. This study demonstrated that nanosilver acts by a different mechanism than ionic silver to cause acute toxicity to *E. fetida*.

Fennell et al. (2017) studied the effect on the biochemical profile in urine of pregnant rats exposed to AgNPs via intravenous or oral administration. In this study, different types of AgNPs were studied: two sizes of AgNP, 20 and 110 nm, and silver acetate (AgAc). NMR metabolomics analysis of urine indicated that AgNP and AgAc exposures impact the carbohydrate and amino acid metabolism. In addition, it was demonstrated that silver crosses the placenta and is transferred to the fetus regardless of the form of silver. Also, Jarak et al. (2017) carried out metabolomics studies together with histological examination, to characterize multiorgan and systemic metabolic responses to AgNPs intravenously administered to mice at 8 mg/kg body weight (a dose not eliciting overt toxicity). The primary target organs of NPs accumulation were liver and spleen. These two organs also showed the most significant metabolic changes in a two-stage response. In particular, the liver of exposed mice was found to switch from glycogenolysis and lipid storage, at 6 h post injection, to glycogenesis and lipolysis, at subsequent times up to 48 h. Besides, metabolites related to antioxidative defense, immunoregulation, and detoxification played a crucial role in hepatic protection. The spleen showed the depletion of several amino acids, possibly reflecting impairment of hemoglobin recycling, while only a few differences remained at 48 h post injection. In the heart, the main metabolic change was tricarboxylic acid (TCA) cycle intensification, and increased ATP production possibly reflecting a beneficial adaptation to the presence of AgNPs. On the other hand, the TCA cycle appeared to be downregulation in the lungs of injected mice with signs of inflammation.

Another relevant piece of works carried out in recent years are examining the transformations by NPs in terrestrial plants. For example, the nontargeted metabolomics changes by foliar exposure of 4-week-old cucumber (*Cucumis sativus*) plants to AgNPs (4 or 40 mg/plant) or Ag^+ (0.04 or 0.4 mg/plant) for 7 days were studied by GC-MS (Zhang et al., 2018a). Multivariate analysis revealed that all the treatments significantly altered the metabolite profile, including activation of antioxidant defense systems by upregulation of phenolic compounds and downregulation of photosynthesis upregulation of phytol. Moreover, exposed plants enhanced respiration, therefore, the upregulation of TCA cycle intermediates was observed, inhibited photorespiration shown by the downregulation of glycine/serine ratio, altered membrane properties (upregulation of pentadecanoic and arachidonic acids, downregulation of linoleic and linolenic acids), and reduced inorganic nitrogen fixation (downregulation of glutamine and asparagine). Although Ag^+ induced some of the same alterations, the levels of lactulose, raffinose, carbazole, citraconic acid, lactamide, acetanilide, and p-benzoquinone were AgNP-specific (Zhang et al., 2018a).

Another type of metal-NPs that has been studied is gold nanoparticles (Au-NPs). In particular, a high number of studies on the safety of Au-NPs have been carried out due to the successful application in drug delivery and in trials of new treatment of diseases. However, only a limited number of works were focused on metabolomics and most of them with in vitro studies using cell cultures (Lindeque et al., 2018), and no studies have been directed to study environmental impacts.

The use of copper nanoparticles (Cu-NPs) is nowadays increasing in agricultural practices. Therefore, it is of particular interest to perform metabolomics studies using plants and crops. The metabolite changes of cucumber plants under nano-Cu stress were possible through the use of both ^{1}H-NMR and GC-MS. The results showed that the metabolite profile was influenced by exposure to Cu-NPs. GC-MS data showed that concentrations of some sugars, organic acids, amino acids, and fatty acids were increased or decreased by Cu-NPs (Zhao et al., 2017). A similar work carried out by the same group of authors investigated by LC-MS/MS studies the antioxidant response of cucumber (*C. sativus*) exposed to Cu-NPs (nanocopper pesticide) (Huang et al., 2018, 2019).

4.2 Metal oxide nanoparticles

Metal oxide NPs are among the most used NMs (Aitken et al., 2006). Bulk materials as titanium dioxide (TiO_2), aluminum, and iron oxides have been used for years, but, during the last decade, nanosized forms started to be used by the industry and are being used in different consumer products. Among them, NMs based of TiO_2 and zinc oxide (ZnO) are widely exploited due to their photolytic properties (Panayotov and Morris, 2016). Nano-Al_2O_3 derivatives are as well some of the most used materials for polymer composites and core-shell NPs for different applications, including catalysis (Hanemann and Szabó, 2010; Yekeen et al., 2019). ZnO and TiO_2 are finding extensive applications in sunscreens, cosmetics, and bottle coatings because of their ultraviolet blocking capability (Mitrano et al., 2015; Subramaniam et al., 2019; Wang and Tooley, 2011). Other relevant metal oxide NPs are based on cerium dioxide (CeO_2-NPs), chromium dioxide (CrO_2-NPs), molybdenum trioxide (MoO_3-NPs), and binary oxides such as indium-tin-oxide (InSnO-NPs). CeO_2 is finding major uses as a combustion catalyst in diesel fuels to improve emission quality (Ratnasamy and Wagner, 2009), and in solar cells, gas sensors, oxygen pumps, and metallurgical and glass or ceramic applications.

The type of metal oxide NPs better studied from the environmental metabolomics point of view is titanium dioxide nanoparticles (TiO_2-NPs), due to their increasing use in industrial and domestic applications. In addition, due to TiO_2-NP use in personal care products, different studies have been carried out to assess human health impacts by metabolic studies using cell-lines approaches (Bo et al., 2014; Garcia-Contreras et al., 2015; Tucci et al., 2013), but they will not be discussed here as they are not the objective of this chapter.

To assess the impact of NMs, several environmental metabolomics studies have been performed using microorganisms. Some of the advantages of metabolomics using microorganisms are the lack of ethical problems, they are easy to manipulate, cost and time-effective, and have crucial roles in the biosphere. Microbial metabolomics integrates biological information into systems microbiology to facilitate the understanding of microbial interactions and cellular functions. But, microorganisms are context-dependent subjects reflecting the overall physiological state of the cell and cannot be dissociated from its host or ecological niche. For example, the study of mold interaction with NPs under the metabolomic point of view can be employed as a rapid and powerful tool to investigate the interaction among organisms and NMs under relevant environmental conditions. GC/MS-based metabolomics was used to investigate the effect of TiO_2-NPs on the slime mold *Physarum polycephalum macroplasmodium* under dark conditions (Zhang et al., 2018c). In this work, at least 60 metabolic biomarkers related to sugar metabolism, amino acid metabolism, nucleotide metabolism, and secondary metabolites pathways were significantly perturbed by TiO_2-NP exposure. Many of them were related to antioxidant mechanisms, suggesting that TiO_2-NPs may induce oxidative stress, even under dark conditions. The authors hypothesize that the oxidative stress might be associated with a TiO_2-NP—induced imbalance of cellular reactive oxygen species (ROS) (Zhang et al., 2018c).

The exposure of *Escherichia coli* to TiO_2-NPs has been shown to induce two main effects on bacterial metabolism (Planchon et al., 2017). The first effects showed the upregulation of proteins and the increase of metabolites related to energy and growth metabolism. The second effects showed the downregulation of other proteins increasing amino acids. Some proteins, such as chaperonin-1 or isocitrate dehydrogenase, as well as some metabolites (e.g., phenylalanine or valine) might be used as biomarkers of cellular stress by NPs. However, in a contradictory manner, in this study it was shown that the ATP content gradually rises proportionally to TiO_2-NP concentration in the medium, indicating the release of ATP by the damaged cells. These results support the hypothesis of heterogeneity of the bacterial population. This heterogeneity is also confirmed by scanning electron microscopy which showed that while some bacteria were covered by TiO_2-NPs, the major part of the bacterial population remains free from NPs, resulting in a difference of proteome and metabolome results. The use of a combined omics approach allowed to understand the bacterial response to TiO_2-NP stress due to heterogeneous interactions under environmental conditions (Planchon et al., 2017).

Considering the characteristics of NMs, another important group of studies are those involving soil organisms. Ratnasekhar et al. (2015) used a method based on GC-MS—based metabolomics to understand the sublethal toxicity at concentrations 7.7 and 38.5 µg/mL of TiO_2-NPs (<25 nm) in a soil nematode *Caenorhabditis elegans*. Multivariate pattern recognition analysis reflected the perturbations in the metabolism of amino acids, organic acids, and sugars. The biological pathways affected were the TCA cycle, arachidonic acid metabolism, and glyoxylate dicarboxylate metabolism. The use of organisms to assess the impact of NMs in aquatic environments is also relevant. In a recent study, the sea urchin *Paracentrotus lividus* was used in a metabolomics study to assess the TiO$_2$-NP effects (Alijagic et al., 2020). The findings of this study highlighted that TiO$_2$-NPs interact with immune cells suppressing the expression of genes encoding for proteins involved in immune response and apoptosis, and boosting the antioxidant metabolic activity, for example, pentose phosphate, cysteine-methionine, and glycine-serine metabolism pathways.

Interesting results were also obtained by Raja et al. (2018) for ecotoxicological information in aquatic environments. In a study, male and female zebrafish (*Danio rerio*) were exposed to sublethal concentrations of TiO$_2$-NPs dosed at 0.1 mg/L (low) and 5.0 mg/L (high), respectively. ^{1}H-NMR metabolomics combined with univariate and multivariate statistics was used to assess the alterations in treated fish. The purine metabolism, including the anabolism/catabolism of ATP, ADP, AMP, inosine, and xanthine, was significantly influenced by TiO$_2$-NP exposure, which may indicate some genotoxicity. Moreover, increasing the dose and time of exposure led to metabolic disturbances in glycerophospholipid metabolism and the Krebs cycle. In order to assess the potential environmental concentrations to produce environmental damages also different studies have been performed with crops and plants. A GC-MS−based metabolomics approach was used to investigate the potential toxicity of TiO$_2$-NPs on hydroponically cultured rice (*Oryza sativa* L.) (Wu et al., 2017). Metabolomics changes were recorded after plant exposure to 0, 100, 250, or 500 mg/L of TiO$_2$-NPs for 14 days. Results showed that the biomass of rice was decreased and the antioxidant system was significantly disturbed. One hundred and five identified metabolites showed significant differences compared to the control. Among them, the concentrations of glucose-6-phosphate, glucose-1-phosphate, succinic acid, and isocitric acid were those that were most increased, while the concentrations of sucrose, isomaltulose, and glyoxylic acid were those that were most decreased. Basic energy-generating ways such as the TCA cycle and the pentose phosphate pathway were significantly increased, while the carbohydrate synthesis metabolism including starch and sucrose metabolism were inhibited. However, the biosynthetic formation of most of the identified fatty acids, amino acids, and secondary metabolites which correlated to crop quality was increased. The results suggest that the metabolism of rice plants is distinctly disturbed after exposure to TiO$_2$-NPs (Wu et al., 2017). In another type of research (Lian et al., 2020), the impact of TiO$_2$-NPs at 0, 100, and 250 mg/L and Cd at 0, 50 μM coexposure on hydroponic maize (*Zea mays* L.) was determined under two exposure modes. Results showed that root coexposure to TiO$_2$-NPs and 100 mg/L Cd significantly enhanced Cd uptake and produced greater phytotoxicity in maize than foliar exposure to TiO$_2$-NPs. Moreover, the chlorophyll content showed a reduction of 45.3% and 50.5%, respectively, when compared with single Cd treatment. By contrast, foliar exposure of TiO_2-NPs could decrease shoot Cd contents and had an influence on superoxide dismutase and glutathione S-transferase activities alleviating Cd toxicity. Besides, several metabolic pathways, including the galactose metabolism and citrate cycle, alanine, aspartate, and glutamate metabolism, as well as glycine, serine, and threonine metabolism, were upregulated. This study revealed the beneficial application of TiO_2-NPs in more safety crop production in Cd-contaminated soils (Lian et al., 2020).

The metabolomics stress of *Enchytraeus crypticus*, was studied after exposure to copper oxide nanoparticles (CuO-NPs) and $CuCl_2$ along 0−7−14 days. Early effects were mainly related to amino acids and later to lysophospholipids (downregulation). Furthermore, the underlying mechanisms CuO-NP toxicity (e.g., neurotransmission, nucleic acids generation, cellular energy, and immune defense) differ from $CuCl_2$, where later metabolomics responses are mostly linked to the metabolism of lipids and fewer to amino acids.

In another study (Kumar Babele, 2019), two-dimensional gel electrophoresis (2DE) and ^{1}H-NMR−based proteomics and metabolomics, respectively, were used to assess the toxicological mechanism of ZnO-NPs in the budding yeast, *Saccharomyces cerevisiae*. Almost 40% of proteins were downregulated in ZnO-NPs (10 mg/L) exposed cells as compared to control. Different metabolites involved in central carbon metabolism, cofactors synthesis, amino acid and fatty acid biosynthesis, purines and pyrimidines, and nucleoside and nucleotide biosynthetic pathways were disrupted in the exposed cells. These metabolic changes may be associated with energy metabolism, antioxidation, DNA and protein damage, and membrane stability (Kumar Babele, 2019).

In another recent analysis (Chavez Soria et al., 2019), MS-based metabolomics has been applied in a nontargeted metabolomics study to assess the effects of CuO-NPs on a model plant, *Arabidopsis thaliana*. Two platforms of HRMS were used LC-QToF-MS and LC Q Exactive Hybrid Quadrupole-Orbitrap-MS (LC-Orbitrap-MS). This double approach was performed to identify specific features (mass-to-charge ratios, m/z) that significantly changed in a reproducible manner. Besides, the total copper concentrations taken up in the plant tissues were quantified using inductively coupled plasma mass spectrometry (ICP-MS), which provided evidence of translocation of CuO-NPs from roots to leaves and flowering shoots. The main results showed that 65 plant metabolites were altered resulting from CuO-NPs exposure of *A. thaliana*. These metabolites belong to the jasmonic acid and glucosinolates pathways, suggesting a stress response induced by CuO-NPs in *Arabidopsis* (Chavez Soria et al., 2019).

The effects on cilantro plants (*Coriandrum sativum*) cultivated for 35 days in soil amended with ZnO nanoparticles (ZnO-NPs), bulk ZnO (bulk-ZnO), and $ZnCl_2$ (ionic-Zn) at 0−400 mg/kg were studied (Reddy Pullagurala et al., 2018). The photosynthetic pigments and the lipid peroxidation were assessed by ^{1}H-NMR, and ICP-MS was used to establish the metabolomics profiles. The results of this study revealed that all Zn compounds increased the chlorophyll content by at least 50%, compared to control. Only ZnO-NPs at 400 mg/kg decreased lipid peroxidation by 70%. ^{1}H-NMR data showed that all compounds significantly changed the carbinolic-based compounds, compared to control. Highest root and shoot uptake of Zn was observed with soils amended with bulk-ZnO at 400 mg/kg and ionic-Zn at 100 mg/kg, respectively. Results of this study corroborate that ZnO-NPs at a concentration <400 mg/kg enhanced the defense response in cilantro plants cultivated in organic soil (Reddy Pullagurala et al., 2018).

4.3 Quantum dots

Quantum dots (QDs) are semiconductor nanocrystals that have a reactive core which controls their optical properties (Farré et al., 2011). These cores are made of semiconductors, as, for example, cadmium selenide (CdSe), cadmium telluride (CdTe), indium phosphide (InP), or zinc selenide (ZnSe). In general, the reactive semiconductor cores are covered by a shell of silica or a ZnSe monolayer that protects semiconductors from oxidation and enhances the photoluminescence yield. QDs are used in medical imaging, solar cells, photovoltaics, photonics, and telecommunications (Farré et al., 2011).

Due to their widespread application, it is expected that residues of QDs can reach natural environments. Metabolomics can offer valuable information in order to assess their potential risk at realistic concentrations. However, very few studies have been carried out until now.

Falanga et al. (2018) studied the effects on the metabolomics profiles of *D. magna* across three generations (F0, F1, and F2) by QDs functionalized with the antimicrobial peptide indolicidin. *D. magna* was exposed to sublethal concentrations of the complex QDs-indolicidin, a normal survival of daphnids was observed from F0 to F2, but a delay of the first brood, fewer broods per female, and a decrease in length of about 50% compared to control was observed. Besides, QD-indolicidin induced the production of ROS significantly higher, and proportional in each generation. An impairment of enzyme's response to oxidative stress such as superoxide dismutase (SOD), catalase (CAT), and glutathione transferase (GST) was measured. These effects were confirmed by metabolomics profiles that pointed out a gradual decrease of metabolomics content over the three generations. The toxic effect of QD-indolicidin was suggested to be related to the higher accumulation of ROS and decreased antioxidant capacity in F1 and F2 generations. Results highlighted the capability of metabolomics to reveal an early metabolic response to stress induced by environmental QDs. In order to assess the response of exposed mammals, Khoshkam and collaborators (Khoshkam et al., 2018) investigated the toxicity of cadmium telluride quantum dots (CdTe-QDs) and bulk Cd_2^+ by metabolomics in mice plasma using ^{1}H-NMR data from four groups of mice at different time intervals. The results showed that significant changes in steroid hormone biosynthesis, lysine biosynthesis, and taurine and hypotaurine metabolism were the most affected pathways by CdTe-QDs especially in estrogenic steroids. Since the pattern of metabolite alteration of CdTe-QDs with equivalent Cd_2^+ was similar to those of $CdCl_2$, it was postulated that besides Cd_2^+ effects, the toxicity of CdTe-QDs was associated with other factors.

Some works have also been focused on investigating the metabolomics response of plants and crops. Zinc selenide quantum dots (ZnSe-QDs) are getting valuable due to extensive industrial and agriculture usage. Their effects on living organisms, including plants, are still unknown. Kolackova et al. (2019) evaluated the antioxidant response to the foliar exposure of *A. thaliana* to 100 and 250 µM ZnSe-QDs. In this study, an increase of the antioxidant response in the leaves was reported. Although QDs induced oxidative stress, the applied treatment dose of ZnSe-QDs did not have a significant toxic effect on the plants and even no morphological changes were observed.

4.4 Carbon-based nanomaterials

Since the C_{60} fullerene discovery by Kroto et al. (1985), subsequent investigations led to rise in a high number of new materials including the synthesis of cylindrical fullerene derivatives, the CNTs, single-walled carbon nanotubes and multi-walled carbon nanotubes with concentric cylinders up to 5—40 nm in diameter. These structures have excellent thermal and electrical conductivities. Due to their inherent hydrophobicity, a lot of research has been devoted to modifying the surface properties of CNTs to improve the stability of their aqueous suspensions (Bosi et al., 2003). The main current uses of these materials are in microelectronics, catalysis, battery and fuel-cell electrodes, supercapacitors, conductive coatings, water-purification systems, plastics, orthopedic implants, adhesives, and sensors. Annual worldwide production of CNTs is estimated to be around 1000 tonnes.

Graphene is another carbon allotrope relatively new made of a single layer of atoms arranged in a honeycomb-shaped lattice. Despite being one atom thick and chemically simple, graphene is extremely strong and highly conductive, making it ideal for high-speed microelectronics and photonics (Farré et al., 2011).

Several works have studied the sublethal toxicity through the metabolic response of diverse species to fullerenes exposure. To assess soil-dwelling organisms, Lankadurai et al. (2015) used ^{1}H-NMR—based metabolomics to investigate the response of *E. fetida* earthworms to sublethal C_{60} NP exposure in both surface contact and soil tests. The results obtained by principal component analysis of ^{1}H-NMR data showed a clear differentiation between the controls and the exposed earthworms after 2 days of exposure. When the exposure time increased, the separation decreased in soil but increased in contact tests suggesting earthworm's adaptation in soil exposure. The amino acids leucine, valine, isoleucine, and phenylalanine, the nucleoside inosine, and the sugars glucose and maltose emerged as potential bioindicators of exposure to C_{60} NPs.

In another work, Du et al. (2017) investigated the effects of nC_{60} on *S. obliquus* using transcriptomics and metabolomics approaches. The cells were exposed to various concentrations of nC_{60} for 7 days. Low dose of nC_{60} was found to have a minor growth inhibitory effect, but at levels over 0.1 mg/L, six metabolites significantly changed, including sucrose, D-glucose, and malic acid. The TCA cycle was the primary target. However, the accumulation of sucrose (end product) could have induced feedback inhibition of photosynthesis in *S. obliquus*, explaining the slight growth inhibition observed (Du et al., 2017).

Sanchís et al. (2018) have studied the metabolomics response of Mediterranean mussels (*Mytilus galloprovincialis*) exposed through the diet to fullerene by HPLC-HRMS. The experiments were conducted in a marine mesocosms emulating natural conditions during 35 days (7 days of acclimatization, 21 days of exposure, and 7 days of depuration). The results showed the bioaccumulation of fullerenes. Metabolomics studies revealed significant differences in the concentrations of seven free amino acids in comparison to the control group. The small nonpolar amino acids (e.g., alanine) and branched-chain amino acids (leucine and isoleucine) increased, which could be associated with hormesis, the glutamine concentrations significantly decreased, suggesting the activation of facultative anaerobic energy metabolism. Other significant differences were observed on the lipid content, such as the general increase of free fatty acids, i.e., long-chain fatty acids (lauric, myristic, and palmitic acids) when the concentrations of exposure increased. These results can be related to hypoxia and oxidative stress (Sanchís et al., 2018). The hormetic response has been also shown by other species exposed to fullerenes. For example, Wang et al. (2019a) evaluated hormetic effect of oxidative stress exerted by fullerene (nC_{60}) on *Daphnia pulex*, employing transcriptomics and metabolomics. *D. pulex* were exposed for 21 days to various concentrations of nC_{60}. After 7 days of exposure, hormetic effect of oxidative stress was evident. An increased L-glutathione (GSH) concentration and superoxide dismutase (SOD) activity was shown at low doses of exposure, and oppositely at high doses. Moreover, the nC_{60} interfered the TCA cycle. The synthesis of L-cysteine and glutamate was affected and the synthesis of GSH further disturbed.

There have been only a few studies on potential effects of carbon NMs on plants. Among the initial investigations by Zhao et al. (2019), 3-week-old cucumber plants were exposed to C_{60}-fullerol [$C_{60}(OH)_{24}$], a water soluble fullerene derivative. The exposure was carried out by foliar application of 1 or 2 mg per plant. Metabolomics revealed that C_{60}-fullerols upregulated antioxidant metabolites including 3-hydroxyflavone, 1,2,4-benzenetriol, and methyl trans-cinnamate, among others, while it downregulated cell membrane metabolites such as linolenic and palmitoleic acid. The global view of the metabolic pathway network suggests that C_{60}-fullerols accelerated electron transport rate, which induced ROS overproduction in chloroplast thylakoids. Plant activated antioxidant and defense pathways to protect the cells from the potential damage resulting from ROS. The revealed benefit (enhance electron transport) and risk (alter membrane composition) suggest a cautious use of C_{60}-fullerol for agricultural application.

Almost no studies have been carried out that focused on assessing subacute ecotoxicology of other carbon NMs by metabolomics studies. However, under natural conditions, the formation of the oxides of carbon-based NMs is a typical behavior. Hu et al. (2015) studied the effect on *Chlorella vulgaris* of graphene oxide (GO) and carboxyl single-walled carbon nanotubes (C-SWCNT). The results showed that cell division of *C. vulgaris* was fostered after 24 h, but inhibited after 96 h of exposure. Metabolomics analysis revealed significant differences in the metabolic profiles among the control, C-SWCNT and GO groups. The metabolisms of alkanes, lysine, octadecadienoic acid, and valine were associated with ROS. The nanotoxicological mechanisms involved the inhibition of fatty acid and amino acid.

4.5 Polymeric nanoparticles

Very few studies have been presented until now on nanoplastics. However, nanoplastics are widely used in cosmetics and cleaning industrial products, and as a consequence of the environmental fractionation. In recent years, there has been an increasing concern over their potential toxic effects on the environment. In this context, the uptake of nanopolystyrene particles has been studied in *C. elegans* (Kim et al., 2019). In this study, nanopolystyrene particles with sizes of 50 and 200 nm were prepared, and the L4 stage of *C. elegans* was exposed to these particles for 24 h. Besides various phenotypic alterations of the exposed nematode, a metabolomics study was performed. Exposure to nanopolystyrene particles caused the perturbation of metabolites related to energy metabolism, such as TCA cycle intermediates, glucose, and lactic acid.

5. Conclusions and future works

Some specific effects have found in most of the metabolomics studies evaluating potential impacts of NMs on plants, aquatic, and terrestrial invertebrates. Some of the effects are on the metabolism of the amino acids, by upregulating or downregulating free amino acids content, cellular ROS effects, and the TCA cycle. However, only a limited number of materials have been studied until now. The central challenges in future studies are the lack of standardized approaches for both metabolomics approaches and manipulation of NMs, that will permit the comparison of different studies and to extend environmental metabolomics studies to a wider number of NMs. Besides, there is a gap of analytical methods to assess NMs residues in complex environmental samples. This limits our capability to quantify their bioaccumulation in some cases. In addition to this, the lack of sensitive analytical approaches limits our abilities to carry out in-field metabolomics studies that are needed to assess subacute effects under more realistic environmental scenarios. An increasing attention needs to be laid on NPs of polymers, and future studies are required in addition to interactive impact studies of mixtures of xenobiotics and NMs. Finally, comparing the different analytical approaches to assess the metabolome of exposed populations, nontargeted metabolomics studies offer broader responses, and in this sense, HRMS offers higher sensitivities being the most suitable technique of choice Table 9.1.

Table 9.1: Examples of metabolomics studies to assess nanomaterial subacute effects on living organisms.

Nanomaterials	Organism	Metabolites changed	Analytical technique	References
AgNPs	*Aspergillus niger* and *Penicillium chrysogenum*	Organic acids (oxalic, citric, malic, succinic).	LDI-TOF-MS	Pietrzak et al. (2016)
AgNPs	*Microcystis aeruginosa*	Upregulation of the amino acids arginine and proline, the indole alkaloid biosynthesis, and phospholipid metabolism. ROS effect.	LC-MS	Zhang et al. (2018b)
AgNPs	*Scenedesmus obliquus*	Carbohydrates D-galactose, sucrose, and D-fructose and the amino acids as glycine. ROS effect.	GC-QTOF-MS	Wang et al. (2019b)
AgNPs	*Daphnia magna*	ROS effect. Increase the levels of lactate.	^{1}H-NMR	Li et al. (2015a)
AgNPs	*Eisenia fetida*	Increase of leucine and arginine and decreased of ATP.	^{1}H-NMR	Li et al. (2015b)
AgNPs	*Cucumis sativus*	Upregulation of TCA cycle intermediates, downregulation of glycine/serine ratio, upregulation of pentadecanoic and arachidonic acids, downregulation of linoleic and linolenic acids, and downregulation of glutamine and asparagine.	GC-MS	Zhang et al. (2018a)
Cu-NPs	*C. sativus*	Sugars (fructose, xylose), organic acids (citric), amino acids (glycine and proline), and fatty acids (caprylic acid, linolenic acid).	^{1}H-NMR and GC-MS	Zhao et al. (2017)

Continued

Table 9.1: Examples of metabolomics studies to assess nanomaterial subacute effects on living organisms.—cont'd

Nanomaterials	Organism	Metabolites changed	Analytical technique	References
TiO$_2$-NPs	*Physarum polycephalum macroplasmodium*	ROS effect. Sugars (including triose phosphates), sugar alcohols, organic acids, amino acids, polyamines, and flavonoids. Glutathione levels exhibited the greatest induced increase.	GC/MS	Zhang et al. (2018c)
TiO$_2$-NPs	*Escherichia coli*	The energy metabolites ATP, GTP, CTP, and UTP. The organic acids (acetate, oxaloacetate and lactate) involved in TCA. They act as intermediates in many metabolic reactions and fatty acids.	^{1}H-NMR and GC-MS	Planchon et al. (2017)
TiO$_2$-NPs	*Caenorhabditis elegans*	Amino acids, organic acids, and sugars. The biological pathways affected are the TCA cycle, arachidonic acid metabolism, and glyoxylate dicarboxylate metabolism.	GC-MS	Ratnasekhar et al. (2015)
TiO$_2$-NPs	*Paracentrotus lividus*	Pentose phosphate, cysteine-methionine, glycine-serine metabolism pathways.		Alijagic et al. (2020)
TiO$_2$-NPs	*Danio rerio*	ATP, ADP, AMP, inosine, and xanthine. Disturbances in glycerophospholipid metabolism and the Krebs cycle.	^{1}H-NMR	Raja et al. (2018)
TiO$_2$-NPs	*Oryza sativa* L	TCA cycle and the pentose phosphate pathway were significantly increased.	GC-MS	Wu et al. (2017)
ZnO-NPs	*Saccharomyces cerevisiae*	Different metabolites involved in central carbon metabolism, cofactors synthesis, amino acid and fatty acid biosynthesis, purines and pyrimidines, nucleoside, and nucleotide biosynthetic pathways were disrupted.	2DE and ^{1}H-NMR	Kumar Babele (2019)
CuO-NPs	*Arabidopsis thaliana*	65 plant metabolites were altered resulting from CuO-NPs exposure of *A. thaliana*. These metabolites belong to the jasmonic acid and glucosinolates pathways.	LC-QToF-MS and LC-orbitrap-ms	Chavez Soria et al. (2019)

Continued

Table 9.1: Examples of metabolomics studies to assess nanomaterial subacute effects on living organisms.—cont'd

Nanomaterials	Organism	Metabolites changed	Analytical technique	References
ZnO-NPs	*Coriandrum sativum*	Compounds significantly changed the carbinolic-based compounds.	^{1}H-NMR and ICP-MS	Reddy Pullagurala et al. (2018).
QDs-indolicidin	*D. magna*	ROS effect. Amino acids alanine and threonine and other metabolites like lactic acid.	^{1}H-NMR	Falanga et al. (2018)
CdTe-QDs	Mice	Significant changes in steroid hormone biosynthesis, lysine biosynthesis, and taurine and hypotaurine metabolism.	^{1}H-NMR	Khoshkam et al. (2018)
C$_{60}$-fullerene	*E. fetida*	Amino acids leucine, valine, isoleucine, and phenylalanine, the nucleoside inosine, and the sugars glucose and maltose	^{1}H-NMR	Lankadurai et al. (2015)
nC$_{60}$-fullerene	*S. obliquus*	Sugars sucrose, D-glucose, and malic acid. The citrate cycle (TCA cycle) was the primary target.	^{1}H-NMR	Du et al. (2017)
Fullerenes	*Mytilus galloprovincialis*	Small nonpolar amino acids (alanine), branched-chain amino acids (leucine and isoleucine) increase, while glutamine concentrations significantly decreased. Other significant differences were the lipid content, such as the general increase of free fatty acids (lauric, myristic, and palmitic).	LC-orbitrap-MS	(Sanchís et al. 2018)
Nanopolystyrene particles	*C. elegans*	Energy metabolism, such as TCA cycle intermediates, glucose, and lactic acid.	GC-MS	Kim et al. (2019)

[1]*H-NMR*, Proton- nuclear magnetic resonance; *2DE*, Two-dimensional gel electrophoresis; *GC-MS*, Gas chromatography coupled to mass spectrometry; *LC-MS*, Liquid chromatography coupled to mass spectrometry; *LDI-TOF-MS*, Laser desorption ionization time-of-flight mass spectrometry.

Acknowledgments

This study was supported by the Spanish Ministry of Economy and Competitiveness, State Research Agency, and by the European Union through the European Regional Development Fund through the project PLAS-MED (CTM2017-89701-C3-2-R).

References

Aitken, R.J., Chaudhry, M.Q., Boxall, A.B.A., Hull, M., 2006. Manufacture and use of nanomaterials: current status in the UK and global trends. Occup. Med. 56, 300−306.

Alijagic, A., Gaglio, D., Napodano, E., Russo, R., Costa, C., Benada, O., Kofroňová, O., Pinsino, A., 2020. Titanium dioxide nanoparticles temporarily influence the sea urchin immunological state suppressing inflammatory-relate gene transcription and boosting antioxidant metabolic activity. J. Hazard Mater. 384.

Aretz, I., Meierhofer, D., 2016. Advantages and pitfalls of mass spectrometry based metabolome profiling in systems biology. Int. J. Mol. Sci. 17.

Bo, Y., Jin, C., Liu, Y., Yu, W., Kang, H., 2014. Metabolomic analysis on the toxicological effects of TiO_2 nanoparticles in mouse fibroblast cells: from the perspective of perturbations in amino acid metabolism. Toxicol. Mech. Methods 24, 461−469.

Bosi, S., Da Ros, T., Spalluto, G., Prato, M., 2003. Fullerene derivatives: an attractive tool for biological applications. Eur. J. Med. Chem. 38, 913−923.

Chavez Soria, N.G., Bisson, M.A., Atilla-Gokcumen, G.E., Aga, D.S., 2019. High-resolution mass spectrometry-based metabolomics reveal the disruption of jasmonic pathway in *Arabidopsis thaliana* upon copper oxide nanoparticle exposure. Sci. Total Environ. 693.

Dastjerdi, R., Montazer, M., 2010. A review on the application of inorganic nano-structured materials in the modification of textiles: focus on anti-microbial properties. Colloids Surf. B Biointerfaces 79, 5−18.

De Hoffmann, E., Charette, J., Stroobant, V., 1996. Mass Spectrometry: Principles and Applications.

De La Luz-Hdez, K., 2012. Metabolomics and mammalian cell culture. Metabolomics 3−18.

Du, C., Zhang, B., He, Y., Hu, C., Ng, Q.X., Zhang, H., Ong, C.N., ZhifenLin, 2017. Biological effect of aqueous C_{60} aggregates on Scenedesmus obliquus revealed by transcriptomics and non-targeted metabolomics. J. Hazard Mater. 324, 221−229.

Emwas, A.H.M., Salek, R.M., Griffin, J.L., Merzaban, J., 2013. NMR-based metabolomics in human disease diagnosis: applications, limitations, and recommendations. Metabolomics 9, 1048−1072.

Falanga, A., Mercurio, F.A., Siciliano, A., Lombardi, L., Galdiero, S., Guida, M., Libralato, G., Leone, M., Galdiero, E., 2018. Metabolomic and oxidative effects of quantum dots-indolicidin on three generations of Daphnia magna. Aquat. Toxicol. 198, 158−164.

Farré, M., Sanchís, J., Barceló, D., 2011. Analysis and assessment of the occurrence, the fate and the behavior of nanomaterials in the environment. TrAC Trends Anal. Chem. 30, 517−527.

Fennell, T.R., Mortensen, N.P., Black, S.R., Snyder, R.W., Levine, K.E., Poitras, E., Harrington, J.M., Wingard, C.J., Holland, N.A., Pathmasiri, W., Sumner, S.C.J., 2017. Disposition of intravenously or orally administered silver nanoparticles in pregnant rats and the effect on the biochemical profile in urine. J. Appl. Toxicol. 37, 530−544.

Garcia-Contreras, R., Sugimoto, M., Umemura, N., Kaneko, M., Hatakeyama, Y., Soga, T., Tomita, M., Scougall-Vilchis, R.J., Contreras-Bulnes, R., Nakajima, H., Sakagami, H., 2015. Alteration of metabolomic profiles by titanium dioxide nanoparticles in human gingivitis model. Biomaterials 57, 33−40.

Gross, J.H., 2011. Isotopic composition and accurate mass. Mass Spectrom. 67−116.

Hanemann, T., Szabó, D.V., 2010. Polymer-nanoparticle composites: from synthesis to modern applications. Materials 3, 3468−3517.

Hu, X., Ouyang, S., Mu, L., An, J., Zhou, Q., 2015. Effects of graphene oxide and oxidized carbon nanotubes on the cellular division, microstructure, uptake, oxidative stress, and metabolic profiles. Environ. Sci. Technol. 49, 10825−10833.

Huang, Y., Adeleye, A.S., Zhao, L., Minakova, A.S., Anumol, T., Keller, A.A., 2019. Antioxidant response of cucumber (*Cucumis sativus*) exposed to nano copper pesticide: quantitative determination via LC-MS/MS. Food Chem. 270, 47−52.

Huang, Y., Li, W., Minakova, A.S., Anumol, T., Keller, A.A., 2018. Quantitative analysis of changes in amino acids levels for cucumber (*Cucumis sativus*) exposed to nano copper. NanoImpact 12, 9−17.

Jarak, I., Carrola, J., Barros, A.S., Gil, A.M., Pereira, M.L., Corvo, M.L., Duarte, I.F., 2017. Metabolism modulation in different organs by silver nanoparticles: an NMR metabolomics study of a mouse model. Toxicol. Sci. 159, 422−435.

Khoshkam, M., Baghdadchi, Y., Arezumand, R., Ramazani, A., 2018. Synthesis, characterization and in vivo evaluation of cadmium telluride quantum dots toxicity in mice by toxicometabolomics approach. Toxicol. Mech. Methods 28, 539−546.

Kim, H.M., Lee, D.K., Long, N.P., Kwon, S.W., Park, J.H., 2019. Uptake of nanopolystyrene particles induces distinct metabolic profiles and toxic effects in *Caenorhabditis elegans*. Environ. Pollut. 246, 578−586.

Kolackova, M., Moulick, A., Kopel, P., Dvorak, M., Adam, V., Klejdus, B., Huska, D., 2019. Antioxidant, gene expression and metabolomics fingerprint analysis of *Arabidopsis thaliana* treated by foliar spraying of ZnSe quantum dots and their growth inhibition of *Agrobacterium tumefaciens*. J. Hazard Mater. 365, 932−941.

Kroto, H.W., Heath, J.R., O'Brien, S.C., Curl, R.F., Smalley, R.E., 1985. C_{60}: buckminsterfullerene. Nature 318, 162−163.

Kuehnbaum, N.L., Britz-Mckibbin, P., 2013. New advances in separation science for metabolomics: resolving chemical diversity in a post-genomic era. Chem. Rev. 113, 2437−2468.

Kumar Babele, P., 2019. Zinc oxide nanoparticles impose metabolic toxicity by de-regulating proteome and metabolome in *Saccharomyces cerevisiae*. Toxicology Reports 6, 64−73.

Lankadurai, B.P., Nagato, E.G., Simpson, A.J., Simpson, M.J., 2015. Analysis of *Eisenia fetida* earthworm responses to sub-lethal C_{60} nanoparticle exposure using 1H-NMR based metabolomics. Ecotoxicol. Environ. Saf. 120, 48−58.

Larive, C.K., Barding, G.A., Dinges, M.M., 2015. NMR spectroscopy for metabolomics and metabolic profiling. Anal. Chem. 87, 133−146.

Li, L., Wu, H., Ji, C., van Gestel, C.A.M., Allen, H.E., Peijnenburg, W.J.G.M., 2015a. A metabolomic study on the responses of daphnia magna exposed to silver nitrate and coated silver nanoparticles. Ecotoxicol. Environ. Saf. 119, 66−73.

Li, L., Wu, H., Peijnenburg, W.J.G.M., Van Gestel, C.A.M., 2015b. Both released silver ions and particulate Ag contribute to the toxicity of AgNPs to earthworm *Eisenia fetida*. Nanotoxicology 9, 792−801.

Lian, J., Zhao, L., Wu, J., Xiong, H., Bao, Y., Zeb, A., Tang, J., Liu, W., 2020. Foliar spray of TiO_2 nanoparticles prevails over root application in reducing Cd accumulation and mitigating Cd-induced phytotoxicity in maize (*Zea mays* L.). Chemosphere 239.

Lindeque, J.Z., Matthyser, A., Mason, S., Louw, R., Taute, C.J.F., 2018. Metabolomics reveals the depletion of intracellular metabolites in $HepG_2$ cells after treatment with gold nanoparticles. Nanotoxicology 12, 251−262.

Markley, J.L., Brüschweiler, R., Edison, A.S., Eghbalnia, H.R., Powers, R., Raftery, D., Wishart, D.S., 2017. The future of NMR-based metabolomics. Curr. Opin. Biotechnol. 43, 34−40.

Meng, X., Martinez, M.A., Raymond-Stintz, M.A., Winter, S.S., Wilson, B.S., 2010. IKK inhibitor bay 11-7082 induces necroptotic cell death in precursor-B acute lymphoblastic leukaemic blasts. Br. J. Haematol. 148, 487−490.

Mitrano, D.M., Motellier, S., Clavaguera, S., Nowack, B., 2015. Review of nanomaterial aging and transformations through the life cycle of nano-enhanced products. Environ. Int. 77, 132−147.

Navarro-Reig, M., Jaumot, J., Baglai, A., Vivó-Truyols, G., Schoenmakers, P.J., Tauler, R., 2017. Untargeted comprehensive two-dimensional liquid chromatography coupled with high-resolution mass spectrometry analysis of rice metabolome using multivariate curve resolution. Anal. Chem. 89, 7675−7683.

Nicholson, J.K., Lindon, J.C., Holmes, E., 1999. 'Metabonomics': understanding the metabolic responses of living systems to pathophysiological stimuli via multivariate statistical analysis of biological NMR spectroscopic data. Xenobiotica 29, 1181−1189.

Oliver, S.G., Winson, M.K., Kell, D.B., Baganz, F., 1998. Systematic functional analysis of the yeast genome. Trends Biotechnol. 16, 373−378.

Panayotov, D.A., Morris, J.R., 2016. Surface chemistry of Au/TiO$_2$: thermally and photolytically activated reactions. Surf. Sci. Rep. 71, 77–271.

Pietrzak, K., Glińska, S., Gapińska, M., Ruman, T., Nowak, A., Aydin, E., Gutarowska, B., 2016. Silver nanoparticles: a mechanism of action on moulds. Metall 8, 1294–1302.

Planchon, M., Leger, T., Spalla, O., Huber, G., Ferrari, R., 2017. Metabolomic and proteomic investigations of impacts of titanium dioxide nanoparticles on *Escherichia coli*. PloS One 12.

Potocnik, J., 2011. Commission recommendation of 18 October 2011 on the definition of nanomaterial. In: Commission Recommendation of 18 October 2011 on the Definition of Nanomaterial.

Raja, G., Kim, S., Yoon, D., Yoon, C., Kim, S., 2018. 1H NMR based metabolomics studies of the toxicity of titanium dioxide nanoparticles in Zebrafish (*Danio rerio*). Bull. Kor. Chem. Soc. 39, 33–39.

Ratnasamy, C., Wagner, J., 2009. Water gas shift catalysis. Catal. Rev. Sci. Eng. 51, 325–440.

Ratnasekhar, C., Sonane, M., Satish, A., Mudiam, M.K.R., 2015. Metabolomics reveals the perturbations in the metabolome of *Caenorhabditis elegans* exposed to titanium dioxide nanoparticles. Nanotoxicology 9, 994–1004.

Reddy Pullagurala, V.L., Adisa, I.O., Rawat, S., Kalagara, S., Hernandez-Viezcas, J.A., Peralta-Videa, J.R., Gardea-Torresdey, J.L., 2018. ZnO nanoparticles increase photosynthetic pigments and decrease lipid peroxidation in soil grown cilantro (Coriandrum sativum). Plant Physiol. Biochem. 132, 120–127.

Sanchís, J., Llorca, M., Olmos, M., Schirinzi, G.F., Bosch-Orea, C., Abad, E., Barceló, D., Farré, M., 2018. Metabolic responses of *Mytilus galloprovincialis* to fullerenes in mesocosm exposure experiments. Environ. Sci. Technol. 52, 1002–1013.

Sharma, V.K., Yngard, R.A., Lin, Y., 2009. Silver nanoparticles: green synthesis and their antimicrobial activities. Adv. Colloid Interface Sci. 145, 83–96.

Shiju, N.R., Guliants, V.V., 2009. Recent developments in catalysis using nanostructured materials. Appl. Catal. Gen. 356, 1–17.

Subramaniam, V.D., Prasad, S.V., Banerjee, A., Gopinath, M., Murugesan, R., Marotta, F., Sun, X.F., Pathak, S., 2019. Health hazards of nanoparticles: understanding the toxicity mechanism of nanosized ZnO in cosmetic products. Drug Chem. Toxicol. 42, 84–93.

Tan, M.L., Choong, P.F.M., Dass, C.R., 2009. Cancer, chitosan nanopartides and catalytic nucleic acids. J. Pharm. Pharmacol. 61, 3–12.

Tucci, P., Porta, G., Agostini, M., Dinsdale, D., Iavicoli, I., Cain, K., Finazzi-Agró, A., Melino, G., Willis, A., 2013. Metabolic effects of TIO$_2$ nanoparticles, a common component of sunscreens and cosmetics, on human keratinocytes. Cell Death Dis. 4.

Viant, M.R., Sommer, U., 2013. Mass spectrometry based environmental metabolomics: a primer and review. Metabolomics 9, 144–158.

Wang, P., Ng, Q., Zhang, B., Wei, Z., Hassan, M., He, Y., Ong, C.N., 2019a. Employing multi-omics to elucidate the hormetic response against oxidative stress exerted by nC$_{60}$ on *Daphnia pulex*. Environ. Pollut. 251, 22–29.

Wang, P., Zhang, B., Zhang, H., He, Y., Ong, C.N., Yang, J., 2019b. Metabolites change of Scenedesmus obliquus exerted by AgNPs. J. Environ. Sci. (China) 76, 310–318.

Wang, S.Q., Tooley, I.R., 2011. Photoprotection in the era of nanotechnology. Semin. Cutan. Med. Surg. 30, 210–213.

Wu, B., Zhu, L., Le, X.C., 2017. Metabolomics analysis of TiO$_2$ nanoparticles induced toxicological effects on rice (*Oryza sativa* L.). Environ. Pollut. 230, 302–310.

Yekeen, N., Padmanabhan, E., Idris, A.K., Chauhan, P.S., 2019. Nanoparticles applications for hydraulic fracturing of unconventional reservoirs: a comprehensive review of recent advances and prospects. J. Petrol. Sci. Eng. 178, 41–73.

Zhang, H., Du, W., Peralta-Videa, J.R., Gardea-Torresdey, J.L., White, J.C., Keller, A., Guo, H., Ji, R., Zhao, L., 2018a. Metabolomics reveals how cucumber (cucumis sativus) reprograms metabolites to cope with silver ions and silver nanoparticle-induced oxidative stress. Environ. Sci. Technol. 52, 8016–8026.

Zhang, J.L., Zhou, Z.P., Pei, Y., Xiang, Q.Q., Chang, X.X., Ling, J., Shea, D., Chen, L.Q., 2018b. Metabolic profiling of silver nanoparticle toxicity in Microcystis aeruginosa. Environ. Sci. J. Integr. Environ. Res.: Nano 5, 2519–2530.

Zhang, Z., Liang, Z.C., Zhang, J.H., Tian, S.L., Le Qu, J., Tang, J.N., De Liu, S., 2018c. Nano-sized TiO_2 ($nTiO_2$) induces metabolic perturbations in Physarum polycephalum macroplasmodium to counter oxidative stress under dark conditions. Ecotoxicol. Environ. Saf. 154, 108–117.

Zhao, L., Hu, J., Huang, Y., Wang, H., Adeleye, A., Ortiz, C., Keller, A.A., 2017. 1H NMR and GC–MS based metabolomics reveal nano-Cu altered cucumber (Cucumis sativus) fruit nutritional supply. Plant Physiol. Biochem. 110, 138–146.

Zhao, L., Zhang, H., Wang, J., Tian, L., Li, F., Liu, S., Peralta-Videa, J.R., Gardea-Torresdey, J.L., White, J.C., Huang, Y., Keller, A., Ji, R., 2019. C60 fullerols enhance copper toxicity and alter the leaf metabolite and protein profile in cucumber. Environ. Sci. Technol. 53, 2171–2180.

CHAPTER 10

Environmental metabolomics and xenometabolomics for the assessment of exposure to contaminant mixtures

Sara Rodríguez-Mozaz[1], Albert Serra-Compte[1], Ruben Gil-Solsona[1], Diana Álvarez-Muñoz[2]

[1]Catalan Institute for Water Research (ICRA), Parc Científic i Tecnològic de la Universitat de Girona, Girona, Spain; [2]Water and Soil Quality Research Group, Department of Environmental Chemistry, IDAEA-CSIC, Barcelona, Spain

Chapter Outline

1. Introduction

Humans and environment are exposed to a wide variety of chemicals. Given that it is unrealistic to assess every possible combination of chemical substances that humans and/ or environment may be exposed to, the major challenge faced nowadays is to develop

Environmental Metabolomics. https://doi.org/10.1016/B978-0-12-818196-6.00010-8

systematic ways of addressing chemical mixtures in environmental assessment, and to identify priority mixtures of potential concern (European Commission, 2012). There is a growing evidence that untargeted metabolomics-based techniques may accomplished both tasks simultaneously: first, by identifying biomarkers of effect and detecting early-stage metabolic dysregulations, and second, by establishing priority mixtures of contaminants accumulated in biological samples (Bonvallot et al., 2018). This chapter attempts to determine the suitability of metabolomics to fulfill both tasks and its potential use as a more routine tool for environmental monitoring.

1.1 Contaminant mixtures in the aquatic environment

Thousands of chemicals are daily used worldwide in different human activities. Industrial, urban, hospital, animal rearing, and agricultural activities (between others) are among the human activities that produce higher chemical release into the aquatic environment worldwide, generating a huge range of contaminants. From each specific source, the contaminants can make their way into the environment through different pathways such as effluents of wastewater treatment plants (WWTPs), surface run-off, atmospheric deposition, direct spillage, etc. One of the most important inputs to the aquatic media is usually through WWTPs. The treatments commonly applied to the wastewater are not specifically designed to eliminate chemical contaminants and, therefore, many chemicals are discharged into the receiving waterbody (Fatta-Kassinos et al., 2011a; Petrie et al., 2014). Therefore, metals, persistent organic pollutants, and emerging organic contaminants are released and commonly found in the aquatic environment (Álvarez-Muñoz et al., 2016). Some of the most ubiquitous groups of contaminants found here comprise polycyclic aromatic hydrocarbons, surfactants, polychlorinated biphenyls, pesticides, flame retardants, pharmaceuticals and personal care products, illicit drugs, fragrances, endocrine-disrupting compounds, plasticizers, nano- and microparticles, per and polyfluorinated alkyl substances, and treatment by-products (Cole et al., 2011; Noguera-Oviedo and Aga, 2016; Petrovic et al., 2004; Rodriguez-Mozaz et al., 2017; Sauvé and Desrosiers, 2014). All of these contaminants make a complex mixture of hazard substances that pollute the different receiving environments, provoking a potential risk for the living organisms and for the human health.

Surface waters (rivers and lakes), coastal areas, and groundwaters are the major contaminant-receiving environments. Once contaminants are discharged into the environment, they can be distributed into the different compartments. Apart from the water phase, contaminants can adsorb to the soil and accumulate in the aquatic organisms exposed to them (Holt, 2000; Petrie et al., 2014; Rodriguez-Mozaz et al., 2017). Moreover, chemicals can be transformed due to biological and physicochemical processes (Fatta-Kassinos et al., 2011b; Picó and Barceló, 2015). These transformations alter the mixture of contaminants as not only the compounds discharged from the WWTPs effluents occur but also their transformation products (TPs) can be found in the aquatic environment. Besides, natural products like marine toxins are also present; new chemicals are continuously developed and used in new applications by the industry, and subsequently they are being discharged as well into the natural media. Therefore, the mixture of contaminants becomes a dynamic system where spatial and temporal variations result in key factors influencing the composition and concentration of contaminants present in a specific environment in a given moment. Consequently, the analysis of contaminant mixtures in the environment becomes a challenging task.

Traditionally, the analytical approach used has been the targeted analysis of different contaminants which generally comprises a concentration step when analyzing water (Gros et al., 2012), whereas for biota and sediment samples, an extraction followed by a clean-up is usually employed. Different instrumental analysis can be performed such as ICP-mass spectrometry (ICP-MS) (for metal analysis), liquid and gas chromatography coupled to different detectors like mass spectrometry (MS), nuclear magnetic resonance (NMR), ultraviolet, or fluorescence detectors, among other analytical techniques (Álvarez-Muñoz et al., 2016). Using these methodologies, huge efforts have been put in developing multiresidue methods (Álvarez-Muñoz et al., 2015; Gros et al., 2012; Huerta et al., 2013). These methods can determine more than 100 compounds in a single analysis within a relatively short run time (about 10 min). However, target approach methods are limited to those contaminants included in the method whose analytical standards are available (most TPs do not have analytical standards). Besides, they need to be continuously adapted to new target contaminants in order to cope with the dynamic of the mixture of contaminants present in the environment, which is time- and money-consuming. In this context, new analytical strategies are needed to overcome these drawbacks when assessing contaminant mixtures in the environment. The European Commission became aware of this problem (European Commission, 2009) and the necessity of adapting the EU chemicals' legislation to address the risk derived from exposure to multiple chemicals in both humans and environment (European Commission, 2012). Currently, the EU chemical legislation, as well as in other parts of the world, is based predominantly on assessments carried out on individual substances according to the Registration, Evaluation, Authorisation and Restriction of Chemicals (REACH) (REGULATION (EC) No 1907/2006). However, as previously reported, humans and environment are exposed to complex mixtures of chemicals, and it is necessary to develop effective tools for addressing them in environmental assessment.

1.2 Study of the toxicity of mixtures: classic and metabolomics approaches

Over the past 100 years, toxicology has been mainly based in the assessment of individual chemicals, studying the effects of exposure to single compounds. However, as previously stated, organisms are exposed to mixtures of chemicals including both anthropogenic and natural occurring compounds. These mixtures can be considered as simple or complex mixtures. According to the United States Environmental Protection Agency (U.S. E.P.A., 2000), simple mixtures contain two or more identifiable components but few enough that the mixture toxicity can be adequately characterized by a combination of the component toxicities and interactions. However, complex mixtures contain many components and any estimation of its toxicity based on its components' toxicities contains too much uncertainty and error to be useful (U.S. E.P.A., 2000). Therefore, the study of mixtures toxicity is a difficult task, especially when dealing with complex mixtures such as in the case of organisms exposed to WWTP effluents.

Two different approaches can be followed in the analysis of mixture toxicity: top-down approach which is based on the analysis of the whole-mixture toxicity and bottom-up approach based on component-interaction analysis (Groten et al., 2001). Usually, the toxicity of the mixtures ranges between the toxicity of the most toxic compound and the toxicity of the less toxic compound. However, it is also commonly observed that even when the individual compounds of a mixture are present at low concentrations and they do not have toxic effects separately, the mixture of the components as a whole has a significant toxicity (Altenburger and Greco, 2009; Faust et al., 2001; Kortenkamp, 2007). This is related to the fact that chemical interactions may emphasize the toxicological outcomes. Chemicals interact in the organism and the resulting toxicity depends on the kinetics and dynamics of the chemicals, together with the biochemical status of the organism (Monosson, 2005). In 1993, Bliss (Bliss, 1993) defined three categories of joint chemical action that are generally accepted: (1) independent joint action, the toxicity of any combination can be predicted from that of the isolated components and the presence of one chemical does not affect the toxicity of another, (2) similar joint action, the components of the mixture act independently but through similar mechanisms and effects, therefore, the presence of one chemical may affect the impact of another, in this case the toxicity can also be predicted based on each chemical, and (3) synergistic action, the toxicity of the mixture may be greater or lower than the one predicted from studies based on isolated components (synergies or antagonizes may occur), therefore the combined toxicity needs to be assessed.

Toxicity can be measured by using either in vivo (with model animals) or in vitro (with cultured bacteria or mammalian cells) experiments. Normally, the response of the model organism or cell is measured at different concentrations of the chemical (or chemical mixture) regarding to a certain endpoint, establishing a dose—effect relationship or the also called concentration—response curve (C-R). A plethora of different endpoints can be used such as lethal dose, effect concentration, inhibitory concentration, no effect concentration, no observed adverse effect concentration, etc.

Since the 1980s with the knowledge gain from "omics" technologies, the endpoints have evolved and mechanisms of toxicity together with biochemical and molecular pathways are investigated. Metabolomics consists on the simultaneous characterization of the metabolites present in an organism and offers a "picture" of the biochemistry of the organism at any one time. It is a high-throughput semiquantitative approach and potentially a fast method for the detection of hundreds of metabolites simultaneously. Traditional metabolomics studies lead to the detection of endogenous metabolites termed metabolome or endometabolome, whose levels are altered due to an external stressor. A similar holistic approach can also be used for profiling the range of chemical xenobiotics and their metabolites in an organism exposed to environmental contaminants, what has been referred to as the exposome or xenometabolome (Al-Salhi et al., 2012; Holmes et al., 2008). Therefore, metabolomics approach offers a major advantage, especially when dealing with complex mixtures, because it offers the possibility of profiling the xenometabolome (approach also known as xenometabolomics) as well as the metabolome simultaneously. By using this approach both the identity of the chemical mixture accumulated in an organism and how they relate to the health status (toxicity of the mixture) can be studied simultaneously. Besides, the measurement of the biochemical responses in the exposed organism can be used to identify the endpoints that are predictive and can be used as biomarkers after appropriate validation. Another advantage is that the pattern of toxicity found may be linked to pathologies and they can be used to screen for similar patterns induced by new xenobiotics (Hartung and Leist, 2008).

2. Application of metabolomics/xenometabolomics for chemical mixtures

A total of 22 studies have been published in the last decade (since 2011 till 2019) where metabolomics has been applied to evaluate the effects of exposure to chemical mixtures in biota from aquatic environments. The articles are listed in Table 10.1 together with specific information about the study conducted, like the type of contaminant mixture (if known) or the specific stressor, if the experiments were performed at lab scale or in the field, if water characterization was carried out, the type of organism, species, and tissues analyzed. The popularity of metabolomics to study the effects of mixtures is rapidly increasing, half of the works listed in Table 10.1 has been published between 2017 and 2019, and it is anticipated to escalate as metabolomics becomes a more routine tool for environmental monitoring. In Fig. 10.1 a scheme of experimental setup and analytical approaches used for metabolomics/xenometabolomics studies of chemical mixtures is presented. This figure will be referred in the next sections as well as Table 10.2 that presents details about the sample pretreatment, analytical approaches, and instrumental analysis followed in the literature (published until 2019) regarding the study of the effects of contaminant mixtures in biota using metabolomics.

Table 10.1: Metabolomics studies conducted to evaluate the effects of exposure to chemical mixtures in biota from aquatic environments. Details about the contaminant mixture used, type of experiment, and target organism are presented.

Author (year)	Contaminant/stressor	Field/lab	Water characterization	Organism	Species	Tissue
	Experiments performed using a mixture of selected compounds					
Serra-Compte (2018)	**Mixture of pharmaceuticals (PhACs)** (ibuprofen, diclofenac, carbamazepine, sulfamethoxazole, erythromycin, metoprolol, atenolol, gemfibrozil, and hydrochlorothiazide)	Lab	Yes	Biofilm	–	Whole community
Melvin (2018)	**Mixture of PhACs** (metformin, diclofenac, and valproic acid)	Lab	Yes (chemical and physicochemical)	Amphibian	*Limnodynastes peronii*	Liver
Wagner (2018)	**Mixture** (carbamazepine, propranolol, and perfluorooctanesulfonic acid (PFOS))	Lab	No	Crustacean	*Daphnia magna*	Whole organism
Jordan (2011)	**Mixture** (bisphenol A, di-(2-ethylhexyl)-phthalate (DEHP), and nonylphenol)	Lab	No	Fish	*Carassius auratus*	Gonads and liver
Søfteland et al. (2014)	**Mixture of polycyclic aromatic hydrocarbons (PAHs) and pesticides in feed** (endosulfan (alfa and beta), chlorpyrifos, phenanthrene, and benzopyrene)	Lab	No	Fish	*Salmo salar L.*	Hepatocyte cells

		Experiments performed using real contaminated waters				
Al-Salhi (2012)	**WWTP effluents**	**Lab**	No	Fish	*Oncorhynchus mykiss*	Plasma and bile
David (2017)	**WWTP effluents**	**Lab**	**Yes** (physicochemical and nonsteroidal antiinflammatory drugs (NSAIDs) analysis)	Fish	*Rutilus rutilus*	Plasma, gonads, gill, liver, and kidney tissues
Mosley (2018)	**WWTP influent and effluent**	**Lab**	**No**	Fish	*Pimephales promelas*	Skin mucus
Wagner (2019)	**Influent** (postsecondary clarification) **and effluent** wastewater (EWW) with PFOSs addition in EWW	**Lab**	**Yes** (concentration of PFOS)	Crustacean	*D. magna*	Whole organism
Huang (2016)	**Exposition to exogenous endocrine compound,** performance chemicals, PhACs, and personal care products (PCPs), petroleum derivative, heavy metals, and EWW	**Lab**	**Yes** (real concentration of contaminants)	Fish	*Danio rerio*	Whole organism
Zhen (2018)	**River + WWTP effluent**	**Lab**	**Yes** (chemical, for over 200 pollutants)	Fish	*D. rerio*	Liver cells
Williams (2014)	Contaminated sediments	**Lab**	**Yes** (sediment chemical analysis)	Fish	*Platichthys flesus*	Muscle, liver, bile, and plasma
Zhang (2014)	**River** WWTP influent and effluent	**Lab**	**Yes** (physicochemical and 22 semivolatile compounds)	Mice	*Mus musculus*	Serum and liver
Campillo (2015)	**Lagoon** (Mar Menor)	**Field (Caged)**	**Yes** (both chemical and physicochemical)	Clams	*Ruditapes decussatus*	Digestive gland
Cappello (2015)	**Petrochemical contaminated area**	**Field (Caged)**	**No**	Mussel	*Mytilus galloprovincialis*	Gill tissue
Ekman (2018)	**River** (South Platte River)	**Field (Caged)**	**Yes** (chemical (118 compounds) and physicochemical)	Fish	*P. promelas*	Liver

Continued

Table 10.1: Metabolomics studies conducted to evaluate the effects of exposure to chemical mixtures in biota from aquatic environments. Details about the contaminant mixture used, type of experiment, and target organism are presented.—cont'd

Author (year)	Contaminant/stressor	Field/lab	Water characterization	Organism	Species	Tissue
Skelton (2014)	**Rivers** impacted by WWTPs	**Field (caged)**	**Yes** (alkylphenols, PhACs, phytochemicals, and hormonal steroids, in transcriptomics experiment)	Fish	*P. promelas*	Liver
Davis (2013)	**Lake** impacted by pulp and paper mill effluent (lake superior)	**Field (Caged)**	Estrogenic activity	Fish	*P. promelas*	Liver
Davis (2016)	**Lakes (5)** impacted by different contaminants	**Field (Caged)**	**Yes** (132 compounds (pesticides, industrial and domestic contaminants, pharmaceuticals, and care products)	Fish	*P. promelas*	Liver
Heffernan (2017)	Contaminated **bays**	**Field (wild animals)**	**Yes** (chemical, 27 compounds)	Turtles	*Chelonia mydas*	Plasma
Glazer (2018)	**Estuary** polychlorinated biphenyl (PCB) contaminated area	**Field (wild animals)**	**No**	Fish	*Fundulus hetroclitus*	Liver
Morris (2019)	General contamination of **Hudson Bay**	**Field (wild animals)**	**No**	Bear	*Ursus maritimus*	Liver, fat, and muscle

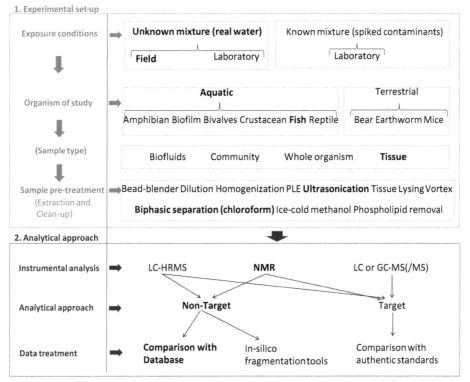

Figure 10.1

Workflow of the metabolomics approach. The different steps considered are defined on the left, whereas the different methodologies for each step are detailed on the right. Bold means the most commonly used approach in each step.

2.1 Exposure conditions

In the following sections, a distinction is made between experiments carried out in the laboratory using a selected mixture of known contaminants and those experiments where organisms are exposed to a real contaminated water with an unknown contaminant mixture.

2.1.1 Experiments performed using a mixture of selected compounds

Among the high amount of contaminants that can occur in a contaminated environment, some of the metabolomics studies have dealt with a selection of pollutants of concern. These contaminants are spiked in water or sediments at a concentration which is expected to exert an effect on the target organisms. That is the case of the study performed by Serra-Compte et al. (2018) where fluvial biofilms were exposed to a mixture of 9 pharmaceuticals at a total concentration of 5000 ng/L (mimicking concentrations and compounds found in polluted aquatic environments). The metabolomics response of

Table 10.2: Sample pretreatment, analytical approaches, and instrumental analysis followed in the literature regarding the study of the effects of contaminant mixtures in biota using metabolomics.

Author (year)	Sample pretreatment	Analytical approach	Instrumental analysis	Other analysis
Experiments performed using a mixture of selected compounds				
Serra-Compte (2018)	Whole community: pressurized liquid extraction (PLE) ((ACN):citric buffer (1:1)) + solid-phase extraction (SPE)	Nontargeted	HRMS	Structural and functional parameters
Melvin (2018)	Liver: ice-cold methanol + ultrasonic extraction (US) + organic and polar phases separation	Targeted/nontargeted	Target HRMS (triglycerides), + nontarget H-NMR (polar compounds)	Morphological parameters
Wagner (2018)	Whole organism: US (phosphate buffer, sodium azide, 2-dimethyl-2-silapentane-5-sulfonate sodium salt (DSS)) + centrifugation	Nontargeted	H-NMR	—
Jordan (2011)	Gonads and liver: US (2:1 methanol chloroform) + centrifugation (water:chloroform)	Target 2D	NMR (H-NMR/C-NMR)	—
Søfteland et al. (2014)	Hepatocyte cells: vortex (chloroform:methanol:water) + separation of polar and nonpolar phases	Nontargeted	H-NMR metabolomics, HRMS (ICR) lipidomics	Transcriptomics, toxicity tests
Experiments performed using real contaminated waters				
Al-Salhi (2012)	Bile: dilution methanol:water. Plasma: dilution ice-cold methanol	Nontargeted	HRMS	
David (2017)	Plasma: Phree plates (phospholipid removal with methanol + formic acid (FA)) + cation-exchange SPE. Tissues: ultrasonic probe + phree plates + SPE	Nontargeted	HRMS	Gene expression
Mosley (2018)	Skin mucus: 96-well captiva (protein removal, ice-cold methanol) + centrifugation	Nontargeted	HRMS	RNA
Wagner (2019)	Whole organism: US (phosphate buffer, sodium azide, DSS) + centrifugation	Targeted	H-NMR and LC-MS/MS	
Huang (2016)	Whole organism: bead-blender (zirconium beads + methanol) + centrifugation + extraction repeating with chloroform	Targeted	HPLC-MS/MS	

Reference	Sample preparation	Targeted/Nontargeted	Analytical method	Additional analysis
Zhen (2018)	Liver cells: quenching + biphasic (separation hydrophilic and lipophilic fractions)	Nontargeted	NMR and GC-MS	
Williams (2014)	Liver: homogenization (methanol/water) + centrifugation(methanol/chloroform/water) x2 polar and nonpolar separation	Nontargeted	H-NMR	Transcriptomics/DNA damage/morphological parameters
Zhang (2014)	Serum: dilution homogenization (phosphate sodium buffer) + centrifugation	Nontargeted	H-NMR	Transcriptomics/microscopic observation of liver cells; Other biomarkers
Campillo (2015)	Digestive gland: incubation (acetonitrile+10mM KH$_2$PO$_4$) + chloroform + centrifugation (biphasic separation)	Targeted	HPLC-MS	
Capello (2015)	Gill tissue: Ultraturrax homogenizer (cold methanol + cold water) + chloroform + water + centrifuge	Targeted	H-NMR	Immunochemical analysis/antibody analysis
Ekman (2018)	Liver: perchloric acid protocol + methanol:chloroform extraction, biphasic extraction: polar and nonpolar (lipophilic) separation	Targeted/nontargeted	Target NMR/nontarget NMR	Estrogenicity of water, transcriptomics
Skelton (2014)	Liver: perchloric acid protocol + methanol:chloroform extraction	Nontargeted	NMR and gc-ms	Transcriptomics
Davis (2013)	Liver: 96-well plate + ice-cold methanol/water + microcentrifuge (ice-cold chloroform). Centrifuge, phase separation	Nontargeted	NMR	Estrogenicity
Davis (2016)	Liver: 96-well plate + ice-cold methanol/water + microcentrifuge (ice-cold chloroform). Centrifuge, phase separation	Nontargeted	NMR	
Heffernan (2017)	Plasma: vortex (acetonitrile/water) + anhydrous magnesium sulfate and sodium chloride + centrifugation + frozen (lipid precipitation	Nontargeted	HRMS	
Glazer (2018)	Liver: vortex (methanol + water) + vortex (chloroform) + centrifugation	Targeted	LC-MS	ELISA for 5-mC and 5-hmC/
Morris (2019)	Lier: bead-blender (zirconium beads + methanol) + centrifugation + extraction repeating with chloroform	Targeted	HPLC-MS/MS	Isotope ratio mass spectrometry/total Hg

biofilm communities was studied with liquid chromatography coupled to high-resolution mass spectrometry (LC-HRMS) using a nontarget approach. In this experiment launched in an artificial fluvial mesocosms, not only chemical but also hydrological stressors were assessed. Melvin et al. (2018) also studied exposure to a mixture of PhACs (diclofenac, metformin, and valproic acid) to frogs (Limnodynastes peronii tadpoles), but in this case the main objective was to evaluate the impact of presence and absence of the carrier solvent methanol using a metabolomics approach. Wagner et al. (2018) examined *Daphnia magna* metabolic responses to acute sublethal water concentrations of binary and tertiary mixtures of the pharmaceuticals propranolol, carbamazepine, and perfluorooctanesulfonic acid (PFOS). The metabolome was measured using ^{1}H nuclear magnetic resonance (H-NMR) through a nontarget approach. A Goldfish model was used to investigate the effects of exposure to a mixture of aquatic pollutant 4,4'-isopropylidenediphenol (Bisphenol-A, BPA), di-(2-ethylhexyl)-phthalate (DEHP), and nonylphenol (NP) by means of H-NMR metabolomics (Jordan et al., 2011). Atlantic salmon primary hepatocytes, selected as experimental model, were exposed for 24 h to single contaminants, i.e., pesticides chlorpyrifos and endosulfan, and polycyclic aromatic hydrocarbons (PAHs) phenanthrene and benzo(a)pyrene, or to simple mixture them according to a factorial experimental design. The objective was to find the most potent mixture using a metabolomics, lipidomics, and transcriptomics approach (Søfteland et al., 2014).

2.1.2 Experiments performed using real contaminated waters

These types of studies with real contaminated water can be performed both in laboratory or directly in the field, where organisms are "caged" for the time of exposure and further processed in the laboratory for the metabolomics and/or xenometabolomics studies. In the case of studies carried out at **laboratory scale** (mesocosms facilities), a bunch of the studies have been performed using real contaminated water (Table 10.1). These exposure experiments under controlled conditions allow maintaining all the organisms under the same general conditions: air and water temperature, pH, hours of light/dark, and oxygen (David et al., 2017). In addition, experimental setup always includes not only the group of organism exposed to contaminated environment but also a reference or control group of organisms in order to compare xenometabolome and/or metabolome alterations between the two groups. Al-Salhi et al. (2012) and David et al. (2017) exposed rainbow trout and roach fish, respectively, to both WWTP-treated effluent and a clean water sample in a laboratory mesocosms. Similar experimental setup was used by Mosley et al. (2018) in their study with fathead minnows, which were exposed to control water or treated wastewater effluent at 5%, 20%, and 100% levels for 21 days, using an on-site, flow-through system. Wagner et al. (2019) studied different stages of wastewater treatment in lab exposure experiments with *D. magna*. The crustacean was exposed to two stages of wastewater, the prechlorinated wastewater and the final effluent, the latest also with a concentration gradient of added PFOS. The impact of a WWTP effluent was assessed through a metabolomics approach by Huang et al. (2016). In a more comprehensive experiment, zebrafish liver cells were exposed 48-h in the laboratory to water samples collected along six critical points of the urban water cycle (Zhen et al., 2018). Namely the WWTP effluent and four points in the river: one point upstream and three points downstream of the discharge of the WWTP. One of the downstream points was located ~14.5 km downstream of the WWTP, where water is taken for the abstraction of the drinking water; finished drinking water was also collected. Finally, flounder fish were exposed for 7 months to a high and low contaminated estuarine sediment at the laboratory mesocosms (Williams et al., 2014), whereas mice (the only study performed with a nonaquatic organism) were fed to untreated and treated wastewater as well as to a control feed (Zhang et al., 2014).

Concerning **field studies**, there is also a series of studies carried out in real environment with different experimental setups. Campillo et al. (2015) performed metabolomics studies in the digestive gland of clams after 7 and 22 days of transplantation in coastal lagoon in Spain affected by the growth of intensive agriculture and urban development in the surrounding area. Exposed clams were compared to control ones caged in a reference more pristine site. The neurotoxicological potential of environmental pollution in marine mussel *Mytilus galloprovincialis* was investigated by Cappello et al. (2015), who caged mussels at a highly polluted petrochemical area and in a reference site along the Mediterranean coastline. At the US Environmental Protection Agency, several metabolomics studies have been carried out with caged fathead minnows in different field conditions: (1) at the outlet of two different WWTPs and in a relatively non-impacted creek as a control site (Ekman et al., 2018), (2) in surface waters upstream and downstream of the effluent point source, as well as to the actual effluent at three different WWTP sites (Skelton et al., 2014), (3) in Great Lakes area near the outflow of a WWTP (receiving 40% of pulp and paper mill effluent), with control fish exposed to reference lake water under laboratory conditions (Davis et al., 2013), and (4) at 18 sites across the Great Lakes basin (Davis et al., 2016). Another example of organisms exposed to natural conditions is the study carried out by Heffernan et al. (2017) who collected blood from wild turtles from a remote offshore control site and two coastal case sites influenced by urban/industrial and agricultural activities in the Great Barrier Reef in Australia. Wild animals (atlantic killifish (*Fundulus hetroclitus* fish)) were also collected and their hepatic metabolite profile compared in another study carried out by Glazer et al. (2018). In this particular case, fish were collected from a PCB-contaminated River Estuary and from a pristine area. Finally, Morris et al. (2019) evaluated the potential influences of contaminant exposure on the hepatic metabolome of male polar bears from two different areas of Canada.

2.2 Model organisms

Fish is usually chosen as a **model organism** (Fig. 10.1) in many ecotoxicological studies, and it is actually the target aquatic organism in more than half of the studies reported on the topic (Table 10.1). Up to 13 manuscripts are devoted to the study of the impact of a mixture of contaminants in goldfish (*Carassius auratus*) (Jordan et al., 2011), roach (*Rutilus rutilus*) (David et al., 2017), salmon (*Salmo salar*) (Søfteland et al., 2014), trout (*Oncorhynchus mykiss*) (Al-Salhi et al., 2012), flounder (*Platichthys flesus*) (Williams et al., 2014), atlantic killifish (*F. hetroclitus*) (Glazer et al., 2018), zebrafish (*Danio rerio*) (Huang et al., 2016; Zhen et al., 2018), and fathead minnow (*Pimephales promelas*) (Davis et al., 2016, 2013; Ekman et al., 2017; Mosley et al., 2018; Skelton et al., 2014). Different type of fish tissue or sample was considered depending on the study: e.g., plasma, gonads, skin mucus, and liver (or liver cells) from fish, as well as the whole organism were processed in the exposure experiments to perform the metabolomics analysis (Table 10.1). Other organisms used in the studies were bivalves like clams (Campillo ct al., 2015) and mussels (Cappello et al., 2015), and crustaceans like *D. magna* (Wagner et al., 2019, 2018), earthworms (McKelvie et al., 2011), turtles (Heffernan et al., 2017), mice (Zhang et al., 2014), and bears (Morris et al., 2019). Bivalves (mainly mussels) are worldwide used as bioindicator organisms of water pollution because they are mobility-limited filter feeders which draw in water and particulates from their surrounding environment, and subsequently bioaccumulate contaminants in their tissues (Dodder et al., 2014). They are included in the Mussel Watch Program of the United States which measures the concentrations of coastal contaminants in bivalves and sediments to determine their spatial distribution and temporal trends, and to provide information to assess the risk posed to marine wildlife and humans through the use of coastal resources (Kimbrough et al., 2008). Therefore, their use in metabolomics studies to evaluate the effects of contaminant mixtures seems a logical step forward. Also of particular interest is the study based on the impact of contaminated water in fluvial biofilm (Serra-Compte et al., 2018), since biofilm is not an organism but a community of (micro)organisms such as bacteria, algae, protozoa, and fungi embedded in a polysaccharide matrix, that lives attached on the submerged substrate in the streambed. Biofilm has been used in ecotoxicological studies since it provides an integrated response at community level to the impact of multiple stressors (e.g., physical, chemical, or biological) in freshwater ecosystems (Sabater et al., 2007). In any case, most of the organisms considered in the studies reported in this chapter were aquatic. They were exposed to contaminated water either real river or seawater, wastewater, or just water spiked with the contaminants of interest. In the case of the experiment performed with earthworms, they were exposed to a contaminant mixture in soil (McKelvie et al., 2011), mice were fed with wastewater (Zhang et al., 2014), and wild bears were exposed to a contaminated environment through different routes like the diet (Morris et al., 2019).

All the model organisms provide information regarding ecological implication when metabolomics studies are performed. However, when assessing contaminant accumulation, the characteristics of each organism may change the type and the amount of contaminants accumulated. For instance, fluvial biofilm possesses a polysaccharide matrix which may attach chemical compounds (Serra-Compte et al., 2018), whereas the fat content, age, and feeding habits of fish determine the type of contaminants that they accumulate. In this sense, studies using organisms with different characteristic or trophic levels provide complementary information regarding ecotoxicological effects and contaminant accumulation, and when applicable, regarding the trophic transfer of contaminants. Actually, the different organisms employed in the articles reported in this chapter belong to different trophic levels, covering primary (fluvial biofilm), secondary (*Daphnia*), and tertiary (fish, turtles, and bears) producers (Table 10.1).

2.3 Sample pretreatment

Sample preparation (Table 10.2) is a key step in the analysis of metabolites or xeno(metabolites), as it determines the amount of compounds that will be recovered from the original matrix. Ideally, the less manipulation of the sample the better, but some minimum preparation is usually required in order to get a sample extract able to be injected in the detector, especially when applying nontargeted approaches because the objective is to recover the highest amount of compounds possible. However, these compounds may highly differ in their physical–chemical properties. Therefore, the sample pretreatment should be as general as possible in order to recover the widest range of metabolites possible. The analysis of organism biofluids such as plasma, serum, bile, or mucus usually requires a simple pretreatment of the sample. A dilution of plasma with methanol and of bile with a mixture of methanol:water prior instrumental analysis can be performed (Al-Salhi et al., 2012), whereas in other experiment (Zhang et al., 2014) the dilution was complemented by homogenization of the serum with phosphate sodium buffer and a centrifugation step. David et al. and Heffernan et al. (David et al., 2017; Heffernan et al., 2017) added a phospholipid removal step prior to injection to plasma samples using Phree filtration plates and plasma frozen for lipid precipitation, respectively. When the whole organism or a specific organ (i.e., liver, gonads, or gills) is analyzed, the extraction can be enhanced using different techniques. Ultrasonication (Jordan et al., 2011; Melvin et al., 2018; Wagner et al., 2019, 2018) and vortex (Glazer et al., 2018; McKelvie et al., 2011; Søfteland et al., 2014) are the most commonly applied. Besides, other extraction techniques have been also applied such as pressurized liquid extraction (Serra-Compte et al., 2018), bead-blender (Huang et al., 2016), tissue lysing and microcentrifuge (Davis et al., 2016, 2013; Ekman et al., 2018; Skelton et al., 2014), and homogenization (Cappello et al., 2015). The extraction solvents employed are in most cases a combination of water with an organic solvent mainly methanol or acetonitrile (Cappello et al., 2015; Glazer et al., 2018; Williams et al., 2014). The utilization of ice-cold methanol for protein removal (Al-Salhi et al., 2012; Melvin et al., 2018; Mosley et al., 2018) or a filtration step for phospholipid elimination (Heffernan et al., 2017) is also commonly applied when dealing with tissues. Furthermore, the application of chloroform combined with the extraction solvent followed by a centrifugation step is extensively used for the separation of hydrophilic and lipophilic compounds (Campillo et al., 2015; Cappello et al., 2015; Davis et al., 2016, 2013; Ekman et al., 2018; Glazer et al., 2018; Huang et al., 2016; Jordan et al., 2011; Skelton et al., 2014; Søfteland et al., 2014; Williams et al., 2014; Zhen et al., 2018). Then, the extracts can be separately analyzed or combined in a final extract prior the instrumental analysis.

2.4 Analytical approach: target versus nontarget/HRMS versus NMR

Two different analytical approaches named target and nontarget analysis are used (Table 10.2). Target analysis offers good sensitivity and reliable identification of the compounds but it has a significant disadvantage as it misses those compounds not included in the method. Nontarget analysis is a powerful tool for the identification of chemicals without a preceding selection of the compounds of interest and its application seems to be the way forward in environmental analysis. Consequently, most of the works published, 15 out of 22 papers, chose to apply nontarget analysis, either alone or in combination with target analysis. The remaining seven studies performed target analysis. Target analysis, in the case of metabolomics approach, consists in analyzing a series of predefined endogenous metabolites in order to evaluate stress related to previously selected biological pathways. This strategy was used by Huang et al. (2016) where 208 metabolites in total were monitored under exposure to different toxicants and effluents of wastewaters, Jordan et al. (2011) with 47 targets, Wagner et al. (2018) with a list of 17 compounds, Campillo et al. (2015) targeted 74 metabolites, Cappello et al. (2015) screened 3 compounds, Glazer et al. (2018) 72 compounds, and Morris et al. (2019) a large list of 219 chemicals.

Two main analytical platforms are generally used: NMR and low- or high-resolution mass spectrometry (MS or HRMS). Their differences, benefits, and drawbacks for metabolomics purposes have been widely discussed in the literature. On one hand, the main benefits of NMR are their high reproducibility, universality, as well as noninvasive and nondestructive way of working. Besides, its elucidation power for unknown compounds is higher than HRMS. Therefore, the majority of the studies working with mixture of contaminants reported in this chapter (13 out of 22) used NMR alone, or in combination with other instrumental technique such as liquid chromatography couple to mass spectrometry (LC-MS) (Wagner et al., 2019), gas chromatography coupled to mass spectrometry (GC-MS) (Zhen et al., 2018), or HRMS (Søfteland et al., 2014). Despite its wide use in metabolomics, NMR is limited by its sensitivity compared to MS and only around 100 metabolites were reported in the 13 scientific papers included in this chapter using NMR. Among them, 83 compounds have been detected both in target and nontarget approaches.

On the contrary, sensitivity is the main advantage of MS and HRMS compared to NMR, since it allows the user to achieve the detection of lower concentration levels for target and nontarget compounds. With these instruments, researchers have detected more than 1500 ions in the biological samples analyzed in nontarget way, although the number of relevant compounds (those features significantly different between two groups) ranged between 41 (David et al., 2017) and 640 ions (Serra-Compte et al., 2018). The main drawback of this nontarget platform is the elucidation process of suspect compounds. This step is usually performed with online spectral databases; notwithstanding, it becomes a time-consuming step and some compounds are not registered in these databases (i.e., TPs).

NMR in combination with MS or HRMS platforms have also been applied for metabolomics studies on chemical mixtures in 5 out of 22 papers published. It provides a wider polarity coverage by analyzing the polar fraction with NMR and the nonpolar one with GC-MS like in the studies performed by Skelton et al. (2014) and Zhen et al. (2018).

3. Biomarkers discovering

3.1 Metabolome

The different endogenous metabolites altered due to exposure to chemical mixtures reported in the literature are shown in Table 10.3 together with the xenobiotics measured (if any). The number of endogenous metabolites (termed metabolome) which levels significantly changed due to exposure to chemical mixtures is very variable. It ranged from only three metabolites in mussels caged for 1 month at a highly polluted petrochemical area in Augusta coastline (Sicily, Italy) (Cappello et al., 2015) to 219 (Morris et al., 2019) in hepatic metabolome of male polar bears from the southern and western Hudson Bay (Canada). These are both targeted studies, but if only untargeted works are considered, the maximum number of metabolites identified decreased up to 124 in Atlantic salmon primary hepatocytes exposed to a mixture of some PAHs and pesticides (Søfteland et al., 2014) (see Table 10.1 for details). In the majority of the papers reported here, the identities of the metabolites are proposed even if it is putatively, but sometimes the metabolites or some of them remained unidentified (Williams et al., 2014; Melvin et al., 2018; Heffernan, 2017). Actually, identification and confirmation of significant markers is considered the bottleneck of the metabolomics workflow. Different kind of molecules can be overexpressed or underexpressed due to exposure to mixtures of contaminants. Amino acids are among the most commonly detected key molecules as well as proteins, lipids, sugars, hormones, and vitamins (Table 10.3). These molecules are involved in biochemical pathways that many times play a crucial role in maintaining cell integrity and viability.

Among the 23 studies covered in this chapter, fish have been selected as a target exposure organism in 13 studies and can be considered as a model organism for pollution monitoring in aquatic systems due to their biological traits. They are ubiquitous and play a

Table 10.3: Endogenous metabolites and xenobiotics identified in the literature.

Author (year)	Endogenous metabolites altered due to exposure	Xenobiotics
	Experiments performed using a mixture of selected compounds	
Serra-Compte (2018)	**(9 compounds)** (azelaic acid, 16-oxohexadecanoic acid, linelenic acid, palmitoleic acid, palmitic acid, behenic acid, lignoceric acid, stearidonic acid, and lysoPA(0:0/16:0))	**Yes** (target analysis, nine xenobiotics: ibuprofen, diclofenac, carbamazepine, sulfamethoxazole, erythromycin, metoprolol, atenolol, gemfibrozil, and hydrochlorothiazide (Corcoll et al., 2015))
Melvin (2018)	**(28 compounds)** (L-isoleucine, L-leucine, L-valine, B-hydroxybutyric acid, L-lactic acid, L-lysine, L-alanine, putrescine, acetate, L-arginine, L-glutamate, L-glutamine, citrate, L-aspartate, creatine, O-phosphocholine, glucose/maltose, myo-inositol, glycogen, UDP, DP, glycine, nucleotides, NAD+, formate, L-tyrosine, and unknown at 6.53)	**No**
Wagner (2018)	**(19 compounds)** (tryptophan, phenylalanine, tyrosine, glucose, lactate, serine, glycine, proline, lysine, asparagine, methionine, glutamine, glutamate, arginine, alanine, threonine, valine, isoleucine, and leucine)	**No**
Jordan (2011)	**(47 compounds)** (2-aminobutyrate, 3-aminoisobutyrate, ADP, AMP, ATP, acetate, acetoacetate, alanine, asparagine, aspartate, choline, citrate, creatine, creatinine, formate, fumarate, glutamate, glutamine, glutathione, glycine, glycolate, inosine, isoleucine, lactate, leucine, malonate, methylsuccinate, NAD+, niacinamide, O-phosphocholine, O-phosphoethanolamine, pantothenate, phenylalanine, proline, propylene, sarcosine, serine, succinate, taurine, threonine, trimethylamine, tryptophan, tyrosine, UDP-glucose, valine, sn-glycero-3-phosphocholine, and p-methylhistidine)	**No**
Søfteland et al. (2014)	**(92 compounds** altered by chlorpyrifos, **22** for endosulfan, compounds not listed) (no differences in 1H-RMN metabolomics)	**No**
	Experiments performed using real contaminated waters	
Al-Salhi (2012)	**(8 compounds)** (taurocholic acid, sphingosine, 2-methylbutryolcarnitine lysoPE(16:0), lysoPE(16:1)/(18:2), and lysoPC(16:1)/(18:2))	**Yes (nontarget analysis, 236 xenobiotics and transformation products (TPs):** surfactants, naphthols, chlorinated xylenols, and phenoxyphenols, chlorophenes, resin acids, mefenamic acid, oxybenzone, and steroidal alkaloids)

Continued

Reference	Compounds	Xenobiotics/TPs
David (2017)	**(10 compounds)** (acetylserotonin, PGF2α, tetranor PGE1 like, dicarboxy LTB4 like, taurodeoxycholic acid, taurolithocholic acid like, tryptophan, indolepyruvate, sphinganine, and palmitoyl serine like)	**Yes (nontarget analysis, 54 xenobiotics and TPs)**: pharmaceuticals, endocrine disrupters, personal care products, pesticides, and antibacterial and human dietary products)
Mosley (2018)	**(30 compounds identified, 50 unidentified)** (3-amino-isobutanoate, choline, 3-hydroxybutanoate, serine, taurine, pyroglutamic acid, isoleucine, threonate, acetylcholine, adipic acid, carnitine, OH-glutamate, phenylalanine, myo-inositol, tyrosine, trans-3-indoleacrylic acid, nonanedioic acid, citric acid, sebacic acid, tryptophan, propionylcarnitine, pantothenic acid, F6P, guanosine, S7P, N-acetyl-α-D-glucosamine 1-phosphate, CMP, UMP, UDP-Glc, and L-glutathione (oxidized))	**Yes (nontarget analysis, 4 xenobiotics TPs:** bisphenol A, 1,7-dimethylxanthine, cotinine, and triclosan)
Wagner (2019)	**(17 compounds H-NMR and 14 compounds LC-MS/MS)** (alanine, asparagine, arginine, glutamate, glutamine, glycine, isoleucine, leucine, lysine, methionine, phenylalanine, proline, serine, threonine, tryptophan, tyrosine, valine, glucose and lactate by NMR) (alpha-ketoglutarate, fumarate, glucose-6-phosphate, malate, pyruvate, succinate, ADP, AMP, ATP, inosine, NAD, and NADP by LC-MS/MS)	No
Huang (2016)	**(208 metabolites)** (21 amino acids, 21 biogenic amines, 4 bile acids, sum of hexose, 17 fatty acids, 40 acylcarnitines, 89 phosphatidylcholines, and 15 sphingomyelins)	No
Zhen (2018)	**(31 compounds NRM and 19 GC-MS)** (leucine/isoleucine, valine, lactate, threonine, alanine, lysine, arginine, acetate, glutamate, glutamine, glutathione, pyruvate, succinate, citrate, creatine/phosphocreatine, choline, phosphocholine, taurine, glycine, myo-inositol, glucose, GXP, UDP-glucose, AXP, histidine, tyrosine, tryptophan, phenylalanine, and NADP+) and (14-demethyllanosterol, lanosterol, 4-methylzymosterol, lathosterol, 7-dehydrocholesterol, desmosterol, cholesterol, glyceryl stearate, glyceryl palmitate, octadecanoic acid (C18:0), octadecenoic acid (C18:1), octadecadienoic acid (C18:2), octadecanoic acid methyl ester, hexadecanoic acid (C16:0), hexadecenoic acid (C16:1), pentadecanoic acid (C15:0), pyridoxine tetradecanoic acid (C14:0), azelaic acid, dodecanol, and nonanoic acid (C9:0))	No
Williams (2014)	**(18 compounds)** (remained unidentified)	**Yes (target analysis, 8 xenobiotics and TPs)**: polycyclic aromatic hydrocarbons (PAHs), chlorinated biphenyls (Leaver et al., 2010)

Table 10.3: Endogenous metabolites and xenobiotics identified in the literature.—cont'd

Author (year)	Endogenous metabolites altered due to exposure	Xenobiotics
Zhang (2014)	**(50 compounds)** (isoleucine, leucine, valine, ethanol, 3-D-hydroxybutyrate, lactate, alanine, arginine, lysine, acetate, threonine, proline, glutamate, methionine, lipids, glutamine, citrate, pyruvate, LDL/VLDL, asparagine, creatine, glutathione, choline, phosphorylcholine, taurine, glycine, N-acetyl-glycoproteins (Nac), O-acetyl-glycoproteins,trimethylamine-N-oxide (TMAO), tyrosine, phenylalanine, tryptamine, scyllo-inositol, indoleacetyl-glycine (IAG), phenylacetyl-glycine(PAG), 3-hydroxy-isovalerate, creatinine, phosphocreatine, isobutyrate, α-ketoisovalerate, α-keto-isocaproate, kynurenine, formate, citrulline, 2-oxoglutarate, dimethylglycine (DMG), trimethylamine(TMA), inosine, glycerol, and succinate)	No
Campillo (2015)	**(74 compounds)** (2-phosphoglyceric acid, 3-phosphoglyceric acid, 8-hydroxy-deoxyguanosine, acetyl-CoA, acetylphosphate, N-acetyl-L-glutamine, adenosine diphosphate, L-alanine, adenosine monophosphate, L-arginine, L-asparagine, L-aspartic acid, adenosine triphosphate, biotin, L-carnitine, cytidine diphosphate, CDP-choline, L-citrulline, cytidine monophosphate, coenzyme A, cytidine triphosphate, L-cysteine, L-cystine, 2'-deoxyadenosine, deoxyadenosine triphosphate, deoxycytidine triphosphate, riboflavin-5'-monophosphate, gamma-butyrobetaine, guanosine diphosphate, L-glutamine, L-glutamic acid, glucosamine 6-phosphate, glucosamine, glycine, guanosine monophosphate, L-glutathione, L-homocysteine, L-homoserine, L-hydroxyproline, hypoxanthine, L-isoleucine, inosine monophosphate, inosine triphosphate, L-leucine, L-lysine, L-methionine, nicotinamide adenine dinucleotide oxidized form, nicotinamide adenine dinucleotide reduced form, nicotinamide adenine dinucleotide phosphate oxidized form, nicotinamide adenine dinucleotide phosphatereduced form, O-phosphorylethanolamine, L-ornithine, L-phenylalanine, L-proline, O-phospho-L-serine, L-serine, taurine, tetrahydrofolic acid, thiamine monophosphate, L-threonine, thymidine, thymine, thymidine monophosphate, L-tryptophan, deoxythymidine triphosphate, L-tyrosine, uridine diphosphate, uridine monophosphate, uridine triphosphate, L-valine, and xanthosine monophosphate)	**Yes (target analysis, 32 xenobiotics:** PAHs, triazines, and organophosphorus pesticides (Campillo et al., 2013)
Cappello (2015)	**(3 compounds)** (tyrosine, serotonin, and acetylcholine)	No
Ekman (2018)	**(4 compounds)** (alanine and glutamate, previously observed, **target**/glutathione and phosphocholine, **nontarget**)	No

Author	Compounds	Xenobiotics
Skelton (2014)	(11 compounds NMR, 1 compound GC-MS) (glycogen, glucose, phosphocholine, choline, glutamate, lysine, branched-chain amino acids (valine, leucine, and isoleucine), taurocholic acid, and unassigned compound by NMR) (cholesterol by GC-MS)	No
Davis (2013)	(18 compounds) (choline, phosphocholine, taurine, creatine, phosphocreatine, glutamate, alanine, glycogen, proline, glucose, NAD+, dimethylamine, adenosine (mono/di/tri)phosphate, dimethylglycine, and histidine)	No
Davis (2016)	(21 compounds) (alanine, aminoisobutyrate, adenosine mono/di/triphosphate, betaine, creatine, dimethylglycine, glucose, glutamate, glycogen, glycerophosphocholine, histidine, lactate, leucine, malonate, nicotinamide adenine dinucleotide, phosphocholine, phosphocreatine, taurine, and taurocholic acid)	No
Heffernan (2017)	(10 compounds) (3-indole propionic acid, vanillylmandelic acid, benzenetriol-sulfate, stearidonic acid, linolenic acid, ecklonialactone a, oxidation products of fatty acid C18:4, C18:3, and C18:2, and an unknown compound)	**Yes (nontarget analysis, 13 xenobiotics, and TPs:** isoquinoline, 4-phenolsulfonic acid, 4-ethoxysulfonic acid, 4-hydroxy-3-methoxy-benzenesulfonic acid, hydroxy(4-methoxyphenyl) methanesulfonic acid, 3-(4-hydroxyphenoxy)-1-propanesulfonic acid, ethiofencarb sulfone, 2,6-dinitrotoluene-4-sulfonic acid, 2-(methylsulfonyl)-1-phenyl ethanesulfonic acid, 4-(phenylsulfonyl)-1-butanesulfonic acid, oleamide, 2-hydroxy-5-[(5-xooxolan-2-yl) methyl]phenyl oxidanesulfonic acid, sphingosine, and docosanamide)
Glazer (2018)	(72 compounds) (altered compounds: inosine monophosphate (imp), pantothenic acid, ornithine, arginine, leucine, tyrosine, sarcosine, ribose-5-phosphate, inosine, choline, adenosine, serine, threonine, glycine betaine, and s-(1,2-carboxyethyl)glutathione)	No
Morris (2019)	(219 compounds) (21 amino acids, 22 biogenic amines, 13 bile acids, total hexose sugars, 18 fatty acids, 40 acylcarnitines, 89 phosphatidylcholines, and 15 sphingomyelins)	**Yes (target analysis, 295 xenobiotics:** polychlorinated biphenyls (PCBs), per- and polyfluoroalkyl substances (PFASs) including perfluorinated carboxylic acids (PFCAs), perfluorinated sulfonic acids (PFSAs), and fluorinated sulfonamides (FASAs), organochlorine pesticides (OCPs), organophosphate esters (OPEs), polybrominated diphenyl ethers (PBDEs), and a suite of alternative halogenated flame retardants (HFRs))

major ecological role, owing to their function as a carrier of energy from lower to higher trophic levels (Huerta et al., 2012). The main biological pathways disrupted in these studies as a result of exposure to chemical mixtures are energy metabolism, phospholipid metabolism, and amino acid metabolism (mainly glutamate). Furthermore, metabolites that are altered in these studies are, in addition, amino acids (branched amino acids, alanine, tryptophan), bile acids, steroids, unsaturated fatty acids, creatine, taurine, cholesterol, carnitine, sphinganine, serotonin, and vitamins as well as liver toxicity markers and oxidative degradation stress pathway markers.

3.2 Xenometabolome

Besides the importance of analyzing changes in the metabolome for understanding the effects of contaminant mixtures in organisms, it is equally important to identify the exogenous compounds or xenobiotics accumulated in the same organism that might be responsible of these alterations. The analysis of xenobiotics bioaccumulated allows to evaluate the exposure to chemical contamination and to link their presence and levels in the organism with metabolic dysregulations. Therefore, their analysis is highly recommended, especially when working with contaminant mixtures. However, out of the 22 studies reported in Table 10.3 only 8 measured the xenobiotics in the respective biota sample, the remaining 14 studies did not carry out any analytical determination of xenobiotics in the exposed organism. The profiling of the xenobiotics present in an organism, also known as the xenometabolome or exposome, is of high relevance in order to connect contaminants and toxic effects. As commented in Section 2.4, there are two different analytical approaches that can be followed for analyzing the xenometabolome. Target analysis of a list of contaminants selected "a priori," and nontarget analysis following a similar holistic approach to the one followed for classic metabolomics studies where a control group is compared with an exposed group. This approach is also known as xenometabolomics. For the so-called xenometabolomics studies, the use of MS instruments is mandatory due to the high sensitivity necessary to measure the low concentrations of contaminants in field biota samples, even in those from lab-scale experiments. In the xenometabolomic approach, the mixture of xenobiotics accumulated in the organism is highlighted by statistics from the data-independent acquisition performed in the HRMS instrument. Among the eight papers published so far (until 2019) where the xenometabolome was explored, only four of them applied xenometabolomics (Al-Salhi, David, Mosley, and Heffernan) (Table 10.3). Actually, studies that include the profile of the xenometabolome are very limited, in spite of the high potential and applicability of the approach. Out of these four studies, three were exposure experiments carried out at lab-scale (Al-Salhi et al., 2012; David et al., 2017; Mosley et al., 2018) and only in one occasion a field study was undertaken (Heffernan et al., 2017). However, xenometabolomics popularity for analyzing biological samples directly from the field is rapidly increasing, and it is anticipated to escalate as metabolomics becomes a more routine tool for environmental monitoring (Southam et al., 2014). Xenometabolomics was initially performed by Al-Salhi et al. in 2012. They discovered that from a total amount of 242 compounds that significantly contributed to the separation of control and exposed fish (to WWTP effluent) only 8 were endogenous metabolites, while the remaining 236 were xenobiotics and TPs. David et al. (2017) followed the same strategy, identifying 54 exogenous compounds and TPs in plasma and tissues of fish exposed to WWTP effluent. Mosley et al. (2018) found four xenobiotics and TPs in skin mucus of fish exposed to WWTP effluent as well. Heffernan et al. (2017) following a xenometabolomic approach with plasma of green sea turtles from the Great Barrier Reef (Australia) found 13 xenobiotics and TPs. These studies point out the strength of the xenometabolomics strategy to cover both xenobiotic compounds and their possible TPs.

Target analysis of the contaminants was performed by Serra-Compte et al. (2018), where the nine pharmaceutically active compounds spiked in the water of the laboratory exposure experiment were analyzed both in water and biofilm samples, Williams et al. (2014) measured eight xenobiotics and some TPs in sediment and fish, Campillo et al. (2015) determined 32 xenobiotics including polycyclic aromatic hydrocarbons (PAHs), triazines, and organophosphorus pesticides in clams caged at Mar Menor lagoon (SE Spain), and Morris et al. (2019) analyzed a broad set of compounds including 295 xenobiotics from different persistent organic groups of pollutants in liver samples of polar bear.

4. Challenges and future research

As reviewed in this chapter, targeted and untargeted metabolomics—based techniques have been used in several studies over the last decade to evaluate biological effects of environmental contaminant mixtures mainly in aquatic ecosystems. The profiling of the metabolome has become a useful tool for the identification of toxic effects and metabolic pathways disrupted due to exposure to multiple chemicals simultaneously. As main drawback, it does not allow to differentiate which component of the mixture produces a specific effect and therefore, for this matter, exposure to single compounds is still needed. However, by carrying out exposure experiments to single compounds and by comparison with the combined effect of the mixture (binary, tertiary, etc.), the joint chemical action can be established as independent, similar, or synergistic action. Besides, it can be concluded that common metabolic pathways are affected by specific groups of chemicals. This has also been observed by Bonvallot et al. (2018) that assessed biological effects of environmental contaminants in humans. Disruption of cellular signaling pathways associated with oxidative stress may be primary drivers for further effects for all studied chemicals (Bonvallot et al., 2018).

Xenometabolomics has proved strength for identifying multiple contaminants accumulated in a certain organism as well as their TPs and it is progressively gaining attention from researchers. The especial interest is its use for environmental biomonitoring that is foreseen to become more popular in the next years for analyzing biological samples directly from the field. Besides, the introduction of more sensitive analytical techniques to perform nontargeted profiling of biological samples seems to be a good option to improve the coverage of the xenometabolome and to provide more relevant information on both chemical exposure and metabolic signatures of environmental contaminants.

As main challenge remains, the lack of standardized approaches and lab procedures, which hinder the comparison of results but it will evolve at the same time that metabolomics and xenometabolomics, becomes more routine techniques in environmental analysis. Sample treatment is also challenging, especially for nontarget approach, where minimum manipulation of the sample is a plus and if an extraction needs to be done is preferable to perform an extraction suitable to get a complete coverage of compounds and TPs.

HRMS is strongly recommended due to its higher sensitivity, it allows the user to achieve the detection of lower concentration levels for target and nontarget compounds. Moreover, it permits a retrospective analysis in order to look for pollutants that earlier were unknown. However, ion suppression and metabolites identification are still bottlenecks of metabolomics workflow.

Acknowledgments

This study was supported by the Spanish Ministry of Economy and Competitiveness, State Research Agency, and by the European Union through the European Regional Development Fund through the projects XENOME-TABOLOMIC (CTM2015-73179-JIN) (AEI/FEDER/UE) and PLAS-MED (CTM2017-89701-C3-2-R). Authors acknowledge the support from the Economy and Knowledge Department of the Catalan Government through Consolidated Research Groups ICRA-ENV 2017 SGR 1124 and 2017 SGR 01404. Albert Serra-Compte acknowledges the FI-DGR research fellowship from the Catalan Government (2018FI_B2_00170). Sara Rodriguez-Mozaz acknowledges the Ramon y Cajal program (RYC-2014-16707).

References

Al-Salhi, R., Abdul-Sada, A., Lange, A., Tyler, C.R., Hill, E.M., 2012. The xenometabolome and novel contaminant markers in fish exposed to a wastewater treatment works effluent. Environ. Sci. Technol. 46, 9080—9088.

Altenburger, R., Greco, W.R., 2009. Extrapolation Concepts for dealing with multiple contamination in environmental risk assessment. Integr. Environ. Assess. Manag. 5, 62—68.

Alvarez-Munoz, D., Huerta, B., Fernandez-Tejedor, M., Rodríguez-Mozaz, S., Barceló, D., 2015. Multi-residue method for the analysis of pharmaceuticals and some of their metabolites in bivalves. Talanta 136, 174—182.

Álvarez-Muñoz, D., Llorca, M., Blasco, D., Barceló, D., 2016. Contaminants in the Marine Environment. Academic Press.

Bliss, C., 1993. The toxicity of poisons applied jointly. Ann. Appl. Biol. 26, 585–615.

Bonvallot, N., David, A., Chalmel, F., Chevrier, C., Cordier, S., Cravedi, J.P., Zalko, D., 2018. Metabolomics as a powerful tool to decipher the biological effects of environmental contaminants in humans. Curr. Opin. Toxicol. 8, 48–56.

Campillo, J.A., Albentosa, M., Valdés, N.J., Moreno-González, R., León, V.M., 2013. Impact assessment of agricultural inputs into a Mediterranean coastal lagoon (Mar Menor, SE Spain) on transplanted clams (*Ruditapes decussatus*) by biochemical and physiological responses. Aquat. Toxicol. 142–143, 365–379.

Campillo, J.A., Sevilla, A., Albentosa, M., Bernal, C., Lozano, A.B., Cánovas, M., León, V.M., 2015. Metabolomic responses in caged clams, Ruditapes decussatus, exposed to agricultural and urban inputs in a Mediterranean coastal lagoon (Mar Menor, SE Spain). Sci. Total Environ. 524–525, 136–147.

Cappello, T., Maisano, M., Giannetto, A., Parrino, V., Mauceri, A., Fasulo, S., 2015. Neurotoxicological effects on marine mussel *Mytilus galloprovincialis* caged at petrochemical contaminated areas (eastern Sicily, Italy): ^{1}H NMR and immunohistochemical assays. Comp. Biochem. Physiol. C Toxicol. Pharmacol. 169, 7–15.

Cole, M., Lindeque, P., Halsband, C., Galloway, T.S., 2011. Microplastics as contaminants in the marine environment: a review. Mar. Pollut. Bull. 62, 2588–2597.

Corcoll, N., Casellas, M., Huerta, B., Guasch, H., Acuña, V., Rodríguez-Mozaz, S., Serra-Compte, A., Barceló, D., Sabater, S., 2015. Effects of flow intermittency and pharmaceutical exposure on the structure and metabolism of stream biofilms. Sci. Total Environ. 503–504, 159–170.

David, A., Lange, A., Abdul-Sada, A., Tyler, C.R., Hill, E.M., 2017. Disruption of the prostaglandin metabolome and characterization of the pharmaceutical exposome in fish exposed to wastewater treatment works effluent as revealed by nanoflow-nanospray mass spectrometry-based metabolomics. Environ. Sci. Technol. 51, 616–624.

Davis, J.M., Collette, T.W., Villeneuve, D.L., Cavallin, J.E., Teng, Q., Jensen, K.M., Kahl, M.D., Mayasich, J.M., Ankley, G.T., Ekman, D.R., 2013. Field-based approach for assessing the impact of treated pulp and paper mill effluent on endogenous metabolites of fathead minnows (*Pimephales promelas*). Environ. Sci. Technol. 47, 10628–10636.

Davis, J.M., Ekman, D.R., Teng, Q., Ankley, G.T., Berninger, J.P., Cavallin, J.E., Jensen, K.M., Kahl, M.D., Schroeder, A.L., Villeneuve, D.L., Jorgenson, Z.G., Lee, K.E., Collette, T.W., 2016. Linking field-based metabolomics and chemical analyses to prioritize contaminants of emerging concern in the Great Lakes basin. Environ. Toxicol. Chem. 35, 2493–2502.

Dodder, N.G., Maruya, K.A., Lee Ferguson, P., Grace, R., Klosterhaus, S., La Guardia, M.J., Lauenstein, G.G., Ramirez, J., 2014. Occurrence of contaminants of emerging concern in mussels (*Mytilus* spp.) along the California coast and the influence of land use, storm water discharge, and treated wastewater effluent. Mar. Pollut. Bull. 81, 340–346.

Ekman, D.R., Ankley, G.T., Villeneuve, D.L., Collette, T.W., Mosley, J.D., Cavallin, J.E., 2017. High-resolution mass spectrometry of skin mucus for monitoring physiological impacts and contaminant biotransformation products in fathead minnows exposed to wastewater effluent. Environ. Toxicol. Chem. 37, 788–796.

Ekman, D.R., Keteles, K., Beihoffer, J., Cavallin, J.E., Dahlin, K., Davis, J.M., Jastrow, A., Lazorchak, J.M., Mills, M.A., Murphy, M., Nguyen, D., Vajda, A.M., Villeneuve, D.L., Winkelman, D.L., Collette, T.W., 2018. Evaluation of targeted and untargeted effects-based monitoring tools to assess impacts of contaminants of emerging concern on fish in the South Platte River, CO. Environ. Pollut. 239, 706–713.

European Commission, 2009. Council conclusions on combination effects of chemicals. In: 2988th ENVIRONMENT Council Meeting, Brussels.

European Commission, 2012. The combination effects of chemicals. In: Chemical Mixtures, Communication from the Commission to the Council, Brussels.

Fatta-Kassinos, D., Meric, S., Nikolaou, A., 2011a. Pharmaceutical residues in environmental waters and wastewater: current state of knowledge and future research. Anal. Bioanal. Chem. 399, 251–275.

Fatta-Kassinos, D., Vasquez, M.I., Kümmerer, K., 2011b. Transformation products of pharmaceuticals in surface waters and wastewater formed during photolysis and advanced oxidation processes – degradation, elucidation of byproducts and assessment of their biological potency. Chemosphere 85, 693–709.

Faust, M., Altenburger, R., Backhaus, T., Blanck, H., Boedeker, W., Gramatica, P., Hamer, V., Scholze, M., Vighi, M., Grimme, L.H., 2001. Predicting the joint algal toxicity of multi-component s-triazine mixtures at low-effect concentrations of individual toxicants. Aquat. Toxicol. 56, 13–32.

Glazer, L., Kido Soule, M.C., Longnecker, K., Kujawinski, E.B., Aluru, N., 2018. Hepatic metabolite profiling of polychlorinated biphenyl (PCB)-resistant and sensitive populations of Atlantic killifish (*Fundulus heteroclitus*). Aquat. Toxicol. 205, 114–122.

Gros, M., Rodrguez-Mozaz, S., Barceló, D., 2012. Fast and comprehensive multi-residue analysis of a broad range of human and veterinary pharmaceuticals and some of their metabolites in surface and treated waters by ultra-high-performance liquid chromatography coupled to quadrupole-linear ion trap tandem. J. Chromatogr. A 1248, 104–121.

Groten, J.P., Feron, V.J., Sühnel, J., 2001. Toxicology of simple and complex mixtures. Trends Pharmacol. Sci. 22, 316–322.

Hartung, T., Leist, M., 2008. Food for thought … on the evolution of toxicology and the phasing out of animal testing. ALTEX 91–96.

Heffernan, A.L., Gómez-Ramos, M.M., Gaus, C., Vijayasarathy, S., Bell, I., Hof, C., Mueller, J.F., Gómez-Ramos, M.J., 2017. Non-targeted, high resolution mass spectrometry strategy for simultaneous monitoring of xenobiotics and endogenous compounds in green sea turtles on the Great Barrier Reef. Sci. Total Environ. 599–600, 1251–1262.

Holmes, E., Ruey, L.L., Cloarec, O., Coen, M., Tang, H., Maibaum, E., Bruce, S., Chan, Q., Elliott, P., Stamler, J., Wilson, I.D., Lindon, J.C., Nicholson, J.K., 2008. Detection of urinary drug metabolite (xenometabolome) signatures in molecular epidemiology studies via statistical total correlation (NMR) spectroscopy (Analytical Chemistry (2007) 79 (2629-2640). Anal. Chem. 80, 6142–6143.

Holt, M.S., 2000. Sources of chemical contaminants and routes into the freshwater environment. Food Chem. Toxicol. 38.

Huang, S.S.Y., Benskin, J.P., Chandramouli, B., Butler, H., Helbing, C.C., Cosgrove, J.R., 2016. Xenobiotics produce distinct metabolomic responses in zebrafish larvae (*Danio rerio*). Environ. Sci. Technol. 50, 6526–6535.

Huerta, B., Jakimska, A., Gros, M., Rodríguez-Mozaz, S., Barceló, D., 2013. Analysis of multi-class pharmaceuticals in fish tissues by ultra-high-performance liquid chromatography tandem mass spectrometry. J. Chromatogr. A 1288, 63–72.

Huerta, B., Rodríguez-Mozaz, S., Barceló, D., 2012. Pharmaceuticals in biota in the aquatic environment: analytical methods and environmental implications. Anal. Bioanal. Chem. 404, 2611–2624.

Jordan, J., Weljie, A.M., Habibi, H.R., Jackson, L.J., Zare, A., 2011. Environmental contaminant mixtures at ambient concentrations invoke a metabolic stress response in goldfish not predicted from exposure to individual compounds alone. J. Proteome Res. 11, 1133–1143.

Kimbrough, K.L., Johnson, W.E., Lauenstein, G.G., Christensen, J.D., Apeti, D.A., 2008. An assessment of two decades of contaminant monitoring in the nation's coastal zone. In: NOAA Technical Memorandum NOS NCCOS 74 Silver Spring, MD, 105 pp.

Kortenkamp, A., 2007. Ten years of mixing cocktails: a review of combination effects of endocrine-disrupting chemicals. Environ. Health Perspect. 115, 98–105.

Leaver, M.J., Diab, A., Boukouvala, E., Williams, T.D., Chipman, J.K., Moffat, C.F., Robinson, C.D., George, S.G., 2010. Hepatic gene expression in flounder chronically exposed to multiply polluted estuarine sediment: absence of classical exposure "biomarker" signals and induction of inflammatory, innate immune and apoptotic pathways. Aquat. Toxicol. 96, 234–245.

McKelvie, J.R., Wolfe, D.M., Celejewski, M.A., Alaee, M., Simpson, A.J., Simpson, M.J., 2011. Metabolic responses of *Eisenia fetida* after sub-lethal exposure to organic contaminants with different toxic modes of action. Environ. Pollut. 159, 3620–3626.

Melvin, S.D., Jones, O.A.H., Carroll, A.R., Leusch, F.D.L., 2018. 1H NMR-based metabolomics reveals interactive effects between the carrier solvent methanol and a pharmaceutical mixture in an amphibian developmental bioassay with *Limnodynastes peronii*. Chemosphere 199, 372–381.

Monosson, E., 2005. Chemical mixtures: considering the evolution of toxicology and chemical assessment. Environ. Health Perspect. 113, 383–390.

Morris, A.D., Letcher, R.J., Dyck, M., Chandramouli, B., Cosgrove, J., 2019. Concentrations of legacy and new contaminants are related to metabolite profiles in Hudson Bay polar bears. Environ. Res. 168, 364—374.

Mosley, J.D., Ekman, D.R., Cavallin, J.E., Villeneuve, D.L., Ankley, G.T., Collette, T.W., 2018. High-resolution mass spectrometry of skin mucus for monitoring physiological impacts and contaminant biotransformation products in fathead minnows exposed to wastewater effluent. Environ. Toxicol. Chem. 37, 788—796.

Noguera-Oviedo, K., Aga, D.S., 2016. Lessons learned from more than two decades of research on emerging contaminants in the environment. J. Hazard Mater. 316, 242—251.

Petrie, B., Barden, R., Kasprzyk-Hordern, B., 2014. A review on emerging contaminants in wastewaters and the environment: current knowledge, understudied areas and recommendations for future monitoring. Water Res. 72, 3—27.

Petrovic, M., Eljarrat, E., Lopez De Alda, M.J., Barceló, D., 2004. Endocrine disrupting compounds and other emerging contaminants in the environment: a survey on new monitoring strategies and occurrence data. Anal. Bioanal. Chem. 378, 549—562.

Picó, Y., Barceló, D., 2015. Transformation products of emerging contaminants in the environment and high-resolution mass spectrometry: a new horizon. Anal. Bioanal. Chem. 407, 6257—6273.

REGULATION (EC) No 1907/, 2006. In: REGULATION (EC) No 1907/2006 of the and Repealing Council Regulation (EEC) No 793/93 and EUROPEAN PARLIAMENT and of the COUNCIL of 18 December 2006 Concerning the Registration, Evaluation, Authorisation and Restriction of Chemicals (REACH), Establishing a 2003R2003, pp. 1—15.

Rodriguez-Mozaz, S., Alvarez-Muñoz, D., Barceló, D., 2017. Pharmaceuticals in marine environment: analytical techniques and applications. In: Environmental Problems in Marine Biology: Methodological Aspects and Applications. Taylor & Francis Publisher, pp. 268—316.

Sabater, S., Guasch, H., Ricart, M., Romaní, A., Vidal, G., Klünder, C., Schmitt-Jansen, M., 2007. Monitoring the effect of chemicals on biological communities. The biofilm as an interface. Anal. Bioanal. Chem. 387, 1425—1434.

Sauvé, S., Desrosiers, M., 2014. A review of what is an emerging contaminant. Chem. Cent. J. 8, 1—7.

Serra-Compte, A., Corcoll, N., Huerta, B., Rodríguez-Mozaz, S., Sabater, S., Barceló, D., Álvarez-Muñoz, D., 2018. Fluvial biofilms exposed to desiccation and pharmaceutical pollution: new insights using metabolomics. Sci. Total Environ. 618, 1382—1388.

Skelton, D.M., Ekman, D.R., Martinović-Weigelt, D., Ankley, G.T., Villeneuve, D.L., Teng, Q., Collette, T.W., 2014. Metabolomics for in situ environmental monitoring of surface waters impacted by contaminants from both point and nonpoint sources. Environ. Sci. Technol. 48, 2395—2403.

Søfteland, L., Kirwan, J.A., Hori, T.S.F., Størseth, T.R., Sommer, U., Berntssen, M.H.G., Viant, M.R., Rise, M.L., Waagbø, R., Torstensen, B.E., Booman, M., Olsvik, P.A., 2014. Toxicological effect of single contaminants and contaminant mixtures associated with plant ingredients in novel salmon feeds. Food Chem. Toxicol. 73, 157—174.

Southam, A.D., Lange, A., Al-Salhi, R., Hill, E.M., Tyler, C.R., Viant, M.R., 2014. Distinguishing between the metabolome and xenobiotic exposome in environmental field samples analysed by direct-infusion mass spectrometry based metabolomics and lipidomics. Metabolomics 10, 1050—1058.

U.S. E.P.A., 2000. Supplementary Guidance for Conducting Health Risk Assessment of Chemical Mixtures. US Environmental Protection Agency. Risk Assess. Forum Tech. Panel. Off. EPA/630/R-00/002.

Wagner, N.D., Helm, P.A., Simpson, A.J., Simpson, M.J., 2019. Metabolomic responses to pre-chlorinated and final effluent wastewater with the addition of a sub-lethal persistent contaminant in Daphnia magna. Environ. Sci. Pollut. Res. 26, 9014—9026.

Wagner, N.D., Simpson, A.J., Simpson, M.J., 2018. Sublethal metabolic responses to contaminant mixture toxicity in Daphnia magna. Environ. Toxicol. Chem. 37, 2448—2457.

Williams, T.D., Davies, I.M., Wu, H., Diab, A.M., Webster, L., Viant, M.R., Chipman, J.K., Leaver, M.J., George, S.G., Moffat, C.F., Robinson, C.D., 2014. Molecular responses of European flounder (*Platichthys flesus*) chronically exposed to contaminated estuarine sediments. Chemosphere 108, 152—158.

Zhang, Y., Deng, Y., Zhao, Y., Ren, H., 2014. Using combined bio-omics methods to evaluate the complicated toxic effects of mixed chemical wastewater and its treated effluent. J. Hazard Mater. 272, 52–58.

Zhen, H., Ekman, D.R., Collette, T.W., Glassmeyer, S.T., Mills, M.A., Furlong, E.T., Kolpin, D.W., Teng, Q., 2018. Assessing the impact of wastewater treatment plant effluent on downstream drinking water-source quality using a zebrafish (*Danio Rerio*) liver cell-based metabolomics approach. Water Res. 145, 198–209.

CHAPTER 11

A snapshot of biomarkers of exposure for environmental monitoring

Diana Álvarez-Muñoz, Marinella Farré

Water and Soil Quality Research Group, Department of Environmental Chemistry, IDAEA-CSIC, Barcelona, Spain

Chapter Outline

1. Introduction

Biomarkers discovery is the last step in the metabolomics workflow after an appropriate experimental design, chemical analysis, data processing, and statistical analysis. According to the World Health Organization (WHO, 1993), the term "biomarker" is used in a broad sense to include almost any measurement reflecting an interaction between a biological system and an environmental agent, which may be chemical, physical, or biological. Particularly this chapter is focused on the study of organic compounds as chemical environmental stressors. Although many other definitions can be found in the literature, from a metabolomics point of view and henceforth, biomarker will be considered as any biological molecule found in blood, body fluids, or tissues of an organism which levels are altered due to an exposure (Lam, 2009). Therefore, biomarkers are used to indicate an exposure to a certain contaminant or the effects of this contaminant in an organism.

Metabolomics is a high-throughput semiquantitative approach that allows simultaneous detection of a set of metabolites. The metabolites' coverage depends, among others, on the type of analysis undertaken. Targeted metabolomics analysis covers a relatively small amount of metabolites, usually a list of predefined metabolites chosen with regard to the

Environmental Metabolomics. https://doi.org/10.1016/B978-0-12-818196-6.00011-X

biological sample analyzed, while nontargeted analysis aims to detect and identify as many metabolites as possible without a prior selection. As mentioned in Chapter 1, around 1500 metabolites are identified using nontargeted analysis and between 200 and 500 using targeted analysis. These figures clearly illustrated the big amount of data available and the higher potential of nontargeted metabolomics analysis for the discovery of biomarkers and profiling of the metabolome. The characterization of the metabolic profile also depends on the instrument, sample pretreatment, and statistical method carried out. Nontargeted analysis is usually undertaken with both nuclear magnetic resonance and high-resolution mass spectrometry (HRMS) coupled to liquid or gas chromatography. Once the metabolites, which levels significantly change due to an exposure to a chemical, are identified, they may be quantified as well, but this only happens in a small proportion compared to the entire set of metabolites identified. Changes in metabolites levels are linked to disruption of metabolic pathways where they are involved and ultimately to pathologies.

In this chapter, a revision of the biomarkers identified using metabolomics that significantly change due to an exposure to organic contaminants is presented. For this purpose, we have selected fish as model organism and liquid chromatography—mass spectrometry as instrumental technique, considering both metabolomics targeted and nontargeted analysis. Exposure under laboratory conditions and to a single contaminant was also chosen in other to narrow the variability of the data. Taken into account these conditions and the information reported along this book a literature research was done. After gathering all the papers published so far (until 2019), an analysis of the reported biomarkers, the effects shown, and the metabolic pathways disrupted was done in other to find out if there was any common trend. The type of molecules most altered due to exposure to organic contaminants and the metabolic pathways affected were studied, and the possibility of finding a potential "universal biomarker" of environmental quality was explored.

1.1 Metabolic profile of fish exposed to organic contaminants

Table 11.1 presents the information gathered in the literature research. A total of 15 articles have been included. They were published in the last 9 years, which indicates the novelty of metabolomics application in environmental field for biomarkers discovery. Besides, an increasing number of papers have been published since 2016, corresponding to an intensification of metabolomics popularity for analyzing biological samples in environmental monitoring. Different fish species have been used by the authors like *Odontesthes bonariensis* (Carriquiriborde et al., 2012), *Oncorhynchus mykiss* (Roszkowska et al., 2018), *Oryzias melastigma* (Lei et al., 2017), *Pimephales promelas* (Ekman et al., 2015), *Rutilus rutilus* (Flores-Valverde et al., 2010), *Solea senegalensis* (Alvarez-Muñoz

Table 11.1: Biomarkers reported in the literature, effects, and pathways disrupted, based on exposure to single compounds, metabolomics analysis with LC-MS, and fish as a model organism.

Type	Contaminant	Biomarkers	Effect observed	Pathway disrupted	Publication
PhAC	17α-Ethinylestradiol	Hydroxyprogesterone	↓	Steroid biosynthesis	Flores-valverde et al. (2010)
		Androstenedione	↓	Steroid biosynthesis	
		11-Hydroxyandrostenedione	↓	Steroid biosynthesis	
		11-Ketotestosterone	↑	Steroid biosynthesis	
		Cortisol	↑	Glucocorticoid biosynthesis	
		Cortisone	↑	Glucocorticoid biosynthesis	
Pesticide	Cypermethrin	Taurocholic acid (TCA) (C24)	↑	Bile acid metabolism	Carriquiriborde et al. (2012)
		Taurotrihydroxycoprostanic acid (TTHCA) (C27)	↑	Bile acid metabolism	
		Taurodeoxycholic acid (TDCA) (C26)	↑	Bile acid metabolism	
		Bilirubin (C33)	↑	Bile acid metabolism	
Surfactant	Alcohol-polyethoxylated	Taurocholic acid (TCA) (c24bile acids)	↑	Bile acid metabolism	Álvarez-Muñoz et al. (2014)
		Hydroxytaurocholic acid (C24 acid)	↑	Bile acid metabolism	
		Scymnol sulfate (C27 bile alcohol)	↑	Bile acid metabolism	
		Cyprinolsulfate (C27 bile alcohol)	↑	Bile acid metabolism	
		Cortisol	↓	Glucocorticoid biosynthesis	
		Tetrahydrocortisone	↓	Glucocorticoid biosynthesis	
		Glycerophosphatidylcholine (PC) (16:0/hydroxy 18:1)	↑	Lipid homeostasis	
		LysoPC (14:0)	↓	Lipid homeostasis	
		PC (16:0/16:0)	↓	Lipid homeostasis	
		PC (18:0/18:1)	↓	Lipid homeostasis	
Plasticizer	Bisphenol A	Palmitoyl carnitine	↑ (M)	Lipid homeostasis	Ekman et al. (2015)
		Guanine	↑ (M)	Purine degradation	
		Xanthine	↑ (M)	Purine degradation	
		Hypoxanthine	↑ (M)	Purine degradation	
		Guanosine	↑ (M)	Purine degradation	
		Inosine	↑ (M)	Purine degradation	
		Uridine	↑ (M)	Pyrimidine metabolism	
		Methionine	↑ (M)	Glutathione metabolism	
		Proline	↑ (F)	Arginine and proline metabolism	
		Uracil	↑ (M)	Pyrimidine metabolism	
		Carnitine	↑ (F)	ABC transporters	
		Glycolic acid	↓ (F)		
		Hydroxynonanoic acid	↓ (F)		
		O-phosphoethanolamine	↓ (F)		

Continued

Table 11.1: Biomarkers reported in the literature, effects, and pathways disrupted, based on exposure to single compounds, metabolomics analysis with LC-MS, and fish as a model organism.—cont'd

Type	Contaminant	Biomarkers	Effect observed	Pathway disrupted	Publication
PAH	Benzo[a]anthracene	Hydroxyphenyllactic acid	↑	Arginine and proline metabolism	Elie et al. (2015)
	Benz[a]anthracene-7,12-dione	Proline	↑		
		Threonine	↑	Glutathione metabolism	
		5-Oxoproline	↑	Glutathione metabolism	
		Isoleucine	↑	Protein degradation (starvation)	
		Lysine	↑	Carnitine metabolism	
		Methionine	↑	Glutathione metabolism	
		Serotonin	↑	Tryptophan metabolism	
		Phenylalanine	↑	Phenylalanine biosynthesis	
		Arginine	↑	Arginine metabolism	
		Tyrosine	↑	Phenylalanine, tyrosine biosynthesis	
		Tryptophan	↑	Phenylalanine metabolism	
		Cystathionine	↑	Glutathione metabolism	
		Serine	↑	Glutathione metabolism	
		N-arachidonoyl taurine	↑		
		S-adenosylhomocysteine	↑	Glutathione metabolism	
		S-adenosylmethionine	↑		
		Indole	↑	Tryptophan metabolism	
		Betaine	↑		
		Tyramine	↑	Phenylalanine, tyrosine biosynthesis	
		Pipecolic acid	↑		
		Indolelactic acid	↑		
		Dimethyl-L-arginine	↑		
		Phosphocreatine	→		
		Propionylcarnitine	↑	Fatty acid metabolism	
		5-L-glutamyl-L-alanine	↑		
		L-threoninyl-L-glutamate	↑		
		Inosine monophosphate (IMP)	↑	Purine metabolism	
		Guanosine monophosphate (GMP)	↑	Purine metabolism	
		Adenosine diphosphate	→	Purine metabolism	
		Phosphoenol pyruvate	↑		
		L-gamma-glutamyl-L-leucine	↑		

Metabolite		Pathway
E-(gamma-glutamyl)-lysine	↑	
Adenosine monophosphate	↓	Purine metabolism
Guanosine 5′diphosphate	↓	Purine metabolism
Adenosine diphosphate ribose	↓	Glycerophospholipid metabolism
2-Hydroxycinnamic acid	↑	
Phosphocholine	↓	Fatty acid metabolism
Docosahexaenoic acid	↓	
D-fructose 1,6-bisphosphate	↓	
D-glycerate-3-phosphate	↑	Purine metabolism
D-ribose-5-phosphate	↑	
6-Phosphogluconic acid	↑	
P-coumaric acid	↑	Purine metabolism
Hypoxanthine	↑	
Xanthine	↑	Purine metabolism
Guanine	↓	Phenylalanine, tyrosine biosynthesis
Dopamine	↑	
Cytosine	↑	Purine metabolism
Cytidine	↑	
Inosine	↑	Purine metabolism
Uridine	↑	
Guanosine	↑	Purine metabolism
Cytidine diphosphate	↓	
Uridine diphosphate	↓	Glycine, serine, and threonine metabolism
Nucleotide adenine dinucleotide	↓	Alanine, aspartate, and glutamate metabolism
Choline	↑	Glutathione metabolism
Citric acid	↑	Purine metabolism
Glutathione	↑	
Uric acid	↑	
Allantoic acid	↑	
Allantoin	↑	
Niacinamide	↑	Nicotinate metabolism

Continued

Table 11.1: Biomarkers reported in the literature, effects, and pathways disrupted, based on exposure to single compounds, metabolomics analysis with LC-MS, and fish as a model organism.—cont'd

Type	Contaminant	Biomarkers	Effect observed	Pathway disrupted	Publication
Pesticide	Chlorpyrifos	L-carnitine	↑	Fatty acid metabolism	Gómez-Canela et al. (2017)
		L-isoleucine	↑	Protein degradation (starvation)	
		L-leucine	↑	Neurotransmitter	
		Taurine	↑	Neurotransmitter	
		γ-Aminobutyric acid (GABA)	↓	Neurotransmitter	
		L-glutamic acid	↓	Protein degradation (starvation)	
		L-valine	↑		
		L-proline	↑	Arginine and proline metabolism	
		B-alanine	↓	Alanine, aspartate, and glutamate metabolism	
		Creatine	↑	Glycine, serine, and threonine metabolism	
		L-threonine	↑	Glutathione metabolism	
		5-Oxoproline	↑	Glutathione metabolism	
		L-glutamine	↓	Purine metabolism	
		L-histidine	↓		
		L-phenylalanine	↑	Phenylalanine metabolism	
		2-Ketobutyric acid	↓		
		N₂-succinyl-L-ornithine	↓		
		Tyramine	↑	Neurotransmitter	
		Propionylcarnitine	↑		
		N-acetylhistidine	↑		
		Phosphocreatine	↑		
		ADP ribose	↑		
		L-acetylcarnitine	↑	Fatty acid metabolism	
		N-acetylornithine	↓		
		Creatinine	↑	Fatty acid metabolism	
		3-Dehydroxicarnitine	↓		
		B-cyprinol sulfate	↑	Fatty acid metabolism	
		Docosahexaenoic acid	↓	Fatty acid metabolism	
		Linoleic acid	↓	Fatty acid metabolism	
		Oleic acid	↓	Fatty acid metabolism	
		Palmitic acid	↓	Fatty acid metabolism	
		Myristoleic acid	↓	Fatty acid metabolism	

BFR compound	Metabolite	Change	Pathway	Reference
2,20,4,40-tetrabromodiphenyl-ether	Glycerol-3-phosphate	↑	Sugar metabolism (starvation)	Lei et al. (2017)
	Hypoxanthine	↓	Purine metabolism	
	Adenine	↓	Purine metabolism	
	Inosine	↓	Purine metabolism	
	Amp	↓	Purine metabolism	
	Inosinic acid	↓	Purine metabolism	
	ADP	↓	Purine metabolism	
	ATP	↑		
	Adenosine	↑		
	L-lactic acid	↓	Sugar metabolism (starvation)	
	Phosphoric acid	←		
	Piperidine	←		
	Glutathione	↓	Glutathione metabolism	
	Threonic acid	↓	Ascorbate and cofactor metabolism	
	Glucose-6-phosphate	↓	Carbon metabolism	
	D-maltose	↓	ABC transporters	
	Maltotriose	↑	ABC transporters	
	Dethiobiotin	↑	Biotin metabolism	
	Niacinamide	↑	Nicotinate metabolism	
	C-aminobutyric acid (GABA)	↑	Neurotransmitter	
	Lysine	↑	Carnitine metabolism	
	Glutamine	↑	Purine metabolism	
	Arginine	↑	Arginine metabolism	
	Glutamate	↑	Alanine, aspartate, and glutamate metabolism	
	Glycine	↑	Neurotransmitter	
	Threonine	↑	Glutathione metabolism	
	Valine	↑	Protein degradation (starvation)	
	Methionine	↑	Glutathione metabolism	
	Spermine	↑	Polyamine metabolism	
	3,4-Dihydroxymandelic acid	↑	Norepinephrine metabolism	
	Vanillylmandelic acid	↑	Norepinephrine and epinephrine metabolism	
	Acetylcholine	↑	Neurotransmitter	

Continued

Table 11.1: Biomarkers reported in the literature, effects, and pathways disrupted, based on exposure to single compounds, metabolomics analysis with LC-MS, and fish as a model organism.—cont'd

Type	Contaminant	Biomarkers	Effect observed	Pathway disrupted	Publication
Plasticizer	Bisphenol A	Guanine	→	Purine metabolism	Ortiz-Villanueva et al. (2018)
		L-proline	→	Arginine and proline metabolism	
		L-threonine	→	Glutathione metabolism	
		Taurine	→	Primary bile acid biosynthesis	
		5-Oxo-D-proline	←	Glycine, serine, and threonine metabolism	
		Creatine	→		
		L-carnitine	→	ABC transporters	
		L-phenylalanine	←	Phenylalanine metabolism	
		Diazepinone riboside	←		
		L-serine	←	Glycine, serine, and threonine metabolism	
		L-valine	←	Protein degradation (starvation)	
		Taurine	→	Primary bile acid biosynthesis	
		L-leucine	←		
		D-glucuronolactone	←	Folate biosynthesis	
		6-Pyruvoyltetrahydropterin	→	Folate biosynthesis	
		6-Lactoyltetrahydropterin	→		
		Adenosine monophosphate (AMP)	→	Purine metabolism	
		Inosine monophosphate (IMP)	→	Purine metabolism	
		5-Aminopentanal	←		
		2,4-Dichlorobenzoate	→		
		5β-Cyprinolsulfate	→	Primary bile acid biosynthesis	
		4-Guanidinobutanoic acid	→		
		L-tyrosine	←	Phenylalanine metabolism	
		Uracil	→	Pyrimidine metabolism	
		4-Hydroxybutanoic acid	←		
		Octanoic acid	←		
		2-Oxo-3-hydroxy-4-phosphobutanoic acid	→		
		Mesaconate	→	Glyoxylate and dicarboxylate metabolism	
		Hypoxanthine	→	Purine metabolism	
		L-glutamine	←	Purine metabolism	

PFC	Metabolite	Change	Pathway	Reference
PFC	Lyso(PC) (22:6)	↑	Glycerophospholipid metabolism	Ortiz-Villanueva et al. (2018)
Perfluorooctanesulfonate	Diacylglycerol (DG) (38:4)	↑		
	Phosphatidylcholine (PC) (32:1)	↑		
	PC(34:1)	↑		
	PC(36:5)	↑		
	PS(44:12)	→		
	Phosphatidylserine (PS) (40:6)	→		
	PS(44:12)	→		
	Glyceryl 1-monostearate	↑		
	Inosine	↑	Purine metabolism	
	Uridine diphosphate-N-acetylglucosamine	→		
	Choline	↑	Glycine, serine, and threonine metabolism	
	L-lactic acid	→		
	Maleamic acid	→		
	Benzoic acid	↑	Phenylalanine metabolism	
	Salicylic acid	→		
	Citric acid	→	Alanine, aspartate, and glutamate metabolism	
	Phosphoric acid	→		
	Glutathione	↑	Glutathione metabolism	
	6-Succinoaminopurine	→		
	Uric acid	→	Purine metabolism	
	Retinal	→		
	L-proline	→	Arginine and proline metabolism	
	Creatine	→	Glycine, serine, and threonine metabolism	
	L-tyrosine	↑	Phenylalanine metabolism	
	L-phenylalanine	↑	Phenylalanine metabolism	
	L-methionine	→	Glutathione metabolism	
	L-asparagine	→	Alanine, aspartate, and glutamate metabolism	
	D-glucuronolactone	↑		
	Guanosine monophosphate (GMP)	→	Purine metabolism	
	Inosine monophosphate (IMP)	→	Purine metabolism	

Table 11.1: Biomarkers reported in the literature, effects, and pathways disrupted, based on exposure to single compounds, metabolomics analysis with LC-MS, and fish as a model organism.—cont'd

Type	Contaminant	Biomarkers	Effect observed	Pathway disrupted	Publication
		Adenosine monophosphate (AMP)	↓	Purine metabolism	
		Hypotaurine	↑	Alanine, aspartate, and glutamate metabolism	
		L-acetylaspartate	↓		
		5β-Cyprinolsulfate	↓	Primary bile acid biosynthesis	
		2-Oxo-3-hydroxy-4-phosphobutanoic acid	↓		
		Oxidized glutathione	↓	Purine metabolism	
		Inosine	↑	Pyrimidine metabolism	
		Uracil	↑	Folate biosynthesis	
		Neopterin	↑	Purine metabolism	
		Guanosine	↑	Phenylalanine metabolism	
		4-Coumarate	↑	Alanine, aspartate, and glutamate metabolism	
		2-Oxoglutarate	↑	Alanine, aspartate, and glutamate metabolism	
		Citric acid	↓	ABC transporters	
		D-maltose	↑	Carbon metabolism	
		D-glucose 6-phosphate	↓	ABC transporters	
		D-maltose	↓		
		Nicotinamide	↑	Alanine, aspartate, and glutamate metabolism	
Biocide	Tributyltin	4-Aminohippuric acid	↑	Protein degradation (starvation)	Ortiz-Villanueva et al. (2018)
		Alanine	↑		
		L-valine	↓	Primary bile acid biosynthesis	
		Taurine	↓		
		5-Oxo-D-proline	↑	Glycine, serine, and threonine metabolism	
		Creatine	↓	Purine metabolism	
		L-glutamine	↑	Glutathione metabolism	
		L-methionine	↑	ABC transporters	
		L-carnitine	↓	Phenylalanine metabolism	
		L-tyrosine	↑	Arginine and proline metabolism	
		L-proline	↓		

Metabolite		Pathway
L-glutamate	→	Alanine, aspartate, and glutamate metabolism
L-acetylcarnitine	←	Fatty acid metabolism
6-Pyruvoyltetrahydropterin	→	
Adenosine monophosphate	→	Purine metabolism
Guanosine monophosphate	→	
Hypotaurine	←	Taurine and hypotaurine metabolism
5-Aminopentanoate	→	Arginine and proline metabolism
B-Nitropropanoate	→	
5β-Cyprinolsulfate	→	Primary bile acid biosynthesis
Methylhexadecanoic acid	←	
4-Hydroxybutanoic acid	←	
2-Oxo-3-hydroxy-4-phosphobutanoic acid	→	
4-Aminobutanoate	→	Alanine, aspartate, and glutamate metabolism
Oxidized glutathione	→	
Dehydrodiconiferyl alcohol	←	
Estrone glucuronide	←	
Hypoxanthine	→	Purine metabolism
LysoPC(18:1)	←	
LysoPC(18:0)	←	
PC(32:1)	←	
PC(36:5)	←	
PS(44:12)	→	
LysoPC(16:0)	←	
LysoPC(22:6)	→	
PS(40:6)	←	
Glyceryl 1-monostearate	←	
Neopterin	←	Folate biosynthesis
Inosine	←	Purine metabolism
Cytidine	→	Pyrimidine metabolism
Uridine	→	Pyrimidine metabolism
Uridine diphosphate-N-acetylglucosamine	←	
Maleamic acid	→	
6-Succinoaminopurine	→	
A-D-glucose	←	Carbon metabolism
Maltotriose	→	ABC transporters

Table 11.1: Biomarkers reported in the literature, effects, and pathways disrupted, based on exposure to single compounds, metabolomics analysis with LC-MS, and fish as a model organism.—cont'd

Type	Contaminant	Biomarkers	Effect observed	Pathway disrupted	Publication
PhAC	Ibuprofen	N-undecanoylglycine	↑		Song et al. (2018)
		Histidinyl-histidine	↓		
		Temocaprilat	↑		
		Cysteineglutathione disulfide	↑	Arginine and proline metabolism	
		Deoxyguanosine	↓	Purine metabolism	
		L-cysteinylglycine disulfide	↑	Arginine and proline metabolism	
		N-a-acetylcitrulline	↓		
		Gamma-glutamyl-beta-aminopropiononitrile	↓		
		3,5-Diiodo-L-tyrosine	↓	Tyrosine metabolism	
		Asparaginyl-alanine	↓		
		(6S)-6-beta-hydroxy-1,4,5,6-tetrahydronicotinamide-adenine dinucleotide	↑		
		Nicotinic acid adenine dinucleotide	↑		
		Benzoyl-coa	↓		
		ADP-ribose 20-phosphate	↓		
		Beta-alanyl-coa	↓	Alanine metabolism	
		Dopamine glucuronide	↓		
		5β-Cyprinolsulfate	↓		
		Alpha-tetrasaccharide	↑		
		Stearoylcarnitine	↓		
		Phosphoribosyl formamido carboxamide	↑	Purine metabolism	
		Nadh	↓		
		(3S)-Hydroxy-tetracosa-6,9,12,15,18,21-all-cis-hexaenoyl-coa	↑		
		Cholesterol ester (15:0)	↓	Glutathione metabolism	
		Eicosadienoic acid	↑		
		Leukotriene B4 ethanolamide	↓		
		Acetyl-coa	↓		
		Oleamide	↓		
		(23S)-23,25-dihdroxy-24-oxovitamins D3 23-(beta-glucuronide)	↑	Alanine metabolism	

Treatment	Metabolite	Change	Pathway	Reference
	2,3-Diaminosalicylic acid	↓		Ziarrusta et al. (2018)[a]
	Ceramide(d18:0/24:0)	↑		
	Phytosphingosine-1-P	↑		
	Secaloside C	↑		
	5,6,7,8-Tetrahydromethanopterin	↑		
	Adenine	↓	Purine metabolism	
	Glutathione	↓		
	Mesoporphyrin IX	↓		
	6-Succinoaminopurine	↓	Purine metabolism	
Oxybenzone	Xanthylic acid	↑		
	D-erythro-Eritadenine	↓	Purine metabolism	
	Deoxycytidine	↑		
	1,4-Beta-D-Glucan	↓		
	Chitobiose	↑		
	3-Carboxy-1-hydroxypropylthiamine diphosphate	↓		
	Ophthalmic acid	↓		
	Vitamin D$_2$ 3-glucuronide	↑		
	D-serine	↓	Glycine, serine, and threonine metabolism	
	5-Oxo-L-proline	↓	Glutathione metabolism	
	Trans-2,3-dihydroxycinnamate	↑	Phenylalanine metabolism	
	3-(4-Hydroxyphenyl)pyruvate/2-hydroxy-3-(4-hydroxyphenyl)propenoate	↑	Phenylalanine metabolism	
	Phenylacetylglycine	↑	Phenylalanine metabolism	
	Hippurate	↑	Phenylalanine metabolism	
	Nervonic acid	↑	Biosynthesis of unsaturated fatty acids	
PCP	Docosahexaenoic acid	↑	Biosynthesis of unsaturated fatty acids	
	Docosadienoic acid	↑	Biosynthesis of unsaturated fatty acids	
	Γ-Linolenic acid/α-Linolenic acid/Crepenynate	↑	Biosynthesis of unsaturated fatty acids	
	3,6-Nonadienal	↓	Biosynthesis of unsaturated fatty acids	
	Sn-glycerol 1-phosphate/sn-glycerol 3-phosphate	↑	Biosynthesis of unsaturated fatty acids	

Continued

Table 11.1: Biomarkers reported in the literature, effects, and pathways disrupted, based on exposure to single compounds, metabolomics analysis with LC-MS, and fish as a model organism.—cont'd

Type	Contaminant	Biomarkers	Effect observed	Pathway disrupted	Publication
PhAC	Amitriptyline	Arginine	↓	Arginine metabolism	Ziarrusta et al. (2019)[a]
		Methionine	↑	Glutathione metabolism	
		Asparagine	↓	Alanine, aspartate, and glutamate metabolism	
		Glutamate	↓		
		N-formimino-l-glutamate	↑		
		C18:0 (stearoylcarnitine)	↑	Lipid metabolism	
		Ceramide (d18:1 /24:1(15Z))	↓	Lipid metabolism	
		Phosphatidylethanolamine	↑	Lipid metabolism	
		Monoacylglyceride (MG)	↓	Lipid metabolism	
		C4	↑	Lipid metabolism	
		PC (c34:1)	↑	Lipid metabolism	
		PC (c36:3)	↑	Lipid metabolism	
		PC (c38:2)	↑	Lipid metabolism	
		LysoPC (16:0)	↓	Lipid metabolism	
		LysoPC (18:1)	↑	Lipid metabolism	
		SM (c18:0)	↓	Lipid metabolism	
		C17:1-cooh	↑	Lipid metabolism	
		C16:1-oh	↑	Lipid metabolism	
		C5	↑	Lipid metabolism	
		LysoPC 20:3	↓	Lipid metabolism	
		LysoPC 24:1	↑	Lipid metabolism	
		Phosphatidylcholine (PC) C30:2	↑	Lipid metabolism	
		PC c32:1	↑	Lipid metabolism	
		PC c32:2	↑	Lipid metabolism	
		PC c32:3	↑	Lipid metabolism	
		Sphingomyelin (SM) C18:1	↑	Lipid metabolism	
		Uric acid	↑	Lipid metabolism	
		Pantothenate	↑	Purine metabolism	
		3-Deoxyvitamin D3	↓		

PhAC				Jiang et al. (2019)[b]
Dydrogesterone	Glutathione oxidized	↑	Bile acid synthesis and metabolism	
	Taurocholic acid	↑	Bile acid synthesis and metabolism	
	Taurohyodeoxycholic acid	↑	Fatty acid metabolism	
	Acetylcarnitine	↑		
	Butyryl carnitine	↑		
	Hexanoylcarnitine	↑		
	Myristoylcarnitine	↑		
	Palmitoyl carnitine	↑	Lipid homeostasis	
	5-Oxo-ETE	↑	Fatty acid metabolism	
	Palmitoleic acid	↑	Fatty acid metabolism	
	A-linolenic Acid	↑	Fatty acid metabolism	
	Stearidonic acid	↑	Fatty acid metabolism	
	5-Hetre	↑	Fatty acid metabolism	
	5-Hete	↑	Fatty acid metabolism	
	Phosphocholine	↑	Lipid metabolism	
	PC(14:0/0:0)	↑	Lipid metabolism	
	PE(16:0/0:0)	↑	Lipid metabolism	
	PE(16:1(9z)/0:0)	↑	Lipid metabolism	
	PE(18:1(9z)/0:0)	↑	Lipid metabolism	
	PG(16:0/0:0)	↑	Lipid metabolism	
	Hypoxanthine	↑	Purine degradation	
	Inosine	↑	Purine degradation	
	Guanine	↑	Purine degradation	
	Adenosine monophosphate (AMP)	↑	Purine degradation	
	Inosine monophosphate (IMP)	↑	Purine degradation	
	Guanosine monophosphate (GMP)	↑	Purine degradation	
	Uridine monophosphate UMP	↑	Purine degradation	
	UDP-glucose	↑		
	UDP-glucuronic acid	↑		
	UDP-N-acetylglucosamine	↑		
	Palmitoyl taurine	↑		
	Pantothenic acid	↑		

Continued

Table 11.1: Biomarkers reported in the literature, effects, and pathways disrupted, based on exposure to single compounds, metabolomics analysis with LC-MS, and fish as a model organism.—cont'd

Type	Contaminant	Biomarkers	Effect observed	Pathway disrupted	Publication
PAH	Benzo[a]pyrene	L-proline	↑	Arginine and proline metabolism	Roszkowska et al. (2018)[c]
		L-tryptophan	↑	Tryptophan metabolism	
		N-methyl-a-aminoisobutyric acid	↑		
		L-valine	↑	Protein degradation (starvation)	
		Taurine	↑	Primary bile acid biosynthesis	
		L-leucine	↑	Protein degradation (starvation)	
		L-isoleucine	↑		
		L-glutamate	↑	Fatty acid metabolism	
		L-acetylcarnitine	↑		
		Cysteine acid	↑		
		Tricosanoylglycine	↑		
		Pristanoylglycine	↑		
		N1-methyl-4-pyridone-3-carboxamide	↑		
		N1-methyl-2-pyridone-5-carboxamide	↑		
		5-Aminopentanoic acid	↑		
		Betaine	↑		
		3-Phenylpropionylglycine	↑		
		N-acetyl-L-phenylalanine	↑		
		Phenylpropionylglycine	↑		
		L-4-hydroxyglutamate semialdehyde	↑		
		O-acetylserine	↑		
		N-acetylserine	↑		
		N-methyl-D-aspartic acid	↑		
		5-Oxoprolinate	↑		
		Leucyl-proline	↑		
		Isoleucyl-proline	↑		
		N-acryloylglycine	↑		
		Pc(38:8)	↑	Lipid metabolism	
		Erythronic acid	↑		
		Pyrroline hydroxycarboxylic acid	↑		
		Pyrrolidonecarboxylic acid	↑		
		Pyroglutamic acid	↑		
		Threonic acid	↑	Ascorbate and cofactor metabolism	
		1-Pyrroline-4-hydroxy-2-carboxylate	↑		

Metabolite		Pathway	PCB95	PCB	Xu et al. (2016)
Aspartate acid	↑	Aspartate metabolism			
Glutamic acid	↑	Neurotransmitter			
Serine	↑	Glycine, serine, and threonine metabolism			
Histidine	↑	Glutathione metabolism			
Threonine	↑	Arginine metabolism			
Arginine	↑	Alanine, aspartate, and glutamate metabolism			
Alanine	↑	Phenylalanine, tyrosine biosynthesis			
Tyrosine	↑	Protein degradation (starvation)			
Valine	↑	Glutathione metabolism			
Methionine	↑	Phenylalanine biosynthesis			
Phenylalanine	↑	Protein degradation (starvation)			
Leucine	↑	Carnitine metabolism			
Isoleucine	↑	Arginine and proline metabolism			
Lysine	↑	Tryptophan metabolism			
Proline	↑	Primary bile acids biosynthesis			
Cysteine	↑	Purine metabolism			
Tryptophan	↑	Neurotransmitter			
Taurine	↑	Purine metabolism			
Glutamine	↑				
Y-Aminobutyrate (GABA)	↑				
Inosine monophosphate (IMP)	↑				
Pipecolic acid	↑				
Betaine	↑	Glycine, serine, and threonine metabolism			
N-acetylornithine					
Choline					
Pyroglutamic acid	↑				

Continued

Table 11.1: Biomarkers reported in the literature, effects, and pathways disrupted, based on exposure to single compounds, metabolomics analysis with LC-MS, and fish as a model organism.—cont'd

Type	Contaminant	Biomarkers	Effect observed	Pathway disrupted	Publication
Pesticides	Permethrin	Phenylalanine	↓	Phenylalanine metabolism	Tufi et al. (2016)
		Tryptophan	↓	Phenylalanine metabolism	
		GABA	↑	Neurotransmitter	
		Acetylcholine	↓	Neurotransmitter	
		Choline	↓	Glycine, serine, and threonine metabolism	
	Diazinon-o-analog	Phenylalanine	↓	Phenylalanine metabolism	
		Tryptophan	↓	Phenylalanine metabolism	
		GABA	↑	Neurotransmitter	
		Acetylcholine	↓	Neurotransmitter	
		Choline	↓	Glycine, serine, and threonine metabolism	
	Aldicarb	Phenylalanine	↑	Phenylalanine metabolism	
		Tryptophan	↓	Phenylalanine metabolism	
		GABA	↑	Neurotransmitter	
		Acetylcholine	↓	Neurotransmitter	
		Choline	↑	Glycine, serine, and threonine metabolism	
	Pirimicarb	Phenylalanine	↑	Phenylalanine metabolism	
		Tryptophan	↑	Phenylalanine metabolism	
		GABA	↓	Neurotransmitter	
		Acetylcholine	↓	Neurotransmitter	
		Choline	↓	Glycine, serine, and threonine metabolism	
	Imidacloprid	Phenylalanine	↑	Phenylalanine metabolism	
		Tryptophan	↑	Phenylalanine metabolism	
		GABA	↓	Neurotransmitter	
		Acetylcholine	↓	Neurotransmitter	
		Choline	↓	Glycine, serine, and threonine metabolism	
	Diclorfos	Phenylalanine	↓	Phenylalanine metabolism	
		Tryptophan	↓	Phenylalanine metabolism	
		GABA	↑	Neurotransmitter	
		Acetylcholine	↓	Neurotransmitter	
		Choline	↓	Glycine, serine, and threonine metabolism	

	Biomarker		Pathway
Chlorpyrifos	Phenylalanine	↑	Phenylalanine metabolism
	Tryptophan	↑	Phenylalanine metabolism
	GABA	→	Neurotransmitter
	Acetylcholine	→	Neurotransmitter
	Choline	→	Glycine, serine, and threonine metabolism
Carbaryl	Phenylalanine	→	Phenylalanine metabolism
	Tryptophan	→	Phenylalanine metabolism
	GABA	→	Neurotransmitter
	Acetylcholine	→	Neurotransmitter
	Choline	→	Glycine, serine, and threonine metabolism

F, female; M, male.

[a]Features with more than one potential identity not reported.

[b]Only biomarkers identified using LC-MS reported.

[c]Features detected with high confidence level reported.

et al., 2014), and *Sparus aurata* (Ziarrusta et al., 2018, 2019), but the preferred one was *Danio rerio* (Elie et al., 2015; Gomez-Canela et al., 2017; Jiang et al., 2019; Ortiz-Villanueva et al., 2017, 2018; Song et al., 2018; Tufi et al., 2016; Xu et al., 2016). *D. rerio* is a popular model organism widely used in environmental research. It is a convenient and cost-effective species and provides conceptual insights into many aspects of vertebrate's biology, genetics, toxicology, and disease, besides its genome is sequenced (Segner, 2009).

Table 11.1 shows contaminant type and name, biomarkers reported, effect observed (increase or decrease of levels), pathway disrupted, and publication. Metabolomics studies carried out with fish exposed to different type of organic contaminants have been performed mainly with pesticides, with 11 compounds tested (cypermethrin, tributyltin, chlorpyrifos, permethrin, diazin-o-analog, aldicarb, pirimicarb, imidacloprid, diclorfos, carbaryl, and tributyltin (used as a pesticide additive in industrial and marine paints)). Followed by pharmaceuticals (PhACs) and polycyclic aromatic hydrocarbons (PAHs) with four and three compounds assessed, respectively (17α-ethinylestradiol, ibuprofen, amitriptyline, and dydrogesterone, and benzo[a]pyrene, benzo[a]anthracene, and benz[a] anthracene-7,12-dione). Compounds belonging to other types of contaminants such as a plasticizer (bisphenol A), a brominated flame retardant (BFR) (2,20,4,40-tetrabromodiphenyl-ether), a polychlorinated biphenyl (PCB) (PCB95), a personal care product (PCP) (oxybenzone), a perfluorinated compound (PFC) (perfluorooctanesulfonate), and a surfactant (alcohol polyethoxylated) have been also evaluated. Therefore, a wide representation of contaminant types of general environmental concern has been studied in fish using metabolomics with the exception of nanomaterials. To the best of our knowledge, there is not any metabolomics work performed yet that aims to profile the metabolome of fish exposed to any kind of nanomaterial using HRMS.

Table 11.1 reports 504 biomarkers identified in fish which levels significantly changed due to a certain organic contaminant exposure. In 62% of them, the trend observed was an increase, while 37% showed a decrease in their levels. Hence, a rise in biomarkers' levels is the most common trend observed when fish is exposed to organic contaminants.

Regarding the type of biomarkers, based on the information presented in Table 11.1, amino acids and related compounds are the most abundant group (58%), followed by lipids (17%) (Fig. 11.1). Bile acids, nutrients, organic acids, sugars, and vitamins accounted with percentages that ranged between 2% and 5%. Molecules such as carbohydrates, acylglycerols, and synthetic metabolites were included in the "others" category (9%).

Amino acids are organic compounds that contain both an amine and a carboxyl group. According to the data shown in Table 11.1 and Fig. 11.1, they are the most sensitive type of molecule in fish, which levels change due to organic contaminant exposure. They have key roles such as building blocks of proteins and polypeptides, as well as precursors of hormones and low-molecular weight nitrogen substances (Wu, 2009). An optimal level of

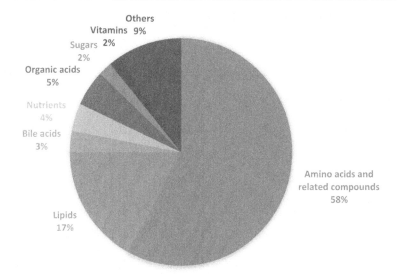

Figure 11.1

Nature of biomarkers most reported in fish exposed to organic contaminants using metabolomics and percentage of hit.

amino acids is crucial for maintaining organism homeostasis. An increase in their levels, as it is the tendency observed after contaminant exposure, may lead to neurological disorders, oxidative stress, and cardiovascular disease.

Lipids are also naturally occurring organic compounds that contain a long hydrocarbon chain (single or multiple) that may be saturated or unsaturated. The have a glycerol group bonded to the hydrocarbon chain and, depending on the type of lipid, to other molecule, for example, to a phosphate group in the case of phospholipids. Fatty acids are also considered a type of lipids; they contain carboxyl groups bonded to the hydrocarbon chain and serve as building blocks for other lipids (i.e., phospholipids). Consequently, they were included in the lipids category in Fig. 11.1. Lipids play important roles in energy storage, signaling, and they are structural components of cell membranes. An increase in their levels may lead to a disruption in cell membrane integrity with potential release of other cell components (Alvarez-Muñoz et al., 2014), oxidative stress (Zhao et al., 2015), and lipid storage disorders such as phospholipidosis (Xia et al., 2000).

2. Biomarkers' identity

Taking into consideration the two types of molecules more sensitive to contaminant exposure, amino acids and lipids, the identity of the specific biomarkers altered within these categories was studied. Table 11.2 presents amino acid and lipid identities together with the number of times that a specific biomarker was reported (considering data shown

Table 11.2: Amino acid and lipid identities together with the number of times that a specific biomarker was reported considering data shown in Table 11.1.

Biomarker	Times detected[a]	Tendency[b]
Phenylalanine	13	↑
Y-Aminobutyric acid (GABA)	10	—
Tryptophan	10	—
Proline	8	↑
Methionine	7	↑
Oxoproline	6	↑
Adenosine monophosphate (AMP)	6	↓
Glutathione	6	—
Hypoxanthine	6	—
Taurine	6	—
Valine	6	↑
Adenosine diphosphate (ADP)	5	↓
Glutamine	5	↑
Inosine monophosphate (IMP)	5	↑
Tyrosine	5	↑
Uridine diphosphate (UDP)	5	↑
Arginine	4	↑
Creatine	4	↓
Glutamate	4	—
Guanosine monophosphate (GMP)	4	—
Inosine	4	↑
Isoleucine	4	↑
Acetylcarnitine	4	↑
6-Succinoaminopurine	3	↓
Alanine	3	↓
Betaine	3	↑
Guanine	3	↑
Guanosine	3	↑
Carnitine	3	↓
Leucine	3	↑
Lysine	3	↑
Threonine	3	↑
Uridine	3	↑
6-Pyruvoyltetrahydropterin	2	↓
Adenine	2	↓
Cytidine	2	—
D-glucuronolactone	2	↑
Hypotaurine	2	↑
N1-methyl-2-pyridone-5-carboxamide	2	↑
N-acetylornithine	2	—
Neopterin	2	↑
Phosphocreatine	2	—
Pipecolic acid	2	↑
Propionylcarnitine	2	↑
Serine	2	↑

Table 11.2: Amino acid and lipid identities together with the number of times that a specific biomarker was reported considering data shown in Table 11.1.—cont'd

Biomarker	Times detected[a]	Tendency[b]
Threonic acid	2	−
Tyramide	2	↑
Xanthine	2	↑
Phosphatidylcholine (PC)	16	↑
Lysophosphatidylcholine (PC)	10	↑
Phosphatidylserine (PS)	5	↓
2-Oxo-3-hydroxy-4-phosphobutanoic acid	3	↓
Docosahexaenoic acid	3	↓
4-Hydroxybutanoic acid	2	↑
Ceramide	2	−
Sphingomyelin	2	−
Linolenic acid	2	↑

The tendency observed in its effect is also shown as increase (↑), decrease (↓), or not clear trend (−).
[a]At least reported 2 times.
[b]Levels increase (↑) or decrease (↓) in >50 times detected, (−) biomarker ↑ 50% of the times reported and ↓ in the other 50%; therefore, a trend was not established.

in Table 11.1), and the tendency observed in its effect. Only biomarkers that were reported at least twice are included in Table 11.2. The tendency was considered as increase (↑) or decrease (↓) when that was the effect shown in more than 50% of the observations. If an increase was observed in half of the observations and a decrease in the other half, a clear trend could not be established and a dash was assigned (−). Amino acids and lipids reported at least two times represented nearly half of the entire fish metabolome (48% according to Table 11.1). Concretely, their identities corresponded to 48 different amino acids and related compounds and 9 lipids (Table 11.2). Regarding the effect observed, the same increasing tendency as previously reported in Table 11.1 was detected (57% of the biomarkers increased, 21% decreased, and 22% not clear trend, Table 11.2).

The top five amino acids were held by phenylalanine, Υ-aminobutyric acid (GABA), tryptophan, proline, and methionine, being the most detected amino acids with 13, 10, 10, 8, and 7 times detected, respectively. Phenylalanine, tryptophan, and methionine are essential amino acids, meaning that they cannot be synthesized from scratch by the organism and they must be supplied in the diet. Proline, although it is generally considered as nonessential, can be essential in some fish (Wu, 2009). GABA chemically speaking is an amino acid but the amino group is attached to the carbon atom at the position gamma instead of alpha; therefore, it is not incorporated into proteins. Regarding their functions, GABA acts as neurotransmitter in the central nervous system and reduces the activity of neurons or nerve cells. Both tryptophan and proline are functional amino acids that can regulate key metabolic pathways necessary for maintenance, growth, reproduction, and immunity in organisms (Wu, 2009). Besides, proline and methionine are related to the antioxidant defense system and to oxidative stress disorders (Zhou et al., 2019). Phenylalanine is the precursor of the catecholamine neurotransmitters L-dopa, dopamine and epinephrine, and methionine a carnitine precursor (Ziarrusta et al., 2019).

In the lipids' groups a top three was established with phosphatidylcholine (PC), lysoPC, and phosphatidylserine (PS) as the most detected biomarkers (16, 10, and 5 times detected, respectively) (Table 11.2). They belong to the phospholipid class that is a major component of cell membranes due to its amphiphilic properties. Two hydrophobic fatty acid "tails" connected to a hydrophilic "head" by a glycerol molecule characterize their molecular structure. The head consists of a phosphate group that can be modified with an organic molecule such as choline, in the case of PC, or serine, for PS. LysoPC is derived from partial hydrolysis of PC by removing one of the fatty acids tails. PC and PS are major constituent of membranes and play an important role in membrane-mediated cell signaling.

3. Pathways disrupted

Once the fish metabolome is profiled and the identity of the biomarkers established, data interpretation, known as secondary analysis, is usually undertaken. The purpose of this analysis is to link the metabolites that significantly change with the metabolic pathways where they are involved in order to draw biological conclusions. This is known as pathway enrichment techniques, and they can be divided into three generations: over representation analysis, functional class scoring, and pathway topology (Khatri et al., 2012). Enrichment tools have been mainly developed for other "omics" technologies, such as genomics or proteomics, and suitable tools for metabolomics are still scarce (Marco-Ramell, 2018). The method identifies biological pathways that are enriched in a metabolites list more than would be expected by chance, by drawing from comprehensive databases like Kyoto Encyclopedia of Genes and Genomes (KEGG), Reactome, WikiPathways, and the Small Molecule Pathway Database (Picart-Armada et al., 2017). MetaboAnalyst is the tool generally used (Elie et al., 2015; Lei et al., 2017; Song et al., 2018; Xu et al., 2016), although manual checking in databases such as KEGG is also popular (Gomez-Canela et al., 2017; Ortiz-Villanueva et al., 2018; Ziarrusta et al., 2018). Manual checking is complex because the same metabolite can be involved in several pathways, for instance, according to KEGG L-Glutamine (C00064) is involved in 20 different pathways. Therefore, software such as MetaboAnalyst or Compound Discoverer, that has recently included this feature, is helpful.

According to the information presented in Table 11.1, 28 pathways were found to be disrupted in fish due to an exposure to organic contaminants. Fig. 11.2 shows the pathways with most compound hits and many of them are closely related. If a "disrupted pathway

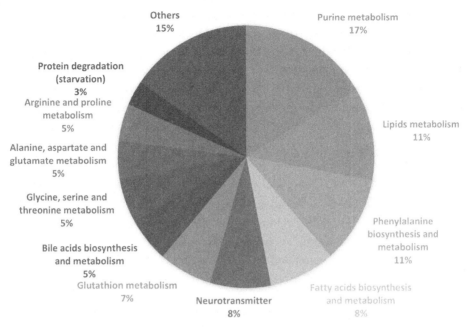

Figure 11.2
Pathways with most compound hits in fish exposed to organic contaminants.

ranking" is established based on the times that they have been pointed out, purine metabolism with 17% of the times holds the first position being the most altered metabolic route due to organic contaminant exposure in fish. In the second place are lipid metabolism and phenylalanine biosynthesis and metabolism with 11% of the times detected. Then, neurotransmitters and fatty acid biosynthesis and metabolism with 8% and glutathione metabolism with 7% hold the third position. In fourth place with 5% bile acid metabolism, glycine, serine, and threonine metabolism, alanine, aspartate, and glutamate metabolism, and arginine and proline metabolism. In fifth position, all the pathways that were pointed out less than a 3% are included, such as the ones related to protein degradation, sugar metabolism, steroid biosynthesis, etc., included in the "others" category.

Purine metabolism, also called purine nucleotide catabolism, may be a component of the homeostatic response of mitochondria to oxidative stress caused by organic contaminants (Elie et al., 2015). Phenylalanine biosynthesis and metabolism disruption in fish may be predictive of developmental neurobehavioral effects associated with contaminant exposure as it has been previously observed in rats and humans (Perera et al., 2012; Xia et al., 2011). Lipids and fatty acids play a major role as sources of energy for fish growing, reproduction, and movement. Besides, some functional lipids such as phospholipids are structural components of the double-layered surface of all cells (lipid bilayer) and disruption of their biosynthesis may affect membrane stability. Glutathione metabolism is also crucial because it plays an important role in the detoxification of contaminants. The enzyme glutathione S-transferase (GST) catalyzes the conjugation of glutathione with xenobiotic compounds containing electrophilic centers. It is important for organisms to deal with this type of contaminants because they can react with macromolecules controlling cell growth such as DNA, RNA, and proteins (Hampel et al., 2016). Besides, many of these chemicals are carcinogenic.

4. Conclusions and future trends

More than 500 metabolites have been reported as potential biomarkers of organic contaminant exposure in fish by using metabolomics techniques. This entails a big amount of data difficult to handle and that generally is not fully exploited. In this chapter, a rational analysis of this information has been performed in order to subtract conclusions that may help to interpret data generated with metabolomics studies and serve as a guideline. The general tendency observed in 62% of the data was an increase in the levels of biomarkers. This suggests that a rise in biomarkers' body burden is the most common effect expected after organic contaminant exposure. Amino acids and lipids were the most sensitive groups of molecules altered, accounting for nearly half of the entire fish metabolome (48%). The top three amino acids and lipids which levels significantly changed due to contaminant exposure were phenylalanine, Υ-aminobutyric acid (GABA), tryptophan, phosphatidylcholine, lysophosphatidylcholine, and phosphatidylserine. They may be proposed as potentials "universal biomarkers" of general organic contamination and environmental quality due to their higher sensitivity and nonspecific response to a certain chemical. To be fair, highly specific biomarkers are less abundant than nonspecific ones, but this may represent an advantage for a first screening of biological status and they can be used as an early warning tool of environmental quality. Later on, a more specific set of biomarkers may be applied in order to characterize exposure or effects of chemicals of concern. Biomarkers were related to metabolic pathways involved in order to draw biological conclusions. Twenty-eight pathways were altered due to contaminant exposure. A "pathway ranking" was established with purine metabolism being the most altered one, followed by lipids and phenylalanine metabolism.

In order to protect biological systems and maintain good environmental quality, it is necessary to study the overall biological effects of organic contaminant exposure. Metabolomics is a useful technique for this because it offers a good coverage of the organism metabolome and a snapshot of its biological status. Its popularity for analyzing biological samples in environmental monitoring is rapidly growing, and it will continue like this until the current challenges that scientist faced are achieved and it will become a routine tool for the assessment of environmental quality.

Acknowledgments

This study was supported by the Spanish Ministry of Economy and Competitiveness, State Research Agency, and by the European Union through the European Regional Development Fund through the project XENOME-TABOLOMIC (CTM2015-73179-JIN) (AEI/FEDER/UE).

References

Alvarez-Muñoz, D., Al-Salhi, R., Abdul-Sada, A., Gonzalez-Mazo, E., Hill, E.M., 2014. Global metabolite profiling reveals transformation pathways and novel metabolomic responses in *Solea senegalensis* after exposure to a non-ionic surfactant. Environ. Sci. Technol. 48, 5203−5210.

Carriquiriborde, P., Marino, D.J., Giachero, G., Castro, E.A., Ronco, A.E., 2012. Global metabolic response in the bile of pejerrey (*Odontesthes bonariensis*, Pisces) sublethally exposed to the pyrethroid cypermethrin. Ecotoxicol. Environ. Saf. 76, 46−54.

Ekman, D.R., Skelton, D.M., Davis, J.M., Villeneuve, D.L., Cavallin, J.E., Schroeder, A., Jensen, K.M., Ankley, G.T., Collette, T.W., 2015. Metabolite profiling of fish skin mucus: a novel approach for minimally-invasive environmental exposure monitoring and surveillance. Environ. Sci. Technol. 49, 3091−3100.

Elie, M.R., Choi, J., Nkrumah-Elie, Y.M., Gonnerman, G.D., Stevens, J.F., Tanguay, R.L., 2015. Metabolomic analysis to define and compare the effects of PAHs and oxygenated PAHs in developing zebrafish. Environ. Res. 140, 502−510.

Flores-Valverde, A.M., Horwood, J., Hill, E.M., 2010. Disruption of the steroid metabolome in fish caused by exposure to the environmental estrogen 17alpha-ethinylestradiol. Environ. Sci. Technol. 44, 3552−3558.

Gomez-Canela, C., Prats, E., Pina, B., Tauler, R., 2017. Assessment of chlorpyrifos toxic effects in zebrafish (*Danio rerio*) metabolism. Environ. Pollut. 220, 1231−1243.

Hampel, M., Blasco, J., Martín-Díaz, M.L., 2016. Chapter 5 Biomarkers and effects. Mar. Ecotoxicol. Curr. Knowl. Future Issues 122−165.

Jiang, Y.X., Shi, W.J., Ma, D.D., Zhang, J.N., Ying, G.G., Zhang, H., Ong, C.N., 2019. Dydrogesterone exposure induces zebrafish ovulation but leads to oocytes over-ripening: an integrated histological and metabolomics study. Environ. Int. 128, 390−398.

Khatri, P., Sirota, M., Butte, A.J., 2012. Ten years of pathway analysis: current approaches and outstanding challenges. PLoS Comput. Biol. 8, e1002375.

Lam, P.K.S., 2009. Use of biomarkers in environmental monitoring. Ocean Coast. Manag. 52, 348−354.

Lei, E.N., Yau, M.S., Yeung, C.C., Murphy, M.B., Wong, K.L., Lam, M.H., 2017. Profiling of selected functional metabolites in the central nervous system of marine medaka (*Oryzias melastigma*) for environmental neurotoxicological assessments. Arch. Environ. Contam. Toxicol. 72, 269−280.

Marco-Ramell, A., Palau-Rodriguez, M., Alay, A., Tulipani, S., Urpi-Sarda, M., Sanchez-Pla, A., Andres-Lacueva, C., 2018. Evaluation and comparison of bioinformatic tools for the enrichment analysis of metabolomics data. BMC Bioinform. 19, 1−11.

Ortiz-Villanueva, E., Jaumot, J., Martinez, R., Navarro-Martin, L., Pina, B., Tauler, R., 2018. Assessment of endocrine disruptors effects on zebrafish (*Danio rerio*) embryos by untargeted LC-HRMS metabolomic analysis. Sci. Total Environ. 635, 156−166.

Ortiz-Villanueva, E., Navarro-Martin, L., Jaumot, J., Benavente, F., Sanz-Nebot, V., Pina, B., Tauler, R., 2017. Metabolic disruption of zebrafish (*Danio rerio*) embryos by bisphenol A. An integrated metabolomic and transcriptomic approach. Environ. Pollut. 231, 22−36.

Perera, F.P., Tang, D., Wang, S., Vishnevetsky, J., Zhang, B., Diaz, D., Camann, D., Rauh, V., 2012. Prenatal polycyclic aromatic hydrocarbon (PAH) exposure and child behavior at age 6−7 years. Environ. Health Perspect. 120, 921−926.

Picart-Armada, S., Fernandez-Albert, F., Vinaixa, M., Rodriguez, M.A., Aivio, S., Stracker, T.H., Yanes, O., Perera-Lluna, A., 2017. Null diffusion-based enrichment for metabolomics data. PLoS One 12, e0189012.

Roszkowska, A., Yu, M., Bessonneau, V., Bragg, L., Servos, M., Pawliszyn, J., 2018. Metabolome profiling of fish muscle tissue exposed to benzo[a]pyrene using in vivo solid-phase microextraction. Environ. Sci. Technol. Lett. 5, 431−435.

Segner, H., 2009. Zebrafish (*Danio rerio*) as a model organism for investigating endocrine disruption. Comp. Biochem. Physiol. C Toxicol. Pharmacol. 149, 187−195.

Song, Y., Chai, T., Yin, Z., Zhang, X., Zhang, W., Qian, Y., Qiu, J., 2018. Stereoselective effects of ibuprofen in adult zebrafish (*Danio rerio*) using UPLC-TOF/MS-based metabolomics. Environ. Pollut. 241, 730−739.

Tufi, S., Leonards, P., Lamoree, M., de Boer, J., Legler, J., Legradi, J., 2016. Changes in neurotransmitter profiles during early zebrafish (*Danio rerio*) development and after pesticide exposure. Environ. Sci. Technol. 50, 3222−3230.

WHO, 1993. Biomarkers and risk assessment: concepts and principles. Environ. Health Criter. 155.

Wu, G., 2009. Amino acids: metabolism, functions, and nutrition. Amino Acids 37, 1−17.

Xia, Y., Cheng, S., He, J., Liu, X., Tang, Y., Yuan, H., He, L., Lu, T., Tu, B., Wang, Y., 2011. Effects of subchronic exposure to benzo[a]pyrene (B[a]P) on learning and memory, and neurotransmitters in male Sprague-Dawley rat. Neurotoxicology 32, 188−198.

Xia, Z., Ying, G., Hansson, A.L., Karlsson, H., Xie, Y., Bergstrand, A., DePierre, J.W., Nassberger, L., 2000. Antidepressant-induced lipidosis with special reference to tricyclic compounds. Prog. Neurobiol. 60, 501−512.

Xu, N., Mu, P., Yin, Z., Jia, Q., Yang, S., Qian, Y., Qiu, J., 2016. Analysis of the enantioselective effects of PCB95 in zebrafish (*Danio rerio*) embryos through targeted metabolomics by UPLC-MS/MS. PLoS One 11, e0160584.

Zhao, Y.Y., Wang, H.L., Cheng, X.L., Wei, F., Bai, X., Lin, R.C., Vaziri, N.D., 2015. Metabolomics analysis reveals the association between lipid abnormalities and oxidative stress, inflammation, fibrosis, and Nrf2 dysfunction in aristolochic acid-induced nephropathy. Sci. Rep. 5, 12936.

Zhou, X., Li, Y., Li, H., Yang, Z., Zuo, C., 2019. Responses in the crucian carp (*Carassius auratus*) exposed to environmentally relevant concentration of 17α-Ethinylestradiol based on metabolomics. Ecotoxicol. Environ. Saf. 183, 109501.

Ziarrusta, H., Mijangos, L., Picart-Armada, S., Irazola, M., Perera-Lluna, A., Usobiaga, A., Prieto, A., Etxebarria, N., Olivares, M., Zuloaga, O., 2018. Non-targeted metabolomics reveals alterations in liver and plasma of gilt-head bream exposed to oxybenzone. Chemosphere 211, 624−631.

Ziarrusta, H., Ribbenstedt, A., Mijangos, L., Picart-Armada, S., Perera-Lluna, A., Prieto, A., Izagirre, U., Benskin, J.P., Olivares, M., Zuloaga, O., Etxebarria, N., 2019. Amitriptyline at an environmentally relevant concentration alters the profile of metabolites beyond monoamines in gilt-head bream. Environ. Toxicol. Chem. 38, 965−977.

CHAPTER 12

Future trends in environmental metabolomics analysis

Diana Álvarez-Muñoz, Marinella Farré
Water and Soil Quality Research Group, Department of Environmental Chemistry, IDAEA-CSIC, Barcelona, Spain

Chapter Outline

The aim of environmental metabolomics is to be used as a bottom-up approach where changes in molecular parameters indicate the physiological function of organisms, which may then be related to the health of a population and ecosystem (Maloney, 2019). Based on its unique capabilities and recent advances, metabolomics could serve as a powerful routine bioanalytical tool for the evaluation of the ecotoxicity and the mode of action (MoA) of environmental contaminants. Even for assessing the MoA of new bioactive compounds before their commercialization, metabolomics can greatly assist this process and reduce the corresponding costs of research and development. The profiling of the metabolome has become a very useful tool for the identification of toxic effects and metabolic pathways disrupted due to exposure to multiple chemicals simultaneously. The discovery of robust, reliable biomarkers or sets of biomarkers of impact through the application of cross-platform analyses is highly anticipated to help out in this direction. However, as the main drawback, the current metabolomics approaches do not allow to differentiate which component of a chemical mixture produces a specific effect and therefore, for this matter, exposure to single compounds is still needed. However, only exposure experiments to individual compounds do not inform on the potential synergistic or antagonistic effects once these single compounds are integrated into real complex mixtures. Nevertheless, it has been shown that specific groups of chemicals affect common metabolic pathways.

The progress in environmental metabolomics has been linked to the advances of analytical chemistry. Initially, the main analytical platforms have been based on nuclear magnetic resonance (NMR) spectroscopy, but the development of sample preparation approaches and the improvement of analytical instrumentation, in terms of sensitivity and selectivity, have been prominent. In particular, the inclusion in metabolomics studies of the new analytical platforms, such as gas chromatography, high-performance liquid chromatography, ultrahigh-performance liquid chromatography coupled to high-resolution mass spectrometry, facilitates separation, detection, characterization, and quantification of new metabolites and related metabolic pathways. In addition, it can be used to detect molecular-level changes in organisms exposed to target contaminants and real mixtures of contaminants at environmentally relevant concentrations in the ng/L to low μg/L range. Combining NMR spectroscopy and mass spectrometry (MS) techniques for nontargeted and targeted analysis can provide a holistic view of the impacted metabolic pathways and the MoA of contaminants. Metabolomics is the only *omics* techniques also providing comprehensive information on environmental exposures. However, taking into consideration the complexity of the metabolome, no single analytical platform can be applied to detect all metabolites in a biological sample. The combined use of modern instrumental analytical approaches is needed to increase the coverage of detected metabolites and propose useful biomarkers for environmental monitoring. As a result of a broader knowledge on biomarkers of environmental stress, a more effective, rapid, and routine track of ecosystem health could be performed. However, the use of metabolomics as an early warning signal for environmental biomonitoring and ecological risk assessment continues being one of the main challenges currently faced by scientist (Bahamonde et al., 2016).

It should be remarked that environmental samples are complex mixtures that include organic materials and toxic metals. For this reason, the application of instrumental analytical approaches based on the use of inductively coupled plasma MS allows tracing the presence of these metals in complex molecules assisted by separation techniques, mainly chromatography, generally in multidimensional arrangements. In some cases, the problem under study can be approached with targeted analysis. Still, in others, nontarget approaches are necessary, usually complementarily supported with organic MS to annotate or identify the metallobiomolecules. Therefore, this binomial methodology (metallomics/metabolomics) offers excellent possibilities to deep insight into detoxification processes in living organisms as a consequence of contamination. Besides, it is a valuable tool in exposure experiments in the laboratory, which can be translated to contamination episodes in the field.

As it has been mentioned before, current metabolomics studies mainly depend on NMR or chromatographic approaches coupled to MS analysis, which still needs to be improved to better identify and quantify metabolites of concerns (biomarkers) within a single sample. Despite the methodological advances achieved until now, one of the primary challenges that should be faced by metabolomics, in general, is the development of standardized methods. The standard approaches involving sample collection and pretreatment, data acquisition, processing, and analysis are still lacking. It should be pointed out that the standardization of analytical conditions and the incorporation of extra parameters complementing the mass spectra and ionization source information, such as type of columns in chromatography, could help to facilitate metabolite identification. Other useful information included in the databases could be the fragments and collision energy if spectra are available. One of the main limitations of metabolomics is the efficiency for recognition of metabolites because in most cases the metabolites detected by NMR and MS have an unknown chemical nature. According to a recent study, reference databases for metabolomics currently provide information for hundreds of thousands of compounds, but only experimental data correspond to barely the 5% of these molecules (Frainay et al., 2018). Moreover, in toxicological studies, the distinction between metabolic changes due to the toxicant or the consequences of damage, and variability between individuals, and for one individual under different conditions is sometimes technically tricky. To summarize, standardized procedures, compatible software, and databases recognized by the scientific community are needed for validation and interpretation of results. They are of primary importance for the development of future environmental metabolomics.

Most of the works performed until now have been focused in aquatic environments. They aim to identify new biomarkers for assessing pollution, and they have been based on whole organism metabolomics using mainly small fish models. However, shortly it is expected that biofluids will gain closer attention since they reflect the health status of an organism at a certain time, they are easy to obtain, and they also require less sample preparation. For example, low protein biofluids, such as urine, can be directly analyzed after a dilution step. This could facilitate automation, which is another crucial challenge for the near future, and in this manner, make easier the analysis of a large sample number in environmental monitoring. On the other hand, environmental metabolomics on plants has strongly started recently, informing about soil contamination and supporting the development of new phytosanitary compounds. This opens a new window to a more comprehensive environmental assessment and involves test organisms that do not present ethical issues.

References

Bahamonde, P.A., Feswick, A., Isaacs, M.A., Munkittrick, K.R., Martyniuk, C.J., 2016. Defining the role of omics in assessing ecosystem health: perspectives from the Canadian environmental monitoring program. Environ. Toxicol. Chem. 35, 20–35.

Frainay, C., Schymanski, E.L., Neumann, S., Merlet, B., Salek, R.M., Jourdan, F., Yanes, O., 2018. Mind the gap: mapping mass spectral databases in genome-scale metabolic networks reveals poorly covered areas. Metabolites 8, 51.

Maloney, E.M., 2019. How do we take the pulse of an aquatic ecosystem? Current and historical approaches to measuring ecosystem integrity. Environ. Toxicol. Chem. 38, 289–301.

Index

Note: 'Page numbers followed by "t" indicate tables, "f" indicate figures and "b" indicate boxes'.

Printed in the United States
By Bookmasters